Tumors in Domestic Animals

Tumors in Domestic Animals

Dr. Pradeep Kumar Choda

RANDOM
RANDOM PUBLICATIONS
NEW DELHI (INDIA)

Tumors in Domestic Animals

ISBN 978-93-5111-887-9

Published in 2016 in India by

RANDOM PUBLICATIONS

4376-A/4B, Gali Murari Lal, Ansari Road
New Delhi-110 002
Phone : +9111-43580356, 011-23289044, 011-43142548
e-mail: sales@randompublications.com,
info@randompublications.com, randomexports@gmail.com

Reprinted 2026

Type Setting by : Friends Media, Delhi-110089

Preface

A tumor is an abnormal growth of cells that serves no purpose. A benign tumor is not a malignant tumor, which is cancer. It does not invade nearby tissue or spread to other parts of the body the way cancer can. In most cases, the outlook with benign tumors is very good. But benign tumors can be serious if they press on vital structures such as blood vessels or nerves. Cancer is common in pet animals and the incidence increases with age. Cancer accounts for almost half of the deaths of pets over 10 years of age. Dogs get cancer at roughly the same rate as humans, while cats get fewer cancers. Strong circumstantial evidence of cancer can be attained from x-rays, blood tests, ultrasonography, the pet's physical examination and medical history. Most cancers, however, will require a biopsy (a removal of a piece of tissue) for confirmation that cancer exists and to grade the level of severity from benign to aggressively malignant.

Many types of skin tumors, both benign (noncancerous) and malignant (cancerous), exist. Approximately 20-40% of primary skin tumors are malignant in dogs and 50-65% aremalignant in cats. Not all forms of skin cancer in cats and dogs are caused by sun exposure, but it can happen occasionally. On dogs, the nose and pads of the feet contain sensitive skin and no fur to protect from the sun. Also, cats and dogs with thin or light-colored coats are at a higher risk of sun damage over their entire bodies. Tumors are abnormal growths of cells. Tumors affecting the skin or the tissue just under the skin are the most commonly seen tumors in dogs. Skin tumors are diagnosed more frequently than other tumors in animals in part because they are the most easily seen tumors and in part because the skin is constantly exposed to many tumor-causing factors in the environment. Chemicals, solar radiation, and viruses are just some of the things that can cause skin tumors. Hormonal abnormalities and genetic factors may also play a role in the development of skin tumors.

This book serves the purpose of a resourceful and authentic information about the tumors of domestic animals including poultry. Every attempt has been made to compile facts about the origin of tumors.

– Author

Contents

1

Lymphoma in Animals

Lymphoma (lymposarcoma) in animals is a type of cancer defined by a proliferation of malignant lymphocytes within solid organs such as the lymph nodes, bone marrow, liver and spleen. The disease also may occur in the eye, skin, and gastrointestinal tract.

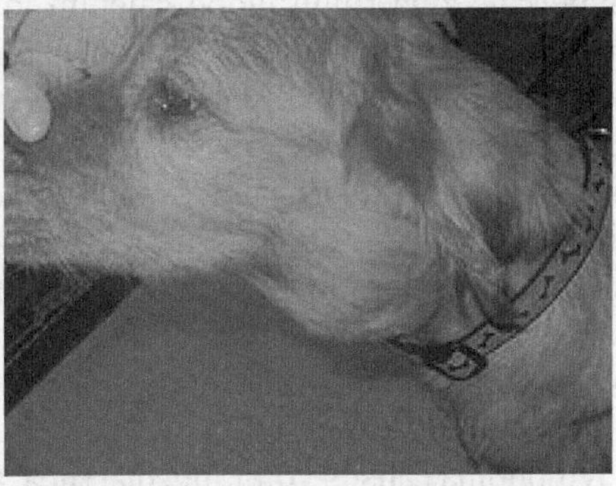

Fig. Lymphoma in a Golden Retriever

LYMPHOMA IN DOGS

Lymphoma is one of the most common malignant tumors to occur in dogs. The cause is genetic, but there are also suspected environmental factors involved, including in one study an increased risk with the use of the herbicide 2,4-D. This risk was not confirmed in another study.

Breeds that are commonly affected include Boxer, Scottish Terrier, Basset Hound, Airedale Terrier, Chow Chow, German Shepherd, Poodle, St. Bernard, Bulldog, Beagle,Rottweiler and Golden Retriever. The Golden Retriever is especially susceptible to developing lymphoma, with a lifetime risk of 1:8.

Classification

The cancer is classified into low and high grade types. Classification is also based on location. The four location types are multicentric, mediastinal, gastrointestinal, and extranodal (involving the kidney, central nervous system, skin, heart, or eye). Multicentric lymphoma, the most common type (by greater than 80 percent), is found in the lymph nodes, with or without involvement in the liver, spleen, or bone marrow. Mediastinal lymphoma occurs in the lymph nodes in the thorax and possibly the thymus. Gastrointestinal lymphoma occurs as either a solitary tumor or diffuse invasion of the stomach or intestines, with or without involvement in the surrounding lymph nodes, liver or spleen. Classification is further based on involvement of B-lymphocytes or T-lymphocytes. Approximately 70 percent are B-cell lymphoma. Cutaneous lymphoma can be classified as epitheliotropic (closely conforming to the epidermis) or non-epitheliotropic. The epitheliotropic form is typically of T-cell origin and is also called mycosis fungoides. The non-epitheliotropic form is typically of B-cell origin.

Signs and symptoms

General signs and symptoms include depression, fever, weight loss, loss of appetite, loss of hair or fur and vomiting. Lymphoma is the most common cancerous cause ofhypercalcemia (high blood calcium levels) in dogs. It can lead to the above signs and symptoms plus increased water drinking, increased urination, and cardiac arrhythmias. Hypercalcemia in these cases is caused by secretion of parathyroid hormone-related protein.

Multicentric lymphoma presents as painless enlargement of the peripheral lymph nodes. This is seen in areas such as under the jaw, the armpits, the groin, and behind the knees. Enlargement of the liver and spleen causes the abdomen to distend. Mediastinal lymphoma can cause fluid to collect around the lungs, leading to coughing and difficulty breathing. Hypercalcemia is most commonly associated with this type.

Gastrointestinal lymphoma causes vomiting, diarrhea, and melena (digested blood in the stool). Low serum albumin levels and hypercalcemia can also occur.

Lymphoma of the skin is an uncommon occurrence. The epitheliotropic form typically appears as itchy inflammation of the skin progressing to nodules and plaques. The non-epitheliotropic form can have a wide variety of appearances, from a single lump to large areas of bruised, ulcerated, hairless skin. The epitheliotropic form must be differentiated from similar appearing conditions such as pemphigus vulgaris, bullous pemphigoid, and lupus erythematosus. Signs for lymphoma in other sites depend on the location. Central nervous system involvement can cause seizures or paralysis. Eye involvement, seen in 20 to 25 percent of cases, can lead to glaucoma, uveitis, bleeding within the eye, retinal detachment, and blindness. Lymphoma in the bone marrow causes anemia, low platelet count, and lowwhite blood cell count.

Diagnosis

Biopsy of affected lymph nodes or organs confirms the diagnosis, although a needle aspiration of an affected lymph node can increase suspicion of the disease. X-rays, ultrasound and bone marrow biopsy reveal other locations of the cancer.

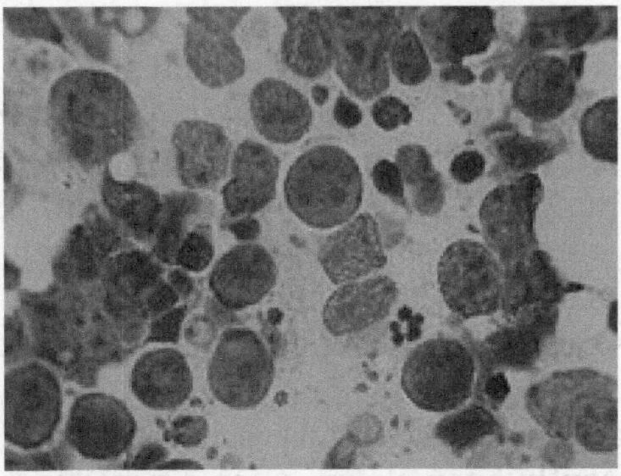

Fig. Cytology of lymphoma in a dog

There are now a range of blood tests that can be utilised to aid in the diagnosis of lymphoma. Flow cytometry detects antibodies linked to tumour cell surface antigens in fluid samples or cell suspensions. Polymerase chain reaction (PCR) for antigen receptor rearrangements (PARR) identifies circulating tumour cells based on unique genetic sequences. The canine Lymphoma Blood Test (cLBT) measures multiple circulating biomarkers and utilises a complex algorithm to diagnose lymphoma. The TK canine cancer panel is an indicator of general neoplastic disease. The stage of the disease is important to treatment and prognosis. Certain blood tests have also been shown to be prognostic.

The stage of the disease is important to treatment and prognosis.

- Stage I - only one lymph node or lymphoid tissue in one organ involved.
- Stage II - lymph nodes in only one area of the body involved.
- Stage III - generalized lymph node involvement.
- Stage IV - any of the above with liver or spleen involvement.
- Stage V - any of the above with blood or bone marrow involvement.

Each stage is divided into either *substage a*, those without systemic symptoms; or *substage b*, those with systemic symptoms such as fever, loss of appetite, weight loss, and fatigue.

Treatment

Due to the high risk of recurrence and ensuing problems, close monitoring of dogs undergoing chemotherapy is important. The same is true for dogs that have entered remission and ceased treatment. Monitoring for disease and remission/recurrence is usually performed by palpation of peripheral lymph nodes. This procedure detects gross changes in peripheral lymph nodes. Some of the blood tests used in diagnosing lymphoma also offer greater objectivity and provide an earlier warning of an animal coming out of remission.

Complete cure is rare with lymphoma and treatment tends to be palliative, but long remission times are possible with chemotherapy. With effective protocols, average first remission times are 6 to 8 months. Second remissions are shorter and harder to accomplish. Average survival is 9 to 12 months. The most common treatment is a combination of cyclophosphamide, vincristine, prednisone, L-asparaginase, and doxorubicin. Other chemotherapy drugs such as chlorambucil, lomustine (CCNU), cytosine arabinoside, and mitoxantrone are sometimes used in the treatment of lymphoma by themselves or in substitution for other drugs. In most cases, appropriate treatment protocols cause few side effects, but white blood cell counts must be monitored.

Allogeneic and autologous stem cell transplantations (as is commonly done in humans) have recently been shown to be a possible treatment option for dogs. Most of the basic research on transplantation biology was generated in dogs. Current cure rates using stem cell therapy in dogs approximates that achieved in humans, 40-50%.

When cost is a factor, prednisone used alone can improve the symptoms dramatically, but it does not significantly affect the survival rate. The average survival times of dogs treated with prednisone and untreated dogs are both one to two months. Using prednisone alone can cause the cancer to become resistant to other chemotherapy agents, so it should only be used if more aggressive treatment is not an option.

Isotretinoin can be used to treat cutaneous lymphoma.

Prognosis

Untreated dogs have an average survival time of 60 days. Lymphoma with a histologic high grade generally respond better to treatment but have shorter survival times than dogs with low grade lymphoma. Dogs with B-lymphocyte tumors have a longer survival time than T-lymphocyte tumors. Mediastinal lymphoma has a poorer prognosis than other types, especially those with hypercalcemia. Clinical stage and substage have some prognostic value, with poorer prognosis associated with Stage V disease, and with substage b (clinical illness at time of presentation).

LYMPHOMA IN CATS

Lymphoma is the most common malignancy diagnosed in cats. Lymphoma in young cats occurs most frequently following infection with feline leukemia virus (FeLV) or to a lesser degree feline immunodeficiency virus (FIV). These cats tend to have involvement of lymph nodes, spine, or mediastinum. Cats with FeLV are 62 times more likely to develop lymphoma, and cats with both FeLV and FIV are 77 times more likely. Younger cats tend to have T-cell lymphoma and older cats tend to have B-cell lymphoma.Older cats tend to have gastrointestinal lymphoma without FeLV infection, although tests more sensitive to low level FeLV infections and replication-defective FeLV have found that many of these cats have been previously exposed. The same forms of lymphoma that are found in dogs also occur in cats, but gastrointestinal is the most common type. Lymphoma of the kidney is the most common kidney tumor in cats, and lymphoma is also the most common heart tumor.

Classification

Gastrointestinal lymphoma is classified into low grade, intermediate grade, and high grade. Low grade types include lymphocytic and small cell lymphoma. High grade types include lymphoblastic, immunoblastic, and large cell lymphoma. Low grade lymphoma is only found in the small intestine, while high grade can commonly be found in the stomach.

Symptoms

Cats that develop lymphoma are much more likely to develop more severe symptoms than dogs. Whereas dogs often appear healthy initially except for swollen lymph nodes, cats will often be physically ill. The symptoms correspond closely to the location of the lymphoma. The most common sites for alimentary (gastrointestinal) lymphoma are, in decreasing frequency, the small intestine, the stomach, the junction of the ileum, cecum, and colon. Cats with the alimentary form of lymphoma often present with weight loss, rough hair coat, loss of appetite, vomiting and diarrhea, although vomiting and diarrhea are commonly absent as symptoms.

The tumor can also cause life-threatening blockage of the intestine. Cats with the mediastinal form often have respiratory distress and fluid in the thoracic cavity. If lymphoma develops in the kidney, the cat may have increased water consumption and increased urination.

Lymphoma of the kidney presents as bilateral kidney enlargement and failure. If the lymphoma is located in the nose, the cat may have discharge from the nose and facial swelling. Lymphoma of the heart causes congestive heart failure, pericardial effusion, and cardiac arrhythmias. Ocular lymphoma in cats often presents as anterior uveitis (inflammation of the inside of the eye). Cats who are also infected with FeLV often present with pale mucous membranes due to anemia. Anemia is a common problem in all cats with lymphoma, but hypercalcemia is rare.

Diagnosis is similar to dogs, except cats should be tested for FeLV and FIV. It is important to differentiate the alimentary form of lymphoma from inflammatory bowel diseasebecause the signs are so similar in cats. A biopsy is necessary to do this. One approach to differentiate inflammatory bowel disease from is to test the infiltrating lymphocytes for their monoclonal origin in lymphomas.

Treatment and prognosis

Chemotherapy is the mainstay of treatment for lymphoma in cats. Most of the drugs used in dogs are used in cats, but the most common protocol uses cyclophosphamide, vincristine, and prednisone. Gastrointestinal lymphoma has also commonly been treated with a combination of prednisolone and high dose pulse chlorambucil with success. The white blood cell count must be monitored. Remission and survival times are comparable to dogs. Lower stage lymphoma has a better prognosis. Multicentric lymphoma has a better response to treatment than the gastrointestinal form, but infection with FeLV worsens the prognosis.

About 75% of cats treated with chemotherapy for lymphoma go into remission. Unfortunately, after an initial remission, most cats experience a relapse, after which they have a median survival of 6 months. However, about one-third of cats treated with chemotherapy will survive more than 2 years after diagnosis; a small number of these cats may be cured of their disease. Untreated, most cats with lymphoma die within 4–6 weeks. Most cats tolerate their chemotherapy well, and fewer than 5% have severe side effects. Cats do not lose their fur from chemotherapy, though loss of whiskers is possible. Other side effects include low white blood cell count, vomiting, loss of appetite, diarrhea, or fatigue. These can typically be controlled well, and most cats have a good quality of life during treatment. If a cat relapses after attaining remission, the cat can be treated with different chemotherapy drugs to try for a second remission. The chances of a second remission are much lower than the chances of obtaining a first, and the second remission is often shorter than the first.

LYMPHOMA IN FERRETS

Lymphoma is common in ferrets and is the most common cancer in young ferrets. There is some evidence that a retrovirus may play a role in the development of lymphoma like in cats. The most commonly affected tissues are the lymph nodes, spleen, liver, intestine, mediastinum, bone marrow, lung, and kidney.

In young ferrets, the disease progresses rapidly. The most common symptom is difficulty breathing caused by enlargement of the thymus. Other symptoms include loss of appetite, weight loss, weakness, depression, and coughing. It can also masquerade as a chronic disease such as an upper respiratory infection or gastrointestinal disease. In older ferrets, lymphoma is usually chronic and can exhibit no symptoms for years. Symptoms seen are the same as in young ferrets, plus splenomegaly, abdominal masses, and peripheral lymph node enlargement.

Diagnosis is through biopsy and x-rays. There may also be an increased lymphocyte count. Treatment includes surgery for solitary tumors, splenectomy (when the spleen is very large), and chemotherapy. The most common protocol uses prednisone, vincristine, and cyclophosphamide. Doxorubicin is used in some cases. Chemotherapy in relatively healthy ferrets is tolerated very well, but possible side effects include loss of appetite, depression, weakness, vomiting, and loss of whiskers. The white blood cell count must be monitored. Prednisone used alone can work very well for weeks to months, but it may cause resistance to other chemotherapy agents. Alternative treatments includevitamin C and Pau d'Arco (a bark extract).

The prognosis for lymphoma in ferrets depends on the their health and the location of the cancer. Lymphoma in the mediastinum, spleen, skin, and peripheral lymph nodes has the best prognosis, while lymphoma in the intestine, liver, abdominal lymph nodes, and bone marrow has the worst.

LYMPHOMA

What does this statement really mean for you and your pet? Presumably you are here because your pet is suspected of having this form of cancer. Perhaps the diagnosis has already been confirmed. In either case, you want to know more and there turns out to be much more to know. Because lymphoma is a common form of cancer in humans, research and new development in this area abounds. Animal patients as well as human patients have benefitted from this research.

WHAT IS THE LYMPH SYSTEM?

The lymph system is represented by a network of lymphatic vessels and lymph nodes through which foreign proteins and disease organisms are circulated. The lymph nodes serve as processing centers where these foreign substances may be presented to the cells of the immune system. There are many different types of immune-related cells; some produce antibodies, some circulate and destroy the foreign materials they encounter, some regulate the activity of other cells.

Lymphocytes are the primary cells of the lymph system and they act in the various ways mentioned. The lymph vessels serve as a circulatory path for lymphocytes in addition to serving as a collection system directing foreign substances toward the lymph nodes. Lymph vessels interface with the blood stream at several areas allowing lymphocytes greater area to patrol.

What is Cancer?

Cancer occurs when a normal cell goes wrong. Its normal regulatory processes disengage and it begins to divide quickly and without control. The organ to which the original cell belonged is destroyed as the cancer cells obliterate its structure. Other local tissues may also become invaded as the tumor cells grow inexorably into them. Cancer cells break off the primary tumor and travel via blood or lymph vessels to new areas of the body. Wherever these cells lodge, they may start new tumors far from the original tumor but just as deadly. This process continues until there is not enough normal tissue left to sustain life. This form of cancer spread is called metastasis.

What is Lymphoma?

When lymphocytes become cancerous within a lymph node, the node swells and hardens. Malignant lymphocytes readily travel through the lymph vessels to nearby lymph nodes. Soon all the nodes are enlarged. Ultimately, the bone marrow (where most blood cells are formed) is affected, the immune system is destroyed, and severe anemia and weakness claim the victim's life.

WITHOUT TREATMENT, ANIMALS WITH LYMPHOMA ARE EXPECTED TO LIVE 4-8 WEEKS FROM THE TIME OF DIAGNOSIS.

Most patients (especially dogs) are not feeling particularly sick at the time of diagnosis. It may be tempting to hold off on treatment until the pet seems more ill. Waiting can drastically reduce the chance for long-term survival; better remission quality is obtained if the patient is treated while he/she still feels healthy.

What is Remission?

Remission is the state in which tumor symptoms have been abated and the patient is as comfortable as and indistinguishable from any normal animal. Prolonged remission is the goal of cancer therapy which, for most lymphoma cases, means chemotherapy.

For any patient, there is an approximately 75% chance of achieving remission regardless of protocol used.

This means that there is an excellent chance of reducing the tumor to undetectable levels. How long a remission lasts depends on what protocol is used and a number of other factors.

What is Cure?

Cure is the permanent removal of all traces of tumor such that no further treatment is needed. In effect, it is a permanent state of remission. While this is a possibility for your pet, it is more constructive and realistic to focus on increasing quality lifetime. With lymphoma, remission is likely but cure is not. Treatment may be thought of as an exchange of only a short time with your pet for a long time with your pet. It is important to keep goals in proper perspective through the treatment of this cancer. The goal is a long remission, not a cure.

LYMPHOCYTE

A lymphocyte is one of the three subtypes of white blood cell in a vertebrate's immune system. They include natural killer cells (NK cells) (which function in cell-mediated, cytotoxic innate immunity), T cells (for cell-mediated, cytotoxic adaptive immunity), and B cells (for humoral, antibody-driven adaptive immunity). They are the main type of cell found in lymph, which prompted the name*lymphocyte*.

TYPES

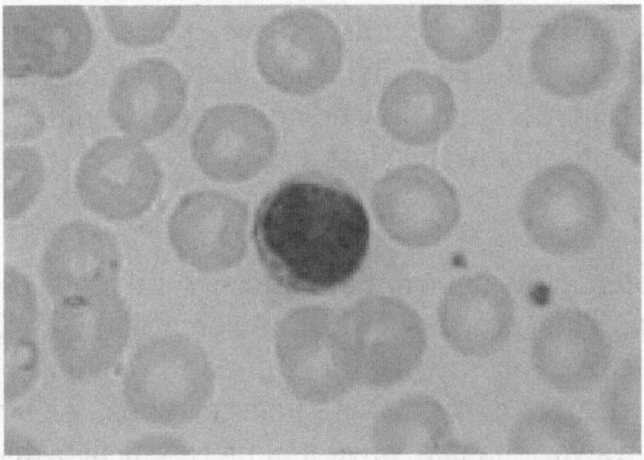

Fig. A stained lymphocyte surrounded by red blood cellsviewed using a light microscope.

The three major types of lymphocyte are T cells, B cells and natural killer (NK) cells. Lymphocytes can be identified by their large nucleus.

T cells and B cells

T cells (thymus cells) and B cells (bone marrow- or bursa-derived cells) are the major cellular components of the adaptive immune response. T cells are involved in cell-mediated immunity, whereas B cells are primarily responsible for humoral immunity (relating toantibodies). The function of T cells and B cells is to recognize specific "non-self" antigens, during a process known as antigen presentation.

Once they have identified an invader, the cells generate specific responses that are tailored to maximally eliminate specific pathogens or pathogen-infected cells. B cells respond to pathogens by producing large quantities of antibodies which then neutralize foreign objects likebacteria and viruses. In response to pathogens some T cells, called *T helper cells*, produce cytokines that direct the immune response, while other T cells, called *cytotoxic T cells*, produce toxic granules that contain powerful enzymes which induce the death of pathogen-infected cells.

Following activation, B cells and T cells leave a lasting legacy of the antigens they have encountered, in the form of *memory cells*. Throughout the lifetime of an animal, these memory cells will "remember" each specific pathogen encountered, and are able to mount a strong and rapid response if the pathogen is detected again.

Natural killer cells

NK cells are a part of the innate immune system and play a major role in defending the host from both tumors and virally infected cells. NK cells distinguish infected cells and tumors from normal and uninfected cells by recognizing changes of a surface molecule called MHC (major histocompatibility complex) class I. NK cells are activated in response to a family of cytokines called interferons. Activated NK cells release cytotoxic (cell-killing) granules which then destroy the altered cells. They were named "natural killer cells" because of the initial notion that they do not require prior activation in order to kill cells which are missing MHC class I.

DEVELOPMENT

Mammalian stem cells differentiate into several kinds of blood cell within the bone marrow. This process is calledhaematopoiesis. All lymphocytes originate, during this process, from a common lymphoid progenitor before differentiating into their distinct lymphocyte types. The differentiation of lymphocytes follows various pathways in a hierarchical fashion as well as in a more plastic fashion. The formation of lymphocytes is known as lymphopoiesis. B cells mature into B lymphocytes in the bursa equivalent, which in humans is the GALT, which is thought to be located in the Peyer's patches of the intestine, while T cells migrate to and mature in a distinct organ, called the thymus. Following maturation, the lymphocytes enter the circulation and peripheral lymphoid organs (e.g. the spleen and lymph nodes) where they survey for invading pathogens and/or tumor cells.

The lymphocytes involved in adaptive immunity (i.e. B and T cells) differentiate further after exposure to an antigen; they form effector and memory lymphocytes. Effector lymphocytes function to eliminate the antigen, either by releasing antibodies (in the case of B cells), cytotoxic granules (cytotoxic T cells) or by signaling to other cells of the immune system (helper T cells). Memory T cells remain in the peripheral tissues and circulation for an extended time ready to respond to the same antigen upon future exposure.

They live weeks to several years to a whole lifetime, which is very long compared to other leukocytes.

CHARACTERISTICS

Microscopically, in a Wright's stained peripheral blood smear, a normal lymphocyte has a large, dark-staining nucleus with little to noeosinophilic cytoplasm. In normal situations, the coarse, dense nucleus of a lymphocyte is approximately the size of a red blood cell (about seven micrometres in diameter). Some lymphocytes show a clear perinuclear zone (or halo) around the nucleus or could exhibit a small clear zone to one side of the nucleus. Polyribosomes are a prominent feature in the lymphocytes and can be viewed with an electron microscope. The ribosomes are involved in protein synthesis, allowing the generation of large quantities of cytokines and immunoglobulins by these cells.

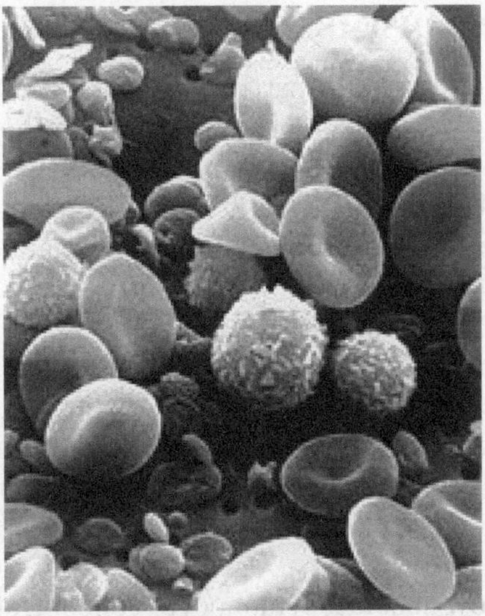

Fig. A scanning electron microscope image of normal circulating human blood showing red blood cells, several types of white blood cells including lymphocytes, amonocyte, a neutrophil and many small disc-shaped platelets.

It is impossible to distinguish between T cells and B cells in a peripheral blood smear. Normally, flow cytometry testing is used for specific lymphocyte population counts. This can be used to specifically determine the percentage of lymphocytes that contain a particular combination of specific cell surface proteins, such as immunoglobulins or cluster of differentiation (CD) markers or that produce particular proteins (for example, cytokines using intracellular cytokine staining (ICCS)). In order to study the function of a lymphocyte by virtue of the proteins it generates, other scientific techniques like the ELISPOT or secretion assay techniques can be used.

Typical recognition markers for lymphocytes			
Class	**Function**	**Proportion**	**Phenotypic Marker(s)**
NK cells	Lysis of virally infected cells and tumour cells	7% (2-13%)	CD16 CD56 but not CD3
Helper T cells	Release cytokines and growth factors that regulate other immune cells	46% (28-59%)	TCRαβ, CD3 and CD4
Cytotoxic T cells	Lysis of virally infected cells, tumour cells and allografts	19% (13-32%)	TCRαβ, CD3 and CD8
γδ T cells	Immunoregulation and cytotoxicity	5% (2%-8%)	TCRγδ and CD3
B cells	Secretion of antibodies	23% (18-47%)	MHC class II, CD19 and CD21

In the circulatory system, they move from lymph node to lymph node. This contrasts with macrophages, which are rather stationary in the nodes.

LYMPHOCYTES AND DISEASE

A lymphocyte count is usually part of a peripheral complete blood cell count and is expressed as the percentage of lymphocytes to the total number of white blood cells counted.

A general increase in the number of lymphocytes is known as lymphocytosis, whereas a decrease is known as lymphocytopenia.

High

An increase in lymphocyte concentration is usually a sign of a viral infection (in some rare case, leukemias are found through an abnormally raised lymphocyte count in an otherwise normal person). A high lymphocyte count with a low neutrophil count might be caused by lymphoma.

Low

A low normal to low absolute lymphocyte concentration is associated with increased rates of infection after surgery or trauma.

One basis for low T cell lymphocytes occurs when the human immunodeficiency virus (HIV) infects and destroys T cells (specifically, the $CD4^+$ subgroup of T lymphocytes). Without the key defense that these T cells provide, the body becomes susceptible to opportunistic infections that otherwise would not affect healthy people. The extent of HIV progression is typically determined by measuring the percentage of $CD4^+$ T cells in the patient's blood. The effects of other viruses or lymphocyte disorders can also often be estimated by counting the numbers of lymphocytes present in the blood.

Tumor-infiltrating lymphocytes

In some cancers, e.g., melanoma and colorectal cancer, lymphocytes can migrate into and attack the tumor. This can sometimes lead to regression of the primary tumor.

LYMPHOMA IN DOGS

The typical canine lymphoma patient is a middle-aged dog who goes to the veterinarian because one or more lumps have been found. The veterinarian rapidly determines that all of the peripheral lymph nodes (those near the skin surface) are enlarged and firm. Usually the dog has not been showing any signs of illness.

The next step is a blood panel and urinalysis to more completely assess the patient's health, and one or more lymph nodes are aspirated or biopsied to confirm the diagnosis of lymphoma.

So here we are.

We have confirmed lymphoma and we know the average life expectancy for a patient with untreated lymphoma is about 2 months from the time of diagnosis. If this is your dog's situation, you probably need some time to absorb the cancer diagnosis. You have many general questions and you know that a decision regarding chemotherapy must be made.

We will attempt to cover commonly asked questions owners have at this point.

HOW DID MY DOG GET LYMPHOMA?

We do not know how dogs or people get cancer most of the time. There are many types of cancer and many possible causes of all those cancers (chemicals in our environment, especially cigarette smoke, sun exposure, assorted viruses and infections). There are important genetic factors as well. Cancer starts with one or a small group of cells that have gone wrong. It appears that such cells arise in our bodies all the time and we have an assortment of natural mechanisms to destroy these cells before they get out of hand. Sometimes these cancer cells escape our natural mechanisms and cancer develops.

It is important to realize that cancer is not contagious and that, as a pet owner, you should not feel that you somehow caused this or brought it on your pet. Many people feel a need to find blame and latch onto the idea that a household cleaner or pesticide was the cause. This is a natural part of grieving but it is important not to focus on cause unduly. Cause is not relevant to treatment; furthermore, there is no way to verify cause. It is best to concentrate on treatment. At this time, there is no way to know what caused lymphoma development in a given patient.

Can my Dog be Cured?

Theoretically, yes, but practically speaking, no. It is best to focus on a realistic outcome that is the longest possible survival with a good quality of life. Different treatment protocols are associated with different disease-free intervals.

Does my Pet Need Further Tests?

Basic blood work and a urinalysis will be needed to assess the patient's ability to take the medications needed to achieve remission.

If a biopsy has not been performed, it may be a good idea to have one done so as to gain the maximum information about the tumor (whether it is slow or fast growing, what type of lymphocytes are involved, etc.) as this information can help predict the response to chemotherapy. Specific staining of the sample, either aspirate or biopsy, is needed to distinguish the type of lymphocyte involved in the tumor. T-cell lymphoma, for example, is less responsive to medication than B-cell lymphoma. Luckily, B-cell lymphoma accounts for 75% of canine lymphoma.

Other tests that may be recommended include a bone marrow aspirate and/or a spleen or liver aspirate. These tests are needed to stage the disease. Lymphoma is classified by stage.

- *Stage I:* Only one lymph node involved
- *Stage II:* Several lymph nodes in the same general area involved
- *Stage III:* All peripheral lymph nodes involved
- *Stage IV:* All peripheral lymph nodes plus the spleen, liver, and/or anterior mediastinum in the chest involved
- *Stage V:* Bone marrow involvement, regardless of any other areas involved

In cases of lymphoma that are not as straightforward as the classical multicentric lymphoma described below, staging may be more important. Staging used to be done regularly after the initial diagnosis of lymphoma but it has since been found that stage of disease does not impact the response to chemotherapy (i.e., it is not true that a stage II will have a better response than a stage IV) and thus there isn't much point to the trouble and expense of staging. The exception is stage V, the most advanced stage. Patients with stage V lymphoma tend to have a poor response to chemotherapy and it may be helpful to stage the patient to rule out whether or not the patient is in stage V.

Another important parameter that should be checked is the blood calcium level. This sounds unrelated to cancer but, in fact, some types of lymphoma produce a hormone called PTH-rp (parathyroid hormone related protein) that is capable of creating dangerous elevations in the blood calcium level. A dog with an elevated blood calcium level tends to have a poorer prognosis. Approximately 15% of dogs with lymphoma (and 40% of dogs with T-cell lymphoma) have elevated blood calcium levels.

How does Lymphoma Cause Death?

Lymphoma is a rapidly-growing malignancy that is able to go and grow anywhere where there is lymph tissue. This is virtually every organ in the body. Eventually, the cancer will infiltrate an organ to such an extent that that organ fails (often this is the bone marrow or the liver). The patient loses appetite, vomits or gets diarrhea, weakens and dies. At some point the tumor becomes resistant to therapy and no further remissions can be obtained.

My Dog does not fit the Above Scenario at all. What are other Forms of Lymphoma?

Lymphoma is classified by the anatomic area affected. By far, the most common form in dogs is the multicentric form, which accounts for 84% of canine lymphoma. In this form, as in the hypothetical case we opened with, all peripheral lymph nodes are large and firm. There are three other forms of lymphoma:

- Gastrointestinal form (affecting only the GI tract) accounts for 5 - 7% of canine lymphoma
- Mediastinal (chest)
- Extranodal (skin, eye and other miscellaneous areas).

Lymphoma can occur anywhere in the body where there is lymph tissue. At this time, we will concentrate on multicentric lymphoma. Eventually further information on these more rare forms will be added.

Chemotherapy FAQ

The word chemotherapy conjures images of people losing their hair and suffering chronic nausea. It is unfortunate that many pets (and probably people, too) do not receive chemotherapy based upon these unpleasant images that do not truly represent the current state of treatment response at least in pets. Chemotherapy simply means therapy using medication (as opposed to surgery or radiation). Decades of research has gone into patient comfort, minimizing side effects and maximizing response so it is important to keep an open mind. The following are common questions pet owners commonly have regarding chemotherapy for their dog.

The median survival time for most dogs on chemotherapy is approximately one year with 25% of dogs surviving two years.

My dog is not acting sick in any way. Shouldn't I wait until she at least feels sick before beginning chemotherapy? This might seem like a reasonable approach at first glance but let us assure you that it is not. One of the most important factors in determining the quality of remission (i.e., how fast we get remission and how long it lasts) is whether or not the patient is feeling sick at the time chemotherapy is started. When lymphoma patients are staged, stages are subcategorized as *a* or *b*, depending on whether or not the patient is feeling ill or not. (For example, a multicentric lymphoma dog who feels well is in stage IIIa compared to a multicentric lymphoma dog who is vomiting or not eating is in stage IIIb). You will have a much better chance for long-term quality survival if lymphoma is treated while the patient is an *a*.

Will chemotherapy make my dog sick? Probably not. Nausea and infection are possibilities but most dogs do not experience any such complications. Only 7% of patients require hospitalization due to side effects of chemotherapy. The bottom line here is to know that animals rarely get sick from chemotherapy but that you should know what to do in case of a problem.

Will chemotherapy make my dog lose his hair or go bald? While whiskers are commonly lost, substantial hair loss is not experienced by most dogs or cats on chemotherapy for cancer. There are some notable exceptions: breeds that have synchronous hair follicle activity. Most breeds have hairs in all different stages of the growth-shed cycle at the same time. A few breeds have all hairs in the same stage of growth-shed at the same time. These are the breeds that can have a baldness issue: the Old English Sheepdog, the poodle, the Lhasa apso, the Shih Tzu.

How will I know when we have achieved remission? A patient in remission is indistinguishable from a completely cancer-free patient. The lymph nodes will go down to normal size and if there were any signs of illness related to the cancer, these should resolve. There is approximately a 75% chance of achieving remission regardless of protocol selected.

How will I know when we have lost remission? The most obvious sign will be that the lymph node enlargement has returned. This means that the cancer is now resistant to the drugs being used and new drugs must be chosen. (This is called a rescue.)

How long will my dog have quality life on chemotherapy? This depends on what protocol you choose, and there are many. There are also many factors that influence how an individual will do relative to the average response. Important parameters to note when reviewing a protocol are:

1. The disease-free interval (i.e., how long the patient is free from illness)
2. Survival time
3. Typical duration of remission
4. Expense
5. Scheduling.

LYMPHOMA IN CATS

Lymphoma is, at this time, the most common malignancy of cats, accounting for as much as 30% of all feline cancers. Luckily, it is just as responsive to medication as it is in dogs and prolonged remissions are common. This article reviews the common forms of lymphoma in the cat including diagnosis and treatment expectations. Lymphoma is a cancer made of lymphocytes and these lymphocytes readily travel throughout the body via the lymph system. Lymphoma is not considered to be a localized disease because of this traveling, so surgery and radiotherapy are not appropriate sole treatments. To reach cancerous lymphocytes in all the places they have gone, medication (chemotherapy) is necessary because medication can be carried all over the body via the circulation. Treatment of lymphoma is going to involve pills and/or injections for your cat regardless of which form of lymphoma has been diagnosed.

CLASSIFICATION BY ANATOMICAL LOCATION

As in dogs, one of the more common ways to classify lymphoma is by the anatomical area where the cancer is mainly located. In dogs, the most common form of lymphoma is the"multicentric form where many (if not all) of the body's lymph nodes seem to enlarge and fill with cancer cells at the same time. Cats are certainly able to develop multicentric lymphoma as well, but by far the most common form of lymphoma in cats is the intestinal form so most of our discussion will center there. Other notable forms of feline lymphoma include: renal (kidney) lymphoma, nasal lymphoma, and mediastinal lymphoma.

INTESTINAL LYMPHOMA

As mentioned earlier, lymphoma can develop anywhere there is lymph tissue and virtually all organs have some bits of scattered lymph tissue associated with them. It is not clear why some areas seem to be especially vulnerable to developing lymphoma over others. In the past, prior to the development of the vaccine for the feline leukemia virus, the intestinal form of lymphoma was unusual but now it accounts for 50-70% of all cases of feline lymphoma.

Cats with intestinal lymphoma tend to be seniors (age 9-13 years on average) with a chronic history of weight loss, vomiting, diarrhea, or all three. Appetite is variable, which means it may be normal, excessive, reduced, or non-existent. Intestinal lymphoma may be a distinct mass or growth or group of growths, or appear as a more subtle infiltration of cancerous lymphocytes into the delicate membranes of the bowel lining. Infiltrative lesions, which are often invisible to the naked eye, generally respond to chemotherapy; surgery is not necessary or even helpful. Distinct masses, however, can cause an acute obstruction and may require surgical removal to relieve this potential emergency. Removing these masses does not control the cancer and chemotherapy is still needed after the surgical site has healed.

DIAGNOSIS

Confirming lymphoma requires a sample of tissue, ideally a small chunk or at least a smear of cells. There are several ways tissue can be harvested for analysis.

Surgical Biopsy

Involves opening the abdomen and removing pieces of tissue for analysis. *Pros:* maximum access to the organs of the abdomen, ability to sample the full thickness of the intestine and see how deep the abnormal cells penetrate, an important piece of information in determining whether cellular infiltration is benign or malignant. *Cons:* invasive. Requires general anesthesia and hospitalization. Also, chemotherapy must be delayed until the surgery site has healed.

Endoscopic Biopsy

Involves the use of a long scope inserted into the mouth and/or anus to view the intestine and stomach from the inside. A small "biter" is inserted through the scope that can bite out little pinches of intestinal tissue. These biopsies are not full thickness but the scope allows for specific areas to be sampled. *Pros*: far less invasive than surgical exploration with patients generally going home the same day as the procedure. Usually produces quality samples for the lab. Chemotherapy need not be delayed after results are obtained. Areas that might not look abnormal when viewed externally may look very abnormal when viewed from inside the intestine. Endoscopy allows for any such specific areas to be biopsied. *Cons*: does not always produce quality samples as readily as surgery. Endoscopic samples are smaller, not full thickness, and may be squashed by the biopsy biter, making interpretation more difficult. Also, only the intestine is accessible; other organs are not.

Needle Aspirate

Involves removing a sample of cells from a larger organ or from a mass with a needle, usually with ultrasound guidance. *Pros:* anesthesia is rarely needed and procedure is less invasive than the other two methods. *Cons*: cells are withdrawn without their tissue architecture, which means some accuracy is sacrificed. Non-diagnostic samples are not unusual. The intestine itself cannot be aspirated, only masses and organs (local lymph nodes, the liver etc.)

ULTRASOUND WITHOUT BIOPSY?

Ultrasound represents a non-invasive means to evaluate the texture of the organs of the abdomen. Intestinal lymphoma tends to have a characteristic disruption of the bowel lining that is visible by ultrasound. Further, lymph node enlargement in the abdomen can be evaluated as well as texture in the liver. Lymphoma has the ability to be very subtle in its manifestation but if the bowel's appearance is classical via ultrasound, it may be reasonable to begin treatment based solely on this information and forego the more expensive/invasive biopsy procedures.

Another way that ultrasound might negate the need for further procedures is by obtaining a diagnostic aspirate. If there is a mass or if the local lymph nodes are enlarged, they may be aspirated as described above. If the cells obtained indicate lymphoma is present, then the diagnosis has been adequately obtained. Furthermore, often the appearance of the intestine and lymph nodes is ambiguous. In this situation, the liver can be aspirated despite its normal appearance. Often lymphoma can be found hiding out there. Again, if a diagnostic aspirate is obtained via ultrasound, it is not necessary to pursue surgery or endoscopy.

High Grade versus Low Grade

One of the advantages of having an actual tissue sample is the ability to grade the malignancy of the lymphoma. Lymphoma is graded by the pathologist reading the biopsy sample as either high-grade, low-grade, or intermediate grade. The grade refers to how rapidly the cells appear to be dividing and how malignant they appear with high grade being the most malignant. The grade of lymphoma bears on its response to chemotherapy. It is may be possible to grade a lymphoma with an aspirate sample but it is vastly easier with an actual chunk of tumor. As a general rule with lymphoma, higher grades tend to be more responsive to chemotherapy drugs. With feline intestinal lymphoma, however, it is the low grade cases that achieve remission easier and for prolonged periods of time.

Treatment

As with most forms of lymphoma, treatment centers on medication rather than surgery or radiation therapy. The use of medication to treat this disease is called chemotherapy, a term fraught with unpleasant connotations. It is important to realize that cats rarely experience nausea, poor appetite, or malaise with their chemotherapy and you should not make decisions thinking the cat will be put through some kind of difficult procedure. Instead, the difficulty centers not around side effects, but around giving daily medication to a cat and regularly bringing the cat to the vet's office for injectable treatments and monitoring tests.

High grade (also called lymphoblastic) lymphoma is bad news. Only 25-50% of cats achieve remission and those that achieve it generally sustain it for only 2-9 months.

Fortunately, most feline lymphoma is the low grade (also called lymphocytic) form of lymphoma. This form behaves much more like inflammatory bowel disease, and, in fact, treatment for severe IBD is largely the same as for low-grade intestinal lymphoma. Approximately 70% of cats with this form of lymphoma will achieve remission and the median survival time is 23-30 months. Treatment for this form of lymphoma generally involves oral medication: prednisolone and chlorambucil. Many people opt to treat with these two medications based on ultrasound appearance of the GI tract, playing the odds that there will be a response. Given the expense of more accurate diagnostics, this may be a reasonable alternative approach.

Renal Lymphoma

Renal lymphoma is lymphoma in the kidneys and is one of the worst forms of lymphoma for a cat to contend with. In most cases, the cat is brought to the veterinarian for symptoms related to kidney failure or insufficiency: poor appetite, excessive water consumption, weight loss, possibly vomiting. Unlike the more typical cases of kidney insufficiency where the kidneys are shrunken and small, the cat with renal lymphoma will have symmetrically large kidneys. A needle aspirate of the kidney usually reveals lymphoma cells.

With chemotherapy, the kidney enlargement reduces and often kidney function improves until remission is lost. Median survival is only 3-6 months, though a recent case report highlighted a cat who lived many years, so long remission times are not impossible. Approximately 50% of cats with renal lymphoma are also positive for the feline leukemia virus, which carries the potential for other complicating issues.

Unlike other forms of lymphoma, renal lymphoma has a strong association with tumor spread to the nervous system (brain or spinal cord). This makes for an unpleasant complication and, unfortunately, this occurs in up to 40% of renal lymphoma cats.

Mediastinal Lymphoma

The mediastinum is not a structure with which most people are familiar. Most people know the chest cavity contains the heart and lungs and some other affiliated structures such as the windpipe (trachea) and esophagus transversing the chest to get to the stomach on the other side of the diaphragm. But these organs are not just loose and flopping around in the chest, and they are not simply held in place by gravity. The mediastinum is a sheet of connective tissue that bisects the chest, contains the heart, esophagus, trachea and other central structures similar to a vacuum-packed plastic bag. The mediastinum divides the chest into right and left halves and stabilizes the location of the organs in the middle.

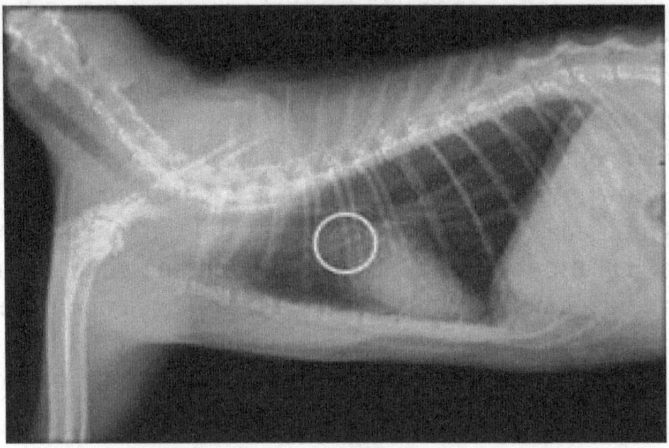

Fig. Chest radiograph of a cat with area of the thymus and mediastinal lymph nodes marked.

Of note in the lymphoma situation are the thymus gland (generally atrophied in adult animals) and the mediastinal lymph nodes. These are lymph system tissues and they become the seat of mediastinal lymphoma.

When these lymph structures fill with lymphoma, the normal fluid drainage of the chest is hampered and fluid begins to back up in the chest, creating what is called a pleural effusion. This fluid takes up space in the chest that would normally go to expanding the lungs and when the lungs cannot expand, the patient cannot breathe. Shallow rapid breaths, possibly with the mouth open, result and the patient must focus on breathing rather than normal activities, including eating. The fluid can be tapped off the chest using a needle and a sample can be sent to the laboratory for analysis. Usually lymphoma cells can be found in the fluid if the cat has mediastinal lymphoma. Treatment is chemotherapy as with the other forms of lymphoma.

Most mediastinal patients are young (less than 5 years old) adults and most (80%) are feline leukemia virus positive. Siamese cats seem to be predisposed to this form of lymphoma. In the days before the vaccine for the feline leukemia virus, mediastinal lymphoma was the most common form of feline lymphoma. Today, we rarely see it. Cats that are positive for the leukemia virus tend not to live as long because of other complications from the infection but remission is readily obtained regardless of the viral infection in most cats.

Nasal Lymphoma

Fig. Cat with nasal lymphoma before chemotherapy, and the same cat two weeks after chemotherapy.

The nasal form of lymphoma is more rare, but bears mentioning because it is the only form that is potentially localized to one area. Cats with nasal lymphoma typically have a nasal swelling, sneezing, and nasal discharge.

Localized disease (i.e. the tumor in the nose) can be treated with radiotherapy or with chemotherapy. In general, as with other forms of lymphoma, the disease is assumed to have traveled to other (distant) areas of the body. Because distant disease is addressed with chemotherapy and it is assumed that lymphoma patients have distant disease with their local disease, most nasal lymphoma patients receive chemotherapy.

Prolonged remission times (median survival times over 500 days) are possible with treatment.

2

Cancer in Animals

SKIN CANCER IN CATS AND DOGS

Many types of skin tumors, both benign (noncancerous) and malignant (cancerous), exist. Approximately 20-40% of primary skin tumors are malignant in dogs and 50-65% aremalignant in cats. Not all forms of skin cancer in cats and dogs are caused by sun exposure, but it can happen occasionally. On dogs, the nose and pads of the feet contain sensitive skin and no fur to protect from the sun. Also, cats and dogs with thin or light-colored coats are at a higher risk of sun damage over their entire bodies.

DIAGNOSIS

Typically, either cytologic or histopathologic analysis of the suspected mass is done prior to initiating treatment. The commonly used diagnostic procedures for skin tumors are fine-needle aspiration cytology and tissue biopsy.

Cytology is an important tool that can help the veterinarian distinguish a tumor from inflammatory lesions. The biopsy technique used will largely depend on the tumor's size and location. Small masses are usually completely excised and sent to the pathology lab to confirm that the surrounding healthy tissues that were excised along with the tumor do not contain any cancer cells. If the tumor is larger, a small sample is removed for analysis and depending on the results, appropriate treatment is chosen. Depending on the tumor type and its level of aggressiveness, additional diagnostic tests can include blood tests to assess the pet's overall health, chest X-rays to check for lung metastasis, and abdominal ultrasound to check for metastasis to other internal organs.

TREATMENT

The specific treatment will depend on the tumor's type, location, size, and whether the cancer has spread to other organs. Surgical removal of the tumor remains the standard treatment of choice, but additional forms of therapy such as radiation therapy, chemotherapy, or immunotherapy exist.

When detected early, skin cancer in cats and dogs can often be treated successfully. In many cases, a biopsy can remove the whole tumor, as long as the healthy tissues removed from just outside the tumor area do not contain any cancer cells.

SKIN CANCER IN HORSES

Skin cancer, or neoplasia, is the most common type of cancer diagnosed in horses, accounting for 45 to 80% of all cancers diagnosed. Sarcoids are the most common type of skin neoplasm and are the most common type of cancer overall in horses. Squamous-cell carcinoma is the second-most prevalent skin cancer, followed by melanoma.Squamous-cell carcinoma and melanoma usually occur in horses greater than 9-years-old, while sarcoids commonly affect horses 3 to 6 years old. Surgical biopsy is the method of choice for diagnosis of most equine skin cancers, but is contraindicated for cases of sarcoids. Prognosis and treatment effectiveness varies based on type of cancer, degree of local tissue destruction, evidence of spread to other organs (metastasis) and location of the tumor. Not all cancers metastasize and some can be cured or mitigated by surgical removal of the cancerous tissue or through use of chemotherapeutic drugs.

SARCOIDS

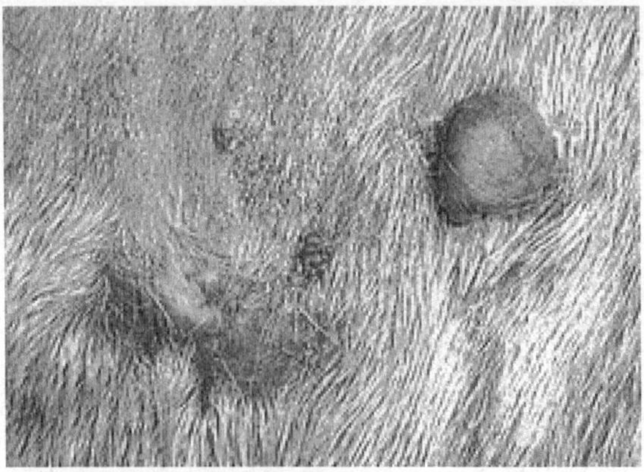

Fig. Occult (hairless area at left) and nodular (large round bump at right) forms of equine sarcoids

Sarcoids account for 39.9% of all equine cancers and are the most common cancer diagnosed in horses. There is no breed predilection for developing sarcoids and they can occur at any age, with horses three to six years old being the most common age group and males being slightly more prone to developing the disease. Sarcoids are also more prevalent in certain familial lines, suggesting that there may be a heritable component. Several studies have found an association between the presence ofBovine papillomavirus-1 and 2 and associated viral growth proteins in skin cells with sarcoid formation, but the exact mechanism that controls or induces epidermal proliferation remains unknown. However, high viral loads within cells are strongly correlated with more severe clinical signs and aggressive lesions.

Clinical signs

The appearance and number of sarcoids can vary, with some horses having single or multiple lesions, usually on the head, legs, ventrum and genitalia or around a wound. The distribution pattern suggests that flies are an important factor in the formation of sarcoids. Sarcoids may resemble warts (verrucous form), small nodules (nodular form), oval hairless or scaly plaques (occult form) or very rarely, large ulcerated masses (fibroblastic form). The occult form usually presents on skin around the mouth, eyes or neck, while nodular and verrucous sarcoids are common on the groin, penile sheath or face. Fibroblastic sarcoids have a predilection for the legs, groin, eyelid and sites of previous injury. Multiple forms may also be present on an individual horse (mixed form). Histologically, sarcoids are composed of fibroblasts (collagen producing cells) that invade and proliferate within the dermis and sometimes the subcutaneous tissue but do not readily metastasize to other organs. Surgical biopsy can definitively diagnose sarcoids, but there is a significant risk of making sarcoids worse. Therefore, diagnosis based solely on clinical signs, fine-needle aspiration or complete excisional biopsy are safer choices.

Treatment

While sarcoids may spontaneously regress regardless of treatment in some instances, course and duration of disease is highly unpredictable and should be considered on a case-by-case basis taking into account cost of the treatment and severity of clinical signs. Surgical removal alone is not effective, with recurrence occurring in 50 to 64% of cases, but removal is often done in conjunction with other treatments. Topical treatment with products containing bloodroot extract (from the plant *Sanguinaria canadensis*) for 7 to 10 days has been reported to be effective in removing small sarcoids, but the salve's caustic nature may cause pain and the sarcoid must be in an area where a bandage can be applied. Freezing sarcoids with liquid nitrogen (cryotherapy) is another affordable method, but may result in scarring or depigmentation.

Topical application of the anti-metabolite 5-fluorouracil has also obtained favorable results, but it usually takes 30 to 90 days of repeated application before any effect can be realized. Injection of small sarcoids (usually around the eyes) with the chemotherapeutic agent cisplatin and the immunomodulator BCG have also achieved some success.In one trial, BCG was 69% effective in treating nodular and small fibroblastic sarcoids around the eye when repeatedly injected into the lesion and injection with cisplatin was 33% effective overall (mostly in horses with nodular sarcoids). However, BCG treatment carries a risk of allergic reaction in some horses and cisplatin has a tendency to leak out of sarcoids during repeated dosing.

External beam radiation can also be used on small sarcoids, but is often impractical. There is a chance of sarcoid recurrence for all modalities even after apparently successful treatment. While sarcoids are not fatal, large aggressive tumors that destroy surrounding tissue can cause discomfort and loss of function and be resistant to treatment, making euthanasia justifiable in some instances. Sarcoids may be the most common skin-related reason for euthanasia.

SQUAMOUS-CELL CARCINOMA

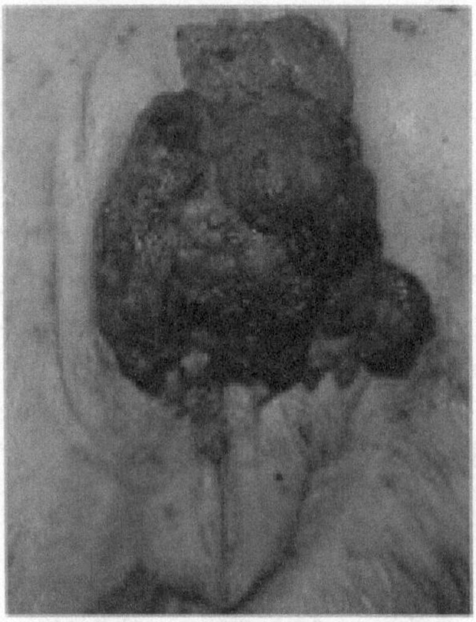

Fig. Squamous-cell carcinoma on the vulva of a gray mare, possibly arising from the clitoris. The tumor is ulcerated and has multiple necrotic areas (black spots).

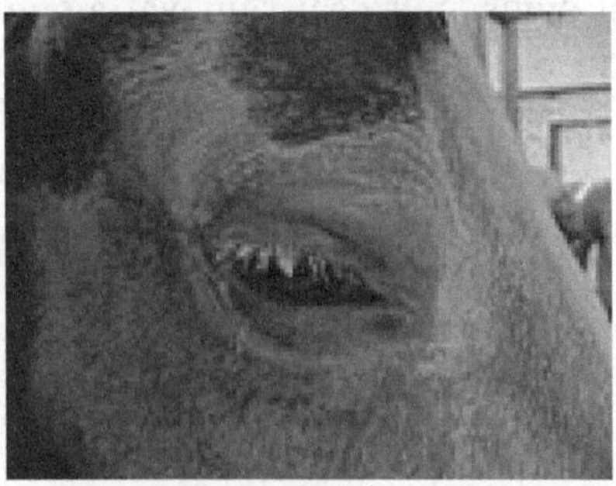

Fig. Smooth, raised plaque on upper eyelid of a Paint horse. This horse developed a carcinoma secondary to sunburn (termed solar keratosis carcinoma in situ).

Squamous-cell carcinoma (SCC) is the most common cancer of the eye, periorbital area and penis, and it is the second most common cancer overall in horses, accounting for 12 to 20% of all cancers diagnosed. While SCC has been reported in horses aged 1 to 29-years, most cases occur in 8 to 15-year-old horses, making it the most common neoplasm reported in older horses.

Carcinomas are tumors derived fromepithelial cells and SCC results from transformation and proliferation of squames, epidermal skin cells that become keratinized. Squamous-cell carcinomas are often solitary, slow-growing tumors that cause extensive local tissue destruction. They can metastasize to other organs, with reported rates as high as 18.6%, primarily to the lymph nodes and lung.

Clinical signs and predisposing factors

Tumors related to squamous-cell carcinoma (SCC) can appear anywhere on the body, but they are most often located in non-pigmented skin near mucocutaneous junctions (where skin meets mucous membranes) such as on the eyelids, around the nostrils, lips, vulva, prepuce, penis or anus. The tumors are raised, fleshy, often ulcerated or infected and may have an irregular surface. Rarely, primary SCC develops in the esophagus, stomach (non-glandular portion), nasal passages and sinuses, the hard palate, gums, guttural pouches and lung. The eyelid is the most common site, accounting for 40-50% of cases, followed by male (25-10% of cases) and female (10% of cases) genitalia. Horses with lightly pigmented skin, such as those with a gray hair coat or white faces, are especially prone to developing SCC, and some breeds, such as Clydesdales, may have a genetic predisposition. Exposure of light-colored skin to UV light has often been cited as a predisposing factor, but lesions can occur in dark skin and in areas that are not usually exposed to sunlight, such as around the anus. Buildup of smegma ("the bean" in horseman's terms) on the penis is also linked to SCC and is thought to be a carcinogen through penile irritation. Pony geldings and work horses are more prone to developing SCC on the penis, due to less frequent penile washing when compared to stallions. Equine papillomavirus-2 has also been found within penile SCCs, but has not been determined to cause SCC.

Treatment and prevention of SCC

Before treatment of squamous-cell carcinoma (SCC) is initiated, evidence of metastasis must be determined either by palpation and aspiration of lymph nodes around the mass or, in smaller horses, radiographs of the thorax. Small tumors found early in the disease process (most frequently on the eyelid) can be treated with cisplatin or radiation with favorable results. For more advanced cases, surgical removal of eye (enucleation), mass orpenile amputation can be curative provided all cancerous cells are removed (wide margins obtained) and there is no metastasis. However, young horses (usually geldings less than 8-years-old) that have a hard or "wooden" texture to SCCs on the glans penis have a very poor prognosis for treatment and recovery.

Regular washing of the penis and prepuce in males as well as cleaning the clitoral fossa (the groove around the clitoris) in mares is recommended to remove smegma buildup, which also gives the opportunity for inspection for suspicious growths on the penis or on the vulva.

MELANOMA

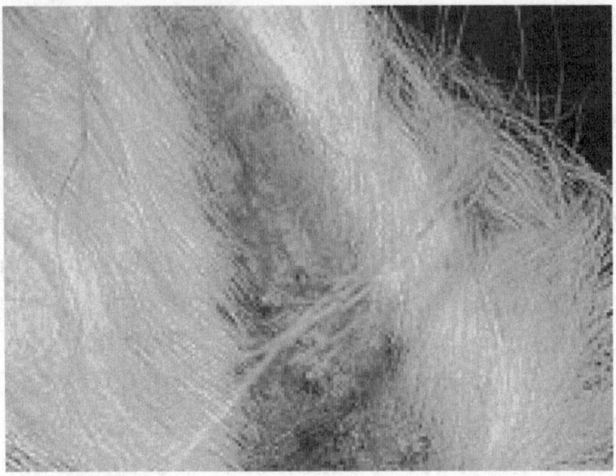

Fig. Multiple nodules at the tail-base.

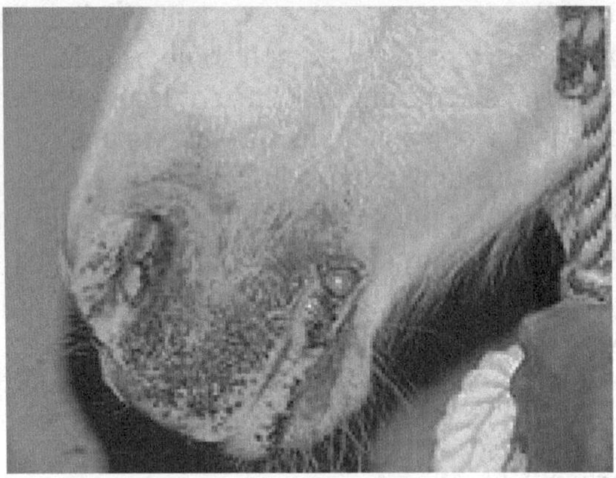

Fig. Small nodules at the lip commisure.

COMMON SITES FOR MELANOMA

Melanoma results from abnormal proliferation and accumulation of melanocytes, pigmented cells within the dermis. Gray horses over 6-years-old are especially prone to developing melanoma. The prevalence of melanoma in gray horses over 15 years old has been estimated at 80%. One survey of Camargue-type horses found an overall population prevalence of 31.4%, with prevalence increasing to 67% in horses over 15 years old. Up to 66% of melanomas in gray horses are benign, but melanotic tumors in horses with darker hair-coats may be more aggressive and are more often malignant. One retrospective study of cases sent to a referral hospital reported a 14% prevalence of metastatic melanoma within the study population. However, the actual prevalence of metastatic melanoma may be lower due to infrequent submission of melanotic tumors for diagnosis. Common sites for metastasis include lymph nodes, the liver, spleen, lung, skeletal muscle, blood vessels and parotid salivary gland.

Clinical signs

The most common sites for melanotic tumors are on the under-side of the tail near the base, on the prepuce, around the mouth or in the skin over the parotid gland (near the base of the ear). Tumors will initially begin as single, small raised areas that may multiply or coalesce into multi-lobed masses (a process called melanomatosis) over time.Horses under 2-years-old can be born with or acquire benign melanotic tumors (called melanocytomas), but these tumors are often located on the legs or trunk, not beneath the tail as in older animals.

Treatment of melanoma

Treatment of small melanomas is often not necessary, but large tumors can cause discomfort and are usually surgically removed. Cisplatin and cryotherapy can be used to treat small tumors less than 3 centimeters, but tumors may reoccur. Cimetidine, a histamine stimulator, can cause tumors to regress in some horses, but may take up to 3 months to produce results and multiple treatments may be needed throughout the horse's life. There are few viable treatment options for horses with metastatic melanoma. However,gene therapy injections utilizing interleukin-12 and 18-encoding DNA plasmids have shown promise in slowing the progression of tumors in patients with metastatic melanoma.

OTHER TYPES OF SKIN CANCER

Lymphoma

Lymphoma is the most common type of blood-related cancer in horses and while it can affect horses of all ages, it typically occurs in horses aged 4–11 years.

BLADDER CANCER IN CATS AND DOGS

Bladder cancer in cats and dogs usually is transitional cell carcinoma (TCC), which arises from the epithelial cells that line the bladder. Less often in these animals, cancer of the urinary bladder is squamous cell carcinoma, adenocarcinoma, or rhabdomyosarcoma.

SIGNS AND SYMPTOMS

The most frequent symptoms of TCC are blood in the urine, painful urination, frequent urination and/or straining to urinate. This can look very similar to an infection of the urinary system.

DIAGNOSIS

Diagnostic tests typically include complete blood tests, urinalysis, urine culture, X-rays of the abdomen and chest, and bladder imaging. The definitive diagnosis of bladder cancer will require a tissue biopsy and subsequent examination of the cells under the microscope.

TREATMENT

Because most bladder cancers are invasive into the bladder wall, surgical removal is usually not possible. The majority of transitional cell carcinomas are treated with 2 types of medications, either traditional chemotherapy or nonsteroidal antiinflammatory drug (NSAID).

BONE CANCER IN CATS AND DOGS

The most common bone tumor is called osteosarcoma, and typically affects middle-age to older dogs of large and giant breeds. Osteosarcoma is less common in cats. Osteosarcoma is an aggressive cancer that can develop in any bone of the body but the majority is seen in the limbs (e.g. long bones such as radius, humerus, femur, and tibia).

SIGNS AND SYMPTOMS

Dogs with limb osteosarcoma typically show lameness and swelling at the affected site. For other sites, dogs may show difficulty to open their mouth (if jaw bone cancer), nasal discharge (if nasal cavity bone cancer) or neurological signs (if spine bone cancer).

Diagnosis

The initial evaluation involves radiographs (X-rays) of the affected site, but the only way to confirm the diagnosis is by sampling the tissue via biopsy or needle aspiration.

Treatment

Depending on the pet's unique condition, there are several treatment options, including surgery, chemotherapy and radiation therapy. Treating the pain adequately is also of crucial importance to improve the pet's quality of life, especially if amputation is not performed.

3

Tumors of the Skin and Soft Tissues

Cutaneous tumors are the most frequently diagnosed neoplastic disorders in domestic animals, in part because they can be identified easily and in part because the constant exposure of the skin to the external environment predisposes this organ to neoplastic transformation. Chemical carcinogens, ionizing radiation, and viruses all have been implicated, but hormonal and genetic factors may also play a role in development of cutaneous neoplasms.

The skin is a complex structure composed of various epithelial (epidermis, adnexa), mesenchymal (fibrous connective tissues, blood vessels, adipose tissue), and neural and neuroectodermal tissues (peripheral nerve, Merkel cells, melanocytes), all with the potential to develop distinctive tumors. Because cutaneous tumors are so diverse, their classification is difficult and often controversial. There is also controversy regarding the criteria used to define whether a lesion that arises in the skin or soft tissues is neoplastic and, if so, whether it is benign or malignant. To avoid confusion, the following terms are used in this discussion: A hamartoma (nevus) is a localized developmental defect associated with enlargement of one or more elements of the skin. A sebaceous hamartoma, for example, refers to a localized region of the skin where sebaceous glands are extremely prominent and sometimes malformed. Although by strict definition hamartomas are present at birth, they may occasionally take a long time to reach a clinically apparent size and may not be diagnosed until an animal is mature. To confuse matters further, some lesions with clinical and histologic features of congenital hamartomas may develop in adult animals. Such "acquired" hamartomas are difficult to separate from benign epithelial and mesenchymal neoplasms. In human medical literature and some veterinary texts, the term "nevus" is used synonymously with hamartoma. A benign neoplasm is localized, noninfiltrative, and because it is surrounded by a capsule, easily excisable. A neoplasm of intermediate malignancy is locally infiltrative and difficult to excise but does not metastasize. A malignant neoplasm is infiltrative with metastatic potential.

Although cutaneous neoplasms characteristically are nodular or papular, they also can occur as localized or generalized alopecic plaques, erythematous and pigmented patches and plaques, wheals, or nonhealing ulcers. The variability in clinical presentation can make distinguishing a neoplasm from an inflammatory disease difficult; furthermore, distinguishing a benign tumor from a malignant tumor is even more subjective because sarcomas or carcinomas early in their development may palpate as discrete, encapsulated masses. To establish a definitive diagnosis, histopathology is generally required. Cytologic evaluation is very useful and should be used to exclude mast cell tumor before surgery so that wide adequate margins are planned as appropriate. For some neoplasms (eg, round cell tumors), cytology can rival or even surpass the value of histologic examination.

Therapy depends largely on the type of tumor, its location and size, and signalment of the animal. For benign neoplasms associated with neither ulceration nor clinical dysfunction, no therapy may be the most prudent option, especially in aged companion animals.

Fine needle aspiration cytology should be performed initially on all tumors to determine the type of tumor and treatment planning. For more aggressive neoplastic diseases or benign tumors that inhibit normal function or are cosmetically unpleasant, there are several therapeutic options. For most, surgical intervention with complete excision provides the best chance of a cure with the least cost and often with the fewest adverse events.

Histologic assessment of margin status is useful to predict local recurrence of cutaneous malignant tumors in dogs and cats treated by means of excision alone. However, method accuracy varies among tumor types and grades. Recurrence times suggest postsurgical follow-up should continue for ≥ 2 yr. Careful postsurgical management is recommended for animals with both infiltrated and close tumor margins.

Lumpectomy is adequate for benign lesions, but if a malignancy is suspected, the lesion should be removed with wide (3 cm) surgical margins. For tumors that cannot be completely excised, partial removal or debulking may prolong the life of the companion animal and increase the effectiveness of radiation or chemotherapy. Cryosurgery is also an option, although it is more effective for benign, superficial lesions than for malignant cutaneous neoplasms. Radiation therapy is of most value for infiltrative neoplasms not surgically resectable, or when surgical intervention would cause unacceptable physical impairment.

Chemotherapy can be used either systemically (and if appropriate, intralesionally) as a primary method for treatment of malignant neoplasms or as an adjunct to surgery or radiation therapy. In the skin, radiation is most commonly used to treat round cell tumors (eg, lymphosarcomas, mast cell tumors, transmissible venereal tumors, etc) or solid tumors that cannot be excised completely. Although generally palliative, long remissions may sometimes be obtained with radiation therapy. Other forms of therapy include hyperthermia, laser therapy, photodynamic therapy, antiangiogenic therapy, metronomic therapy, gene therapy, immunotherapy, and multimodal therapy using a combination or sequencing of various therapies.

BENIGN SOFT TISSUE TUMORS

Benign tumors of soft tissue are more common than benign tumors of bone. They can occur at almost any site, both within and between muscles, ligaments, nerves, and blood vessels. These tumors vary widely in appearance and behavior. Some tumors can be quite aggressive. Invasion of nearby tissues increases the chance of an incomplete excision and the possibility that the tumor will come back. Your surgeon may use specialized techniques and even radiation therapy to reduce the risk of recurrence.

Among the most common tumors which can be classified as benign soft tissue tumors are lipoma, angiolipoma, fibroma, benign fibrous histiocytoma, neurofibroma, schwannoma, neurilemmona, hemangioma, giant cell tumor of tendon sheath, and myxoma. Some conditions, like nodular fasciitis, are not tumors, but may require similar treatment. A small number of these tumors may be related to an underlying inherited condition.

The majority of the tumors in this category are benign tumors of fat called lipomas. Lipomas frequently occur just beneath the skin as a soft painless bulge which grows slowly over a period of months to years. These tumors represent very little risk to the patient and can be adequately treated by a "marginal excision" which is commonly described as "shelling out the tumor." This type of treatment has a high rate of success. However, not all of these lumps need to be removed. Your physician may determine that the lipoma needs only to be observed. Some of these lipomas are deep below the fascia or with the muscle group. Care needs to be taken to distinguish these deep "lipomas" from cancers and lower grade cancers.

In settings where the tumor is not a benign lipoma, the strategy of "shelling out the tumor" can spread a tumor, which might otherwise be easily treated by a more careful surgical procedure. Malignant tumors that are discovered unexpectedly can often be spread in this way.

Therefore, if the lump does not feel like a lipoma, if it is seen to grow, or causes pain, it should be assessed carefully and thoughtfully. Often the use of MRI scans can assist in determining both the location and the internal characteristics of the tumor.

A soft tissue mass, which does not have the appearance of a lipoma on MRI, needs to be approached more cautiously. Often, a minimally invasive biopsy is desirable, prior to commitment to any larger surgical procedure. The vast majority of benign soft tissue tumors can be treated with a single clean surgical procedure and minimal risk to surrounding normal blood vessels, nerves, muscle or bone. This provides the patient with the greatest possible functional recovery and the least chance of local recurrence of the tumor. In general, the orthopaedic surgeon should consider a biopsy before proceeding with an excision of a tumor. Only on rare occasions should the orthopaedic surgeon consider resection without a biopsy.

SOFT TISSUE SARCOMA SUBTYPE INFORMATION

LIPOSARCOMA

Liposarcoma is the most commonly diagnosed soft tissue tumor. They are also among the largest category of sarcomas reported. These tumors usually develop in the deep fatty tissue. They most commonly occur in the thigh, behind the knee, the groin, the gluteal area or behind the abdominal cavity (retroperitoneum). Liposarcomas are usually malignant (cancerous) and rarely are from a pre-existing lipoma, which is a non-cancerous tumor. They are most commonly found in adults between 30 and 60 years old and are slightly more common in men than women.

Tumors are usually firm and bumpy and invade surrounding tissue aggressively. Metastases (spreading) to the lymph nodes occur in approximately 10% of patients with liposarcoma. They are very different than lipomas, which are superficial and typically soft and smooth and are mobile.

FIBROSARCOMA (FIBROUS TISSUE)

Fibrosarcomas usually occur in the arms or legs or on the trunk but can occur in any of the soft tissues. They can occur around scars, muscles, nerves, tendons, and around the lining of the bone (periosteum). Local recurrence after surgery is common. Fibrosarcomas typically invade local tissues and can metastasize through the bloodstream to the lungs. Less than 5% of patients will experience metastases to the lymph nodes.

DERMATOFIBROSARCOMA PROTUBERANS

This lesion develops as a tumor sticking out of the skin usually found on the back or abdomen. In the early stages the tumor remains under the skin and the skin surface appears dark red-violet. If untreated the tumor can break through the skin forming an open wound. It has a tendency to recur but rarely metastasizes. Treatment consists of wide surgical excision.

MALIGNANT FIBROUS HISTIOCYTOMA (MFH)

Malignant fibrous histiocytomas are the most commonly diagnosed soft tissue sarcoma in patients between the ages of 50 and 70. It is twice as common in men than women.

There are four variations of MFH: storiform pleomorphic, myxoid malignant fibrous histiocytoma, malignant giant cell tumor of soft parts, and inflammatory malignant fibrous histiocytoma. Storiform and Myxoid MFH types are typically high grade (fast growing and invasive) tumors, whereas, malignant giant cell MFH and inflammatory MFH tend to be lower grade (slower growing and less invasive) tumors. While the Myxoid type is lower grade they tend to have local recurrence rate and therefore treatment usually consists of radical resection or amputation. Storiform pleomorphic MFH is typically a deep-seated large tumor of the thigh. The deeper and larger tumors found in MFH have a higher rate of metastases than the more superficial smaller tumors.

SYNOVIAL CELL SARCOMA

Synovial Cell sarcoma usually occurs in young adults. They are most commonly found in the arms or legs next to a joint. They are usually found around the joint capsule but rarely invade the joint itself. The most common site is adjacent to the knee. They are also commonly found near the foot, ankle and hand. Unlike other soft tissue sarcomas, synovial cell sarcomas are often painful. Treatment usually consists of radical excision with radiation and chemotherapy or amputation combined with chemotherapy.

EPITHELIOID SARCOMAS

Epithelioid sarcomas typically occur in the hand or foot of young adults. They appear like small nodules, which sometimes merge together. Spreading to lymph nodes occurs in approximately 20% of patients. Because of their proximity to the joints, surgical excision can be difficult and often amputation is the surgery of choice for long-term survival.

RHABDOMYOSARCOMAS (STRIATED MUSCLE TUMOR)

Cancerous tumors in the striated or skeletal muscle are one of the most common types of soft tissue sarcoma, and account for about half the soft tissue sarcomas diagnosed in children. There are several different types, including pleomorphic, alveolar, embryonal and botryoid.

Embryonal Rhabdomyosarcoma

This is the most common type of rhabdomyosarcoma, and is diagnosed most frequently in children under the age of 10. It can be found anywhere in the body, butoften occurs in the head and neck especially around the eye. Chemotherapy is very effective in these tumors. There is a high cure rate in these tumors when combination therapy is used (i.e. surgery and/or radiation combined with chemotherapy).

Alveolar Rhabdomyosarcoma

This tumor is extremely aggressive and typically occurs in adolescents and young adults. Tumors metastasize widely at an early stage. Alveolar rhabdomyosarcoma almost always is associated with a translocation between two genes (PAX3 or PAX7 and FOX01). Treatment includes radiation and chemotherapy as well as surgery for some cases. Localized alveolar rhabdo responds fairly well to treatment.

- *Botryoid Rhabdomyosarcoma:* These lesions usually occur in children at the average age of 7. They generally occur in the genital region and urinary tract. They appear as a swollen lumpy mass.
- *Pleomorphic Rhabdomyosarcoma:* This tumor occurs most commonly in people over 30 years of age. It commonly affects males on their extremities but not exclusively. Because it grows in the skeletal muscle, it usually is deep in the limb. Tumors can arise at many sites within the same muscle group and grow very rapidly. The tumor spreads through the bloodstream although in approximately 15% of cases the tumor spreads through the lymphatic system. Rhabdomyosarcoma is often widely spread at the time of diagnosis. Treatment generally consists of surgery and combination chemotherapy (multiple drugs used).

Leiomyosarcoma (Smooth Muscle Tumor) and Uterine SarcomaLeiomyosarcomas are cancerous tumors of the smooth muscle. They most commonly occur in the organs (e.g., gastrointestinal tract and the uterus). The average age of patients is 60 years. Of the tumors occurring in the GI tract 61% occur in the stomach, 29% in the small bowel, and 10% in the colon. Symptoms of GI or uterine leiomyosarcomas are significant bleeding and pain. Metastases occur in more than half of patients.

Leiomyosarcomas of the retroperitoneum and the vena cava are found mostly in women. The growth of the tumors can accelerate during pregnancy. Metastases usually occur in the lungs except in GI tumors, which often metastasize in the liver. Treatment for uterine leiomyosarcoma is total abdominal hysterectomy.

Gastrointestinal Sarcoma, otherwise known as Gastrointestinal Stromal Tumor (GIST). GIST develops in the stroma, the supporting connective tissue, of the stomach and intestines. It is now often treated with the biological agent Gleevec. Gleevec is a biologic agent that works by shutting down the activities of gtenes that apparently are needed for the tumor's growth. In particular, it targets a gene called c-kit that produces inappropriate amounts of an enzyme that promotes the tumor growth.

EXTRAOSSEOUS TUMORS

There are three types of Extraosseous tumors. They are giant cell tumors, osteosarcomas, and Ewing's sarcoma. These are bone sarcomas, which are found in the soft tissue. They have very similar characteristics as their bone sarcoma variation, Extraosseous Ewing's sarcomas respond well to radiation therapy.

MYXOMA

Myxomas occur in men and women averaging the age of 50. They are most often found in the arms and legs. Tumors can be small nodules or very large tumors. They do not metastasize; however, they do spread locally. Treatment is usually wide local excision.

MESENCHYMOMAS

Mesenchymomas can be distributed throughout the body and approximately 75% are malignant. The tumors are invasive especially when they are located in the skeletal muscle. Treatment is wide excision sometimes combined with radiation and or chemotherapy.

Vascular Sarcomas (containing many blood vessels)

Hemangioendothelioma

Hemangioendotheliomas are a very rare vascular (containing many blood vessels) tumor, which can be malignant. They occur in men and women but rarely in children. One variety of this tumor occurs in younger males and typically arises on the hand and tends to recur locally.

Angiosarcomas

Angiosarcomas represent 1% of all sarcomas. One-third of the cases occur on the skin, one-fourth of the cases occur in the soft tissue or organs such as breast, liver, heart and lungs. Tumors arising in the liver are seen in adults exposed to angiography, insecticides or plastics. Angiosarcomas of the breast typically occur in young or middle aged women. Additionally some angiosarcomas occur at the site of previous radiation therapy. Tumors metastasize widely. The most common form of treatment is radical amputation.

Hemangiopericytoma

Hemangiopericytomas are typically found in the thigh, retroperitoneum or near the kidneys. Intracranial hemangiopericytoma grow along the sinuses, tend to locally recur and can metastasize.

MALIGNANT NEURILEMOMA (MALIGNANT SCHWANNOMA & NEUROSARCOMA)

Malignant Neurilemomas usually develop in young to middle aged adults, most commonly in males. They occur in the peripheral nerves (not brain or spinal cord nerves). Half of these tumors occur in people with Von Recklinghausen's disease or multiple neurofibromatosis. The spreading of these tumors occurs in the surrounding soft tissue forming a nodular tumor. Metastases can occur through the bloodstream. Tumors can be painful, tender and sometimes nerve function is affected. Treatment is generally wide resection of the nerve in either direction. Radiation therapy and chemotherapy can be used as well.

ALVEOLAR SOFT-PARTS SARCOMA

This is a very rare tumor that typically occurs in female adolescents and young adults. It is usually a slow growing tumor found in the extremities and commonly metastasizes to the lung. The typical form of treatment is wide resection of the lesion.

KAPOSI'S SARCOMA

Kaposi's Sarcoma is a disease in which cancerous cells are found in the tissues under the skin or mucous membranes that line the mouth, nose and anus. There are three groups of patients for Kaposi's sarcoma. The first group typically includes older men of Jewish, Italian or Mediterranean heritage. This type of Kaposi's usually progresses slowly over 10-15 years. Patients commonly develop a bluish lesion on the front of the lower leg, which typically spreads to multiple lesions. After some time the disease can spread to other organs.

The second group of Kaposi's sarcoma occurs in patients who have received organ transplant. Due to the Immunosuppressive treatment following a transplant, patients immune systems are weakened thus are more susceptible to infection.

The third group of Kaposi's sarcoma is found in AIDS patients. This group is referred to as epidemic Kaposi's sarcoma. Due to the weakened immune system cause by the HIV virus, Infections and other diseases such as Kaposi's can invade the body. Kaposi's sarcoma in people with AIDS usually spreads more quickly and can be found in many parts of the body.

Radiation therapy is usually the treatment for Kaposi's; however, when lesions have spread to the organs, chemotherapy is often used as well.

TUMORS OF MELANOCYTIC ORIGIN

These tumors are most common in dogs, gray horses, and miniature pigs; uncommon in goats and cattle; and rare in cats and sheep. The terminology used to describe melanocytic lesions in veterinary medicine is different from that used in human dermatology. In animals, the terms melanocytoma and malignant melanoma are used to describe benign and malignant melanocytic proliferations, respectively. In people, a benign melanocytic proliferation (whether congenital or acquired) is called a nevus, and the term melanoma by definition refers to a malignancy (ie, in people, there are no benign melanomas). In addition, although solar injury is a common cause of melanocytic tumors in people, actinic damage is seldom associated with development of analogous tumors in domestic animals.

DOGS

Melanocytomas of the skin are diagnosed much more frequently than malignant melanomas. They most commonly develop on the head and forelimbs in middle-aged or older dogs. There may be a predilection for males. Miniature and Standard Schnauzers, Doberman Pinschers, Golden Retrievers, Irish Setters, and Vizslas are the breeds in which these tumors are most commonly recognized. They can appear as macules or patches; as papules or plaques; or as elevated, occasionally pedunculated masses. Most have a pigmented surface. Although generally solitary, lesions may be multiple, especially in the breeds at risk. These tumors are benign, and complete excision is curative.

Malignant melanomas most commonly develop in dogs somewhat older than those that develop melanocytomas. Miniature and Standard Schnauzers and Scottish Terriers are most at risk. The mucocutaneous junctions of the lips, in the oral cavity, and in the nail beds are the most common sites of development. Malignant melanomas of haired skin are rare, and most arise on the ventral abdomen and the scrotum. Males are affected more commonly than females. Most malignant melanomas appear as raised, generally ulcerated nodules that are variably pigmented depending on the melanin content; some malignant melanomas are amelanotic. When present on the mucocutaneous regions of the lip, the tumors may be pedunculated with a papillated surface; when present in the nail bed, they appear as swellings of the digit, often with loss of the nail and destruction of underlying bone, mimicking osteomyelitis. Whenever a toe is festering in an older dog, radiographs and a deep punch biopsy that includes bone are indicated for diagnosis. Canine malignant melanomas are locally aggressive and have a high potential for metastasis; however, there is considerable variability in reported survival times. Reported median survival times for dogs with oral malignant melanomas treated with definitive surgery alone are 511 days for WHO stage I disease, 324 days for stage II, and 336 days for stage III disease. A comprehensive study found that up to 32% of oral melanocytic tumors in dogs exhibited less aggressive biological behavior, which emphasizes the need to accurately classify them in clinical trials.

Treatment generally consists of complete excision; however, the infiltrative nature of the tumor may make achieving adequate margins difficult. When present on the digits, amputation is indicated; when present on the mandible, a hemimandibulectomy may allow for complete excision with adequate margins and an acceptable postsurgical cosmetic appearance and survival. Lesions at rostral locations on the maxilla or mandible allowed the best survival times of 10.9 mo in one study. Melanomas are generally considered insensitive to radiation therapy, and there is no established chemotherapeutic protocol shown to be highly effective. Survival times range from 1–36 mo, indicating that individual variations in host defense mechanisms and the aggressiveness of the tumor may play a role in establishing a prognosis. The nuclear protein Ki67 has been established as a highly sensitive and specific immunohistochemical marker for predicting the outcome of oral malignant melanoma, with indices >19.5 corresponding to lower survival times. A mitotic index of >4/10 high-power field and a nuclear atypia score of >30% also yield poor survival times. In one study of 117 dogs with digit masses, 24 had melanoma and a median survival time of 12 mo, with 42% alive at 1 yr and 13% alive at 2 yr.

A series of novel xenogeneic gene therapy vaccinations using plasmid DNA encoding human tyrosinase can induce an immune response in dogs, resulting in antibody and cytotoxic T-cell responses that may shrink their melanomas. A canine melanoma vaccine was shown to have potential therapeutic value for dogs as an adjuvant therapy for completely removed oral melanoma and for some advanced malignant melanoma cases in clinical trials in the New York area. It was conditionally licensed in the USA in 2007; however, no benefit was found in a study that followed 23 vaccinated dogs and 17 nonvaccinated dogs in Michigan. A clinical trial using masitinib in combination with the canine melanoma vaccine in dogs with melanomas expressing mutated c-KIT and another study that adds doxorubicin for nonmutated c-KIT are being evaluated for efficacy. Chondroitin sulfate proteoglycan-4 is an immunohistochemical biomarker and a potential immunotherapeutic target for canine malignant melanoma, as in people.

CATS

Cutaneous melanocytic neoplasms are uncommon and most often identified on the head (especially the pinnae), neck, and distal extremities in middle-aged or older cats. An association with the oral cavity and subungual regions is less well defined than in dogs, and a higher percentage are malignant. Excision is the treatment of choice.

HORSES

Most melanocytic neoplasms are in gray horses, in which the coat turns gray (or white) with age. They are especially common in Lipizzaners, Arabians, and Percherons, and 80% of gray or white horses of these breeds may be affected. They are generally recognized in older horses but usually begin their development when animals are 3–4 yr old. The perineum and the base of the tail are the most common sites of development, but these tumors may develop in any location, including the parotid area. The tumors are often multiple and may appear as coalescent, frequently pedunculated nodules that often extend in a linear arrangement up the tail base. They increase in size and number over time. Although most are benign, invasive variants, some with metastatic potential, can develop. Most are black on cross-section. Many gray horses have evidence of lymph node involvement; however, there is debate as to whether this represents metastasis or whether the intranodal melanocytes and melanophages represent a stimulation of extracutaneous melanocytes that are normally present in the lymph node. Treatment consists of surgical or cryosurgical removal; however, affected horses are predisposed to develop additional tumors over time. Little is known about the use of radiation or chemotherapy for the treatment of equine melanocytoma or malignant melanoma. Cimetidine was reported to be of value in controlling recurrence, but followup studies did not find benefit. Electrochemotherapy or intralesional chemotherapy with cisplatin or carboplatin after surgical debulking may be of benefit in the treatment of large or inoperable masses. Recurrent tumors do not develop resistance to cisplatin, and they can be treated a second time, sometimes with good results.

Melanocytic neoplasms of nongray horses are rare tumors usually found on the trunk and extremities of young (often <2 yr old) horses. They may represent expansion of a congenital lesion. Masses characteristically appear as solitary nodules. Most are benign; however, congenital malignant melanomas may infrequently develop. Such tumors are invasive, with little metastatic potential. Surgical excision or cryosurgery is the treatment of choice. If the tumors are benign and surgically extirpated, the prognosis is excellent. For invasive tumors, the prognosis is guarded.

PIGS

Melanocytic neoplasms of pigs are seen as congenital lesions and sporadically in adults of the Sinclair (Hormel) miniature pigs and Duroc and Duroc crosses. Selective breeding in these strains has increased the prevalence of tumors. These tumors develop both pre- and postnatally, anywhere on the body. Generally multiple, they can appear as pigmented macules or patches with smooth borders; as raised, often ulcerated pigmented lesions; or as deeper, slightly raised, blue masses. Deeply invasive melanomas are often associated with metastatic disease. The lymph nodes and lungs are the most common sites of metastasis. Not all of these tumors become invasive, and many undergo spontaneous regression associated with an intense lymphocytic infiltrate. Melanocytic lesions in pigs are not treated; because of the heritable nature of the disease, prevention by selective breeding is recommended if lesions are frequently recognized in a herd.

CATTLE

Melanocytic neoplasms in cattle develop infrequently anywhere on the body. They can be found at any age but are most commonly recognized in young animals; congenital forms have been recognized. Angus cattle appear to be predisposed. Most commonly, the tumors are large nodular masses, densely pigmented on cut surface, and benign. Excision is curative for most; however, rare malignant variants have been recognized with distant metastasis.

SHEEP

Melanocytic neoplasms in sheep are most common in middle-aged or older animals but have been recognized in neonates. They are most common in Suffolks and Angoras, in which they appear as multiple, densely pigmented dermal or subcutaneous masses. They should be considered malignant; metastasis is common.

GOATS

Melanocytic tumors in goats are rare. They are most common in middle-aged or older animals and possibly in Angoras. There may be a site predilection for the coronary band and udder. Lesions are seen as solitary or multiple masses with variable pigmentation on cut surface. Most tend to grow rapidly, and metastasis is common.

METASTATIC TUMORS

The spread of a primary neoplasm to the skin is unusual in domestic animals. It is occasionally identified in dogs; less commonly in cats; and rarely in horses, cows, sheep, goats, and pigs. Although all malignant neoplasms are capable of secondary cutaneous involvement, metastatic potential is greatest in mammary gland adenocarcinomas, squamous cell carcinomas, transitional cell carcinomas, transmissible venereal tumors, pulmonary adenocarcinomas, and angiosarcomas. Although appearance is variable, the lesions most commonly are multiple, ulcerated papulonodules. Early cutaneous metastasis is characterized by aggregates of neoplastic cells within superficial and deep dermal vessels. As these lesions evolve, they extend into the dermis and are associated with effacement of adnexa. Generally, it is difficult to distinguish the primary neoplasm based on the morphologic features of a metastatic site. This is because only a small population of cells in the primary tumor have the potential for metastasis, and these cells may have different microscopic features. In cats, pulmonary adenocarcinomas appear to preferentially metastasize to the distal extremities, and when carcinomas are diagnosed on multiple feet, radiographic examination for a primary lung tumor should be performed. Cutaneous metastasis is usually a feature of aggressive tumors and is associated with a guarded to poor prognosis.

EPIDERMAL AND HAIR FOLLICLE TUMORS

BENIGN, NONVIRUS-ASSOCIATED PAPILLOMATOUS LESIONS

Epidermal hamartomas (nevi) are rare proliferations identified only in dogs, most often in the young. The disease may be heritable in Cocker Spaniels. Grossly, epidermal nevi appear as pigmented, hyperkeratotic, vaguely papillated papules and plaques that are occasionally arranged in a linear pattern. Some forms are associated with pustules and acantholytic cells. They are benign, but their appearance is unpleasant, and the extensive hyperkeratosis is prone to secondary bacterial infection. Localized lesions can be excised; dogs with multiple lesions or lesions too large to be surgically removed may be responsive to isotretinoin or etretinate. Hyperkeratosis may be transiently controlled by use of topical keratolytic shampoos and emollients.

Congenital papillomas of foals are rare and probably a developmental defect rather than a result of papilloma virus infection. They are found anywhere on the body but most commonly on the head. Thoroughbreds may be predisposed. Present at birth, the lesions are often several centimeters in diameter, hairless, pedunculated, and exophytic, with a papillated surface reminiscent of a cauliflower. They are benign, and excision is curative.

Canine warty dyskeratomas are rare, benign neoplasms of uncertain derivation but with histologic features of follicular or apocrine neoplasms (or both). They appear grossly as verrucous papules or nodules with a keratotic, umbilicated center. Excision is curative.

BASAL CELL TUMORS AND BASAL CELL CARCINOMAS

Basal cell tumors represent a heterogeneous group of cutaneous epithelial neoplasms recognized most commonly in cats, less commonly in dogs, occasionally in horses and sheep, and seldom in other domestic animals. These neoplasms are composed of a proliferation of small basophilic cells that exhibit morphology reminiscent of the progenitor cells of the epidermis and adnexa. As these tumors have been examined more closely, evidence of differentiation (follicular, sebaceous, etc) has been discovered, giving justification for reclassification. For example, in dogs, what in the past was called a basal cell tumor is best characterized as a trichoblastoma, a tumor of hair bulb (the site of the follicle that produces the hair shaft) origin.

Some reclassification schemes have suggested that the use of the term basal cell tumor be restricted to a benign neoplasm in cats (the derivation of which has yet to be defined). Because this revised terminology is being adopted slowly, traditional terminology will be used herein. That is, a benign proliferation of basal cells will be called a basal cell tumor; a malignant proliferation will be called a basal cell carcinoma. In domestic animals, most basal cell tumors are benign and originate in the mid to deep dermis, indicating probable adnexal derivation. These features distinguish basal cell tumors in domestic animals from those in people, the latter being locally invasive (ie, they are true carcinomas) and originating in the epidermis. In addition, solar injury is a common cause of neoplasms derived from basal cells in people, but its role in inducing basal cell tumors of other animals is unknown.

Canine basal cell tumors generally develop in middle-aged to older dogs of predisposed breeds such as Wirehaired Pointing Griffons, Kerry Blue Terriers, and Wheaten Terriers. These tumors are found most commonly on the head (especially the ears), the neck, and forelimbs. Older domestic longhair, Himalayan, and Persian cats are the breeds most at risk. Feline basal cell tumors may develop almost anywhere on the body. In both dogs and cats, these tumors generally appear as firm, solitary, encapsulated, and often hairless or ulcerated nodules that may be pedunculated; they vary in size from <1 cm to >10 cm in diameter. In cats more often than dogs, these tumors are often densely pigmented, and on cut section, they can be difficult to distinguish from dermal melanocytomas. Cystic variants are also more common in cats. Although basal cell tumors are benign, they are expansive neoplasms and may be associated with extensive ulceration and secondary inflammation. Complete excision is curative.

Basal cell carcinomas are more frequently recognized in aged cats than in dogs. Persian-type breeds are predisposed. They often appear as ulcerated plaques on the head, extremities, or neck. Unlike benign basal cell tumors, these carcinomas generally have continuity with the epidermis, are locally invasive, and may be multicentric. Although evidence of vascular invasion may be identified on histologic sections, local or systemic metastasis rarely occurs. Consequently, surgical excision is the treatment of choice.

In dogs, most basal cell carcinomas have histologic evidence of cornification, a feature they have in common with squamous cell carcinomas. Therefore, they are generally called basosquamous cell carcinomas. These tumors are generally recognized in older dogs. Saint Bernards, Scottish Terriers, and Norwegian Elkhounds are most at risk. Unlike canine basal cell tumors, basosquamous cell carcinomas do not tend to develop on the head and can be found almost anywhere on the body where they have continuity with the epidermis and appear as exoendophytic nodules or plaques. These tumors are locally invasive but seldom metastasize. Surgical excision is the treatment of choice.

INTRACUTANEOUS CORNIFYING EPITHELIOMAS

Intracutaneous cornifying epitheliomas are benign neoplasms of dogs and possibly cats. As in human keratoacanthomas, these lesions most likely arise from the hair follicle and not from the interfollicular epidermis. They can develop anywhere on the body, with the back, tail, and extremities the most common sites affecting middle-aged dogs. Norwegian Elkhounds, Belgian Sheepdogs, Lhasa Apsos, and Bearded Collies are most likely to develop these tumors, with Norwegian Elkhounds and Lhasa Apsos at risk of developing generalized lesions. The most characteristic presentation is a papule or nodule with a central cornified pore that may protrude above the epidermal surface, giving the appearance of a horn; however, many of these tumors never have continuity with the epidermis and may appear solely as cornified cysts. These tumors are benign and treatment is optional, provided a definitive diagnosis has been established and there is no self-trauma, ulceration, or secondary infection. Tumor wall rupture releases keratin into surrounding tissues, evoking a pyogranulomatous and granulomatous inflammatory response. Excision is curative; however, dogs are prone to develop additional tumors over time. For animals with a generalized form of the disease, oral retinoids (eg, isotretinoin or etretinate) may be of therapeutic benefit.

SQUAMOUS CELL CARCINOMAS

Thought to arise from either the epidermis or the epithelium of the superficial (infundibular) regions of the outer root sheath of the hair follicle, squamous cell carcinomas have been recognized in all domestic animals. Although most arise without prior cause, in many species, especially in the face and ear tips of white cats and in the ventral nonpigmented skin of dogs, prolonged exposure to sunlight on sparsely haired, minimally pigmented areas of the body is a major predisposing factor. The grooming habits of cats also expose them to particulate carcinogens from cigarette smoke and flea collars. In addition, a unique form of feline squamous cell carcinoma associated with papilloma virus infection has been described.

In dogs, these are the most frequently diagnosed carcinomas arising in the skin. Two forms are recognized: cutaneous and subungual.Cutaneous squamous cell carcinomas are tumors of older dogs, with Bloodhounds, Basset Hounds, and Standard Poodles at greatest risk. Lesions commonly arise on the head, distal extremities, ventral abdomen, and perineum. Most cutaneous squamous cell carcinomas appear as firm, raised, frequently ulcerated plaques and nodules; sometimes they can be extremely exophytic and have a wart-like surface. The etiology of most of these tumors is undefined; however, some are induced by prolonged solar injury. These usually develop on ventral abdominal, preputial, scrotal, and inguinal skin in white-skinned, shorthaired breeds such as Dalmatians, Bull Terriers, Pit Bulls, and Beagles. Lesions develop in a ventral location because the poorly haired skin has minimal shielding from ultraviolet radiation, many animals sun themselves lying on their backs, and perhaps because solar radiation reflects from the ground. Before a carcinoma develops, animals acquire focal zones of lichenification, hyperkeratosis, and erythema known as solar keratosis(solar dermatosis, actinic keratosis, senile keratosis).

Subungual squamous cell carcinomas are most commonly found in older Giant and Standard Schnauzers, Gordon Setters, Briards, Kerry Blue Terriers, Standard Poodles, and Scottish Terriers. Generally, all are darkhaired breeds, and a dark coat color has been associated with development of subungual squamous cell carcinomas that arise from the nailbed epithelium and invade the medullary and cortical bone of P3. They may arise on multiple digits, often appear on different extremities, and may metastasize via lymphatics to regional lymph nodes and lungs in as many as 13% of cases. Females have a slight predilection, and both fore- and hindlimbs are equally predisposed to tumor development. In cats, cutaneous squamous cell carcinomas most commonly develop in conjunction with chronic solar injury. Consequently, they usually develop on the pinnae, frontal ridges, eyelids, nose, or lips of cats that have white skin in these regions. There is no breed or sex predilection. As in dogs, solar keratosis or carcinoma in situ (early superficial stage) often precedes development of a malignant tumor. Recently, coat-associated particulate carcinogens from exposure to cigarette smoke and flea collars have been identified as risk factors for cats with oral squamous cell carcinoma. Lesions not caused by sun exposure may develop on the digits and are presumed to be metastatic from primary squamous cell carcinomas of the lung, but primary subungual forms are uncommon in cats.

Cutaneous squamous cell carcinomas are the most common malignant neoplasm in horses. They generally develop in adult or aged horses with white or part-white coats; breeds at risk include Appaloosa, Belgian, American Paint, and Pinto.

Although they can arise anywhere on the body, these tumors most commonly arise in nonpigmented, poorly haired areas near mucous membranes. Thus, the periorbital regions, lips, nose, anus, and external genitalia (especially the penile sheath) are sites most likely to be affected.

In cattle, these tumors are most common in breeds with white hair and poorly pigmented skin (especially Holsteins and Ayrshires) and, as in horses, develop around the mucous membranes, usually at the mucocutaneous junctions, particularly the periocular and vulvar regions. In India, squamous cell carcinomas of the horn core are common in aged bullocks. The most common cause is actinic injury. Solar keratoses often precede development of an invasive tumor; genetic factors, immunodeficiency, and viruses may also play a role.

In sheep, squamous cell carcinomas are of economic significance in some parts of the world. In a study in Australia, they were responsible for more than one-third of all condemnations before slaughter. The Merino breed is most at risk, and females more so than males. The most common sites are the poorly haired skin of the ears, lips, muzzle, and the vulvar lips after they have been externalized by a Mules operation to prevent fly strike. Tumors at these sites develop in conjunction with solar injury, which is heightened when animals ingest photosensitizing plants. Tumors of the ears also occur more frequently after a procedure such as ear tagging. Squamous cell carcinomas can develop from follicular cysts on sites not commonly exposed to sunlight.

In goats, squamous cell carcinomas develop most frequently in females, in which tumors develop on the perineal and vulvar regions and on the skin of the teats and udders. Both males and females can develop sun-induced tumors on the ears. Although Angoras are most at risk, Saanan goats occasionally develop squamous cell carcinomas on the udder in association with papillomas. The role papilloma viruses play in tumor progression is considered contributory.

Squamous cell carcinomas are extremely uncommon in swine.

Most squamous cell carcinomas are solitary lesions; however, multiple tumors may develop in conjunction with solar injury in the "field cancerization" model. They appear as endophytic or exoendophytic lesions, the former as raised, irregular dermal masses with an ulcerated surface, and the latter as raised, irregular dermal masses covered by a papillated epidermis. Cats initially exhibit small crusting facial sores that do not heal. The lesions often persist for months before defects appear on the ear tips, nares, and eyelids. Subungual squamous cell carcinomas of dogs are first identified by lameness or nail malformation, an infection that mimics chronic osteomyelitis, or loss of the claw of the affected digit. In cattle with involvement of the horn, the first sign is distorted growth.

Squamous cell carcinomas are characteristically invasive into adjacent soft and bony tissues. Infrequently, in cattle, they regress spontaneously. In small animals, longterm survival and the likelihood of metastasis are correlated with histologic differentiation. Well-differentiated tumors are slowly progressive or remain localized; undifferentiated tumors are more likely to metastasize or recur within 20 wk of excision. In general, failure of treatment is due to late diagnosis and lack of control of local disease rather than metastasis.

For dogs and cats, surgical excision, such as amputation of the involved digit or pinnae or nosectomy, is the treatment of choice, and margins of at least 2 cm are recommended. One review of 117 digit masses in dogs found that 25% of the lesions were squamous cell carcinomas and 66% were subungual lesions. These dogs had a 95% 1-yr survival after amputation; however, if the lesion originated in other parts of the digit, 1-yr survival was 60%. Excision may be combined with radiation or chemotherapy. Feline squamous cell carcinomas are more radiosensitive than their canine counterparts. Still, the 1-yr survival rate is <10% for invasive neoplasms.

Cryosurgery and hyperthermia can be very helpful for local therapy especially in early (carcinoma in situ) lesions, but controlled studies have not been done to determine their effectiveness. Intralesional implant chemotherapy with 5-fluorouracil (dogs only), cisplatin (dogs only), or carboplatin along with retinoids and photodynamic therapy has been used with variable success. Intratumoral injection of nasal planum squamous cell carcinomas in cats using carboplatin in a water-sesame seed oil emulsion resulted in a 70% general response with 1-yr progression-free survival rate of ~50%. In dogs with multiple ventral actinic keratoses, topical dinitrochlorbenzene or 5-fluorouracil (5%) may be of benefit. Cats should not be given any form of 5-fluorouracil therapy. More recently, the use of topical immunomodulating antineoplastics such as imiquimod cream applied topically on lesions twice daily for 2 wk creates a local application site inflammatory reaction and often resolves lesions. Imiquimod stimulates Toll-like receptor 7 and dendritic cells, modifying immune responses to destroy the targeted carcinoma cells. The response may be enhanced by the application of 6 joules of cold laser therapy to the lesions every 2 wk until regression. Limiting exposure to ultraviolet radiation may help prevent solar-induced squamous cell carcinomas in dogs and cats. This may be accomplished by using UV window screens, sunscreen, and keeping companion animals indoors during hours of peak sunlight between 10 AM and 20 PM. Tattoos, magic markers, and sunscreen are used with variable success.

In horses, radiotherapy using surface or interstitial brachytherapy is the treatment of choice for squamous cell carcinomas. Other options include Sr orIr implants, wide surgical excision (especially for neoplasms of the third eyelid, penis, and prepuce), and cryosurgery. Immunotherapy, with either an autogenous vaccine made from the tumor tissue suspended in Freund's adjuvant, or nonspecific immunomodulation using *Corynebacterium parvum*, has had some success in treating ocular or horn core squamous cell carcinomas in cattle.

Feline multicentric squamous cell carcinoma in situ (feline Bowen disease) is a disease of aged (>10 yr old) cats and may be associated with immunosuppression. There is no defined breed or sex predilection. Clinically, lesions appear as multiple discrete, erythematous, black or brown hyperkeratotic plaques and papules. Lesions are nonpruritic, and ulceration is uncommon. Their development is associated with the presence of a papilloma virus. The term in situ refers to a malignant proliferation of epidermal and follicular outer sheath cells that are not invasive into the underlying dermis. Unfortunately, lesions may progress over time into an invasive carcinoma. Metastasis is extremely uncommon. These lesions usually develop in systemically ill or immunosuppressed cats and are believed to be virally induced.

KERATINIZED CUTANEOUS CYSTS

Most keratinized cutaneous cysts are malformations of the hair follicle. They are common in dogs; occasionally identified in cats, horses, goats, and sheep; and rare in cattle and pigs. Excision is the treatment of choice. Vigorous squeezing of these lesions is contraindicated, because it often incites a severe foreign body inflammatory response due to the release of keratin into surrounding tissues.

Infundibular follicular cysts (epidermoid cysts, epidermal inclusion cysts, erroneously called sebaceous cysts) are the most common. They are a cystic dilatation of the upper portion of the outer sheath of the hair follicle (the infundibulum) lined by a layer of stratified cornifying epithelial cells indistinguishable from the epidermis. These cysts vary in size from 2 mm to >5 cm (lesions <5 mm in diameter are often called milia). The only domestic animals identified at risk are Merino sheep, in which these cysts are often multiple and may progress to squamous cell carcinomas. As with all follicular cysts, these are usually solitary, papular to nodular lesions that are freely movable. They are generally partially compressible on palpation and occasionally have a small opening through the epidermis from which the cystic contents can be extruded. On cut surface, they are filled with a gray, brown, or yellowish, granular, "cheesy" material that is lumenal keratin.

Isthmus catagen cysts (trichilemmal cysts, pilar cysts, cystic intracutaneous cornifying epithelioma) are follicular cysts that have the keratinization pattern of the lower portion of the outer root sheath. They have been definitively identified only in dogs and rarely in cats.

Matrix cysts are follicular cysts in which the wall resembles the epithelium of the hair bulb (the matrix portion of the hair follicle) and the inner root sheath. They occur predominantly in dogs and cats. Many progress to pilomatricomas.

Hybrid cysts (panfollicular cysts) are follicular cysts that have a combination of the characteristics of epidermal inclusion, trichilemmal, and matrix cysts and that are found predominantly in dogs and cats. Many progress to trichoepitheliomas.

Dermoid cysts are congenital malformations found most commonly on the dorsal midline of the head or along the vertebral column. They are most commonly identified in Boxers, Kerry Blue Terriers, and Rhodesian Ridgeback dogs; Thoroughbred horses; and possibly Suffolk sheep. Typically multiple, they differ from other follicular cysts in that on cut surface they contain fully formed hair shafts. They are arguably the only true epidermal inclusion cysts, because they most likely represent an embryonal invagination of the epidermis with associated adnexa. These adnexa are responsible for the hair shafts within the cyst lumens.

Keratomas are cystic lesions in the hoof wall of the toe or, less frequently, the quarter or heel in simple or cloven-hoofed animals. They often occur secondary to a traumatic injury. Although often asymptomatic, they commonly induce lameness and deformity of the hoof wall or sole and may be associated with distal phalangeal lysis. Keratomas are seldom >5 cm in diameter and contain white to brown laminated keratin, often with a necrotic center associated with secondary inflammation. When lameness is present, surgical excision and curettage of the underlying bone, if affected, is the treatment of choice.

Dilated pores of Winer are rare, hair-follicle neoplasms recognized only in aged cats. Males may be predisposed. These lesions most often develop on the head. Clinically, they appear as solitary, dome-shaped lesions with the appearance of a giant comedo. Compact keratin may protrude through (above) the surface, giving them the appearance of a cutaneous horn. These lesions are benign, and complete excision is curative.

TUMORS OF THE HAIR FOLLICLE

The hair follicle is a complex structure composed of eight different epithelial layers. Hair-follicle tumors display a similar complexity, and much work needs to be done to characterize them further. They are most common in dogs, less frequent in cats, and rare in other domestic animals.

Tricholemmomas are rare, benign, hair-follicle neoplasms of dogs, most commonly found on the head. Poodles may be predisposed. These tumors are derived from the lower portion of the outer root sheath and often have areas of transition into basal cell tumors. They have little in common with a tumor of the same name in people that represents an old wart. They appear as firm, ovoid masses, 1–7 cm in diameter, that are encapsulated but expand over time. Excision is curative.

Trichofolliculomas are extremely rare follicular tumors of dogs composed of the inferior and isthmic regions of multiple abortive follicles that extrude their lumenal contents into a dilated abnormal cystic infundibulum. Too few have been recognized to determine age, breed, or sex predilection. Considered by some to be more a hamartoma than a true neoplasm, these tumors are benign, and complete surgical excision is curative.

Trichoepitheliomas are cystic hair follicle neoplasms of dogs and, less commonly, cats, in which all elements of the hair follicle (infundibulum, isthmus, and inferior portions) and the patterns of cornification they produce are represented. The epithelium and cornification of the infundibular and isthmic portions predominate. Benign and malignant forms are recognized. In dogs, these lesions can be seen at any age but are found most commonly during late middle age.

Many breeds are predisposed, including Basset Hounds, Bull Mastiffs, Irish Setters, Standard Poodles, English Springer Spaniels, and Golden Retrievers. There is no defined sex predilection. Tumors can develop anywhere on the body but are most common on the trunk in dogs and on the head, tail, and extremities in cats. Benign forms appear as palpably encapsulated cystic nodules (1–5 cm in diameter) in the dermis and subcutaneous fat. Expansion of cysts or self-trauma may induce ulceration associated with extrusion of lumenal keratin that appears as a condensed, yellow, granular, "cheesy" material. Excision is curative; however, animals that develop one such tumor are prone to develop additional lesions at other sites. This is especially true for Basset Hounds and English Springer Spaniels.

Malignant trichoepitheliomas are much less common than benign trichoepitheliomas and are differentiated by their local invasiveness; continuity with the epidermis; and association with extensive inflammation, necrosis, and fibrosis. Metastasis is uncommon. Wide surgical excision is the treatment of choice and is often curative in those tumors that are invasive but have minimal metastatic potential.

Pilomatricomas (hair matrix tumors, calcifying epitheliomas of Malherbe) are cystic hair follicle neoplasms recognized almost exclusively in dogs. Unlike trichoepitheliomas, in which all elements of the follicle are represented, in pilomatricomas only the cells of the matrix region of the inferior part of the hair follicle and the cornification patterns they produce (hair shaft and inner root sheath) are present. Benign and malignant forms are recognized. Benign tumors are most common on the trunk of middle-aged dogs. Kerry Blue and Wheaten Terriers, Bouviers des Flandres, Bichons Frises, and Standard Poodles are most at risk. Grossly, these tumors are indistinguishable from trichoepitheliomas, but their cystic contents are often gritty because of mineralization. Excision is the treatment of choice. As in trichoepitheliomas, when one such lesion develops, additional lesions often develop over time.

Malignant pilomatricomas (malignant hair matrix tumor, matrical carcinoma) are rare and have been identified most often in dogs. They are a tumor of old dogs and grossly characterized as solitary or multinodular, variably cystic tumors that are often firmly attached to subjacent soft tissues. Because they are invasive, they are difficult to excise, and recurrence is common after attempts at surgical excision. They often metastasize to draining lymph nodes and internal organs, especially the lungs. Aggressive surgery is recommended. It is unknown whether they respond to radiation or chemotherapy.

CUTANEOUS APOCRINE GLAND TUMORS

Sweat glands are of two types: apocrine and eccrine. Apocrine glands are tubular glands with a coiled secretory portion and a long, straight duct that flows into the follicular infundibulum. In domestic animals, all hair follicles have apocrine glands. Apocrine glands in dogs and cats are also present in association with the anal sac, and modified apocrine glands, known as ceruminous glands, are present in the external auditory meatus. In most mammals, apocrine glands produce an odiferous, oily compound that is a sexual attractant, a territorial marker, and a warning signal. In horses and cattle, these glands play a role in thermoregulation by producing sweat.

Apocrine gland tumors and malformations are most common in dogs and cats. Three diseases of apocrine glands of haired skin have been characterized.

Cystic apocrine gland dilations (apocrine gland cysts, cystic apocrine gland hyperplasia, apocrine cystomatosis) are best characterized as hamartomas. Two forms exist: a cystic form in which one or more cysts develop in the mid to upper dermis with a poor association with hair follicles, and a more diffuse form characterized by cystically dilated apocrine glands associated with multiple hair follicles in nontraumatized skin. Both are found in middle-age or older dogs and, less commonly, cats. The head and neck are the most common sites where these lesions develop. In both species, lesions appear as fluctuant dermal cysts or as translucent bullae. Complete excision is curative; however, this may be difficult to accomplish in the more diffuse form.

Apocrine gland adenomas are diagnosed almost exclusively in dogs, cats, and rarely horses. Two types are recognized based on whether their histologic appearance primarily resembles the secretory or ductular portion of the apocrine gland. Apocrine adenomas resemble the secretory region of the apocrine glands. They are found in older dogs and cats. Great Pyrenees, Chow Chows, and Alaskan Malamutes are the most commonly affected breeds. The head, neck, and extremities are the most frequent sites of development. In cats, apocrine adenomas are more likely to be seen in males, and no breed appears at greater risk than any other. The vast majority of apocrine adenomas occur on the head, especially the pinnae. In horses, no age, sex, or breed association is known. The pinnae and vulva are the most likely regions to develop these tumors. In all species, these tumors appear as firm to fluctuant cysts, seldom >4 cm in diameter. They contain varying amounts of clear to brownish fluid. In cats, the luminal fluid may be darkly pigmented, and apocrine cysts can be confused clinically with melanocytomas, especially when present on the inner aspect of the ears. Apocrine ductular adenomas are less common. They are found in older dogs and cats and are putatively derived from or show differentiation toward apocrine ducts. In dogs, these tumors are most commonly recognized in Peekapoos, Old English Sheepdogs, and English Springer Spaniels. They are often smaller, firmer, and less cystic than apocrine adenomas. Because they often consist of a large population of basal cells and because evidence of ductular differentiation can be extremely subtle, these tumors are often diagnosed histologically as basal cell tumors. Apocrine adenomas and apocrine ductular adenomas are benign, and complete surgical excision is curative.

Apocrine gland adenocarcinomas of haired skin are rare in all domestic animals but most frequently identified in older dogs and cats. In dogs, Treeing Walker Coonhounds, Norwegian Elkhounds, German Shepherds, and mixed-breed dogs are most at risk; in cats, Siamese may be predisposed. In both species, this tumor most commonly arises in axillary and inguinal regions—sites that allow it to be easily confused clinically and histologically with mammary gland ductular adenocarcinomas. Apocrine gland adenocarcinomas generally are larger than adenomas and have a variable clinical appearance ranging from fibrotic dermal nodules to ulcerated plaques. They are locally invasive and frequently metastasize to draining lymph nodes. Less commonly, skin and lung metastasis may occur. Complete surgical excision is the treatment of choice. Little is known about response to adjunct chemotherapy.

APOCRINE GLAND TUMORS OF ANAL SAC ORIGIN

Apocrine gland tumors of anal sac origin have been definitively identified only in dogs, although anecdotal reports suggest they may also occur in cats.

Older English Cocker and Springer Spaniels, Dachshunds, Alaskan Malamutes, German Shepherds, and mixed-breed dogs are most at risk. Unlike hepatoid gland tumors, there is no sex predilection. They most commonly appear as deep, firm, nodular masses near the anal sac. As these lesions grow, they may compress the rectum and induce constipation. Some of these tumors are associated with paraneoplastic hypercalcemia, which causes anorexia, weight loss, polyuria and polydipsia, and mineralization of renal tissue with increased BUN and creatinine values. These tumors are often highly infiltrative into the pelvic canal and commonly (90%) metastasize to the sublumbar lymph nodes or to distant internal organs (40%). Wide surgical excision, including involved lymph nodes, is the treatment of choice. Even if the tumor cannot be totally resected, debulking can be of value in dogs with pseudohyperparathyroidism, because the hypercalcemia is related to the production of parathormone-like hormone from the total tumor volume. Adjunct chemotherapy along with tyrosine kinase inhibitors, metronomic chemotherapy, and radiation therapy may also be of benefit to increase time to progression, but few dogs are reported to live >1 yr after the tumor has been recognized.

ECCRINE GLAND TUMORS

Eccrine glands are the coiled, tubular, sweat glands present on the footpads of carnivores, the frog of ungulates, the carpus of pigs, and the nasolabial region of ruminants. Tumors derived from these glands are extremely rare and have been identified only on the footpads of dogs and cats. Most are malignant and invasive. These tumors are reported to have a high potential to metastasize to draining lymph nodes.

SEBACEOUS GLAND TUMORS

Tumors and tumor-like conditions of sebaceous glands are common in dogs, infrequent in cats, and rare in other domestic animals. Based on morphologic more than on behavioral features, four categories of benign sebaceous gland proliferations have been described. In people, in which a roughly similar classification scheme is traditionally used, it has been proposed that all benign sebaceous gland tumors be called sebaceomas.

Sebaceous gland hamartomas are solitary lesions reported only in dogs. These lesions are distinguished from sebaceous gland hyperplasias and adenomas because they are linear or circumscribed, several centimeters in length or diameter, and usually identified shortly after birth.

Sebaceous gland hyperplasias (senile sebaceous hyperplasias) represent a senile change in dogs and cats. In dogs, Manchester, Wheaten, and Welsh Terriers are at greatest risk. In cats, there is no breed predilection, but females develop these lesions more frequently than males. In both species, the skin of the head and abdomen are affected most commonly. Sebaceous hyperplasias commonly appear as papillated masses seldom >1 cm in diameter, often with a shiny, keratotic surface.

Sebaceous gland adenomas are seen in all domestic animals but are so common in older dogs and cats they can be considered primarily a small animal neoplasm. Coonhounds, English Cocker Spaniels, Cocker Spaniels, Huskies, Samoyeds, and Alaskan Malamutes are the canine breeds most likely to develop these tumors; Persians are the feline breed most predisposed. In dogs, these tumors frequently are clinically indistinguishable from sebaceous hyperplasias, but they tend to be larger (typically >1 cm). They are often multiple and may develop anywhere on the body but are commonly found on the head. Sebaceous adenomas may be covered with a serocellular crust and exhibit pleocellular inflammation and superficial pyoderma. Sebaceous gland epitheliomas are a variant of sebaceous adenoma distinguished by lobules composed primarily of basal progenitor cells rather than mature sebocytes. Because they often have irregular lobules that extend into the deep dermis, they can occasionally be confused with sebaceous carcinomas. These tumors are found in older dogs and rarely in cats. They appear as ulcerated nodules that may be several centimeters in diameter. A papillated epidermal surface and pigmentation are variable findings.

Sebaceous gland adenocarcinomas are rare in domestic animals. They are recognized almost exclusively in dogs and cats, generally in middle-aged or older animals. Cavalier King Charles Spaniels; Cocker Spaniels; and Scottish, Cairn, and West Highland White Terriers are most at risk. Male dogs and female cats may be predisposed. These lesions are often ulcerated and may be indistinguishable from sebaceous epitheliomas or other cutaneous carcinomas. They are locally infiltrative and may metastasize to regional lymph nodes late in the disease.

Once a diagnosis is established, treatment is optional for benign sebaceous gland tumors unless they are secondarily inflamed and infected. For malignant adenocarcinomas, excision is the treatment of choice, but complete removal can be difficult because of the infiltrative nature of this tumor; adjunct radiotherapy may be required. Even benign sebaceous gland growths recur if remnants are left at the surgical site. In addition, animals that develop one sebaceous gland hyperplasia or adenoma often develop new lesions at other sites over time. No established protocol of chemotherapy for any of these lesions has been defined. Oral retinoids may prevent recurrence of sebaceous hyperplasia, but their use remains poorly defined, and consultation with a veterinary oncologist or dermatologist is strongly recommended.

HEPATOID GLAND TUMORS

These common neoplasms arise from modified sebaceous glands that are most abundant in the cutaneous tissues around the anus but may also be present along the ventral midline from the perineum to the base of the skull, the dorsal and ventral tail, and in the skin of the lumbar and sacral regions. Because androgens stimulate the development of hepatoid glands, the incidence of proliferative lesions of hepatoid glands in intact, male dogs is three times that in females.

Benign hepatoid gland tumors are divided into hepatoid gland hyperplasias and adenomas; however, as with benign sebaceous gland tumors, there is a continuum from hyperplasia to adenoma. Here, they will be considered as a single entity. Hepatoid gland adenomas are most common in aged dogs. Siberian Huskies, Samoyeds, Pekingese, and Cocker Spaniels are most commonly affected. Tumors may develop at any site where hepatoid glands are present, but 90% are found in the perianal region. Grossly, they appear as one or (more commonly) multiple intradermal nodules 0.5–10 cm in diameter. Larger lesions commonly ulcerate, and hemorrhagic, pertinacious material can often be extruded with local pressure. Large tumors can compress the anal canal and make defecation difficult. Up to 95% of male dogs respond completely to castration; in those that do not, the pituitary-adrenal axis should be evaluated and, if no abnormality is detected, the dog should be reevaluated for the presence of a low-grade hepatoid gland adenocarcinoma. Excision may be used concurrently to remove extremely large or ulcerated tumors that have become secondarily infected. Surgery is the treatment of choice for females with hepatoid gland adenomas but may need to be repeated, because recurrence is common. Radiation therapy is also an option and has a 2-yr cure rate of 69% for benign tumors. Cryosurgery is another therapeutic alternative, but because of the complication of fecal incontinence, should be used only when tumors are not amenable to surgical intervention. Diethylstilbestrol has been used in the past as an alternative to castration, but because of severe adverse effects, including aplastic anemia and cystic prostatic hyperplasia, it should be used with extreme caution, if at all. Antiandrogens may have a role as an alternative to castration.

Hepatoid gland adenocarcinomas are uncommon canine neoplasms that generally appear as nodular lesions affecting the perianal region. These tumors are found in male dogs 10 times more commonly than in females. Siberian Huskies, Alaskan Malamutes, and Bulldogs are most likely to develop this tumor. Histologic evaluation is the best means of diagnosis; however, there is debate about how to distinguish low-grade malignant tumors from hepatoid adenomas, because well-differentiated forms can be confused with adenomas, and anaplastic forms can be confused with apocrine gland adenocarcinomas of anal sac origin. These tumors have metastatic potential and often spread to regional lymph nodes. Treatment consists of wide surgical excision including involved lymph nodes and, possibly, subsequent radiation. These tumors are generally not responsive to castration or to estrogen therapy; however, some studies show that the use of masitinib and other tyrosine kinase inhibitors may overcome chemoresistance, inhibit the proliferation of tumor cells, and prevent the emergence of metastasis. These findings may increase the benefit of combining targeted agents with various forms of chemotherapy such as oral piroxicam and capecitabine or with metronomic chemotherapy to prevent or treat local recurrence and metastatic disease. The overall prognosis is guarded.

PRIMARY CUTANEOUS NEUROENDOCRINE TUMORS

In veterinary medicine, the diagnosis of tumors derived from Merkel cells (tactile, neurosecretory cells of epithelial derivation present in the basal cell layer of the epidermis) has fallen in disfavor, and most pathologists consider this tumor to be an extramedullary plasmacytoma. Merkel cell tumors most likely develop in animals but are not recognized as such.

PAPILLOMAS

Papillomatosis, young Staffordshire pit bull

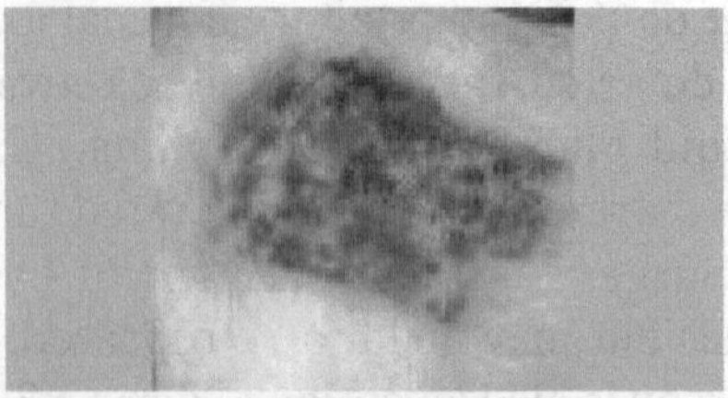

Viral-induced squamous papilloma, shoulder, 12-yr-old Bulldog

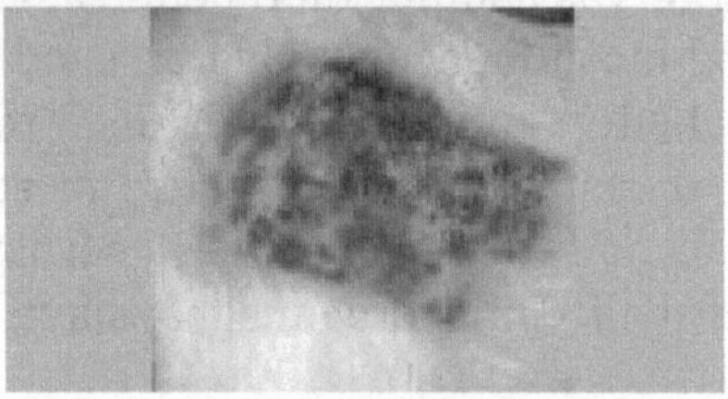

Papilloma viruses are small, double-stranded DNA viruses of the Papovaviridae family. Some mammals have several distinct papilloma viruses—people have >20; cattle, 6; dogs, 3; and rabbits, 2. Different papilloma viruses often have considerable species, site, and histologic specificity. The virus is transmitted by direct contact, fomites, and possibly by insects. Papillomas have been reported in all domestic animals, birds, and fish. Multiple papillomas (papillomatosis) of skin or mucosal surfaces generally are seen in younger animals and are usually caused by viruses. Papillomatosis is most common in cattle, horses, and dogs. Single papillomas are more frequent in older animals, but they may not always be caused by viral infection.

When lesions are multiple, they may be sufficiently characteristic to confirm the diagnosis; however, there are many simulants of warts, and a definitive diagnosis requires identification of the virus or its cytopathic effects on individual cells—a change known as koilocytic atypia or koilocytosis.

In cattle, warts commonly are found on the head, neck, and shoulders, and occasionally on the back and abdomen. The extent and duration of the lesions depend on the type of virus, area affected, and degree of susceptibility. Warts appear ~2 mo after exposure and may last e"1 yr. Papillomatosis becomes a herd problem when a large group of young, susceptible cattle becomes infected. Immunity usually develops 3–4 wk after initial infection, but papillomatosis occasionally recurs, probably due to loss of immunity.

Although most warts appear as epidermal proliferations that have a keratotic surface resembling a cauliflower (verruca vulgaris), some bovine papilloma viruses (bovine papilloma types 1 and 2) involve dermal fibroblasts and keratinocytes and appear as a papulonodule with a warty surface. Such fibropapillomas may involve the venereal regions, where they can cause pain, disfigurement, infection of the penis of young bulls, and dystocia when the vaginal mucosa of heifers is affected.

A form of persistent cutaneous papillomatosis with smaller numbers of papillomas may be seen in herds of older cattle. A bovine papilloma virus has been demonstrated in bladder tumors associated with bracken fern ingestion and in upper GI tract papillomas of cattle in Scotland. It is believed that the papilloma virus acts as a co-carcinogen. When bovine papilloma virus type 1 or 2 is injected into the skin of horses, a dermal tumor similar to equine sarcoid develops.

Papillomas, horse

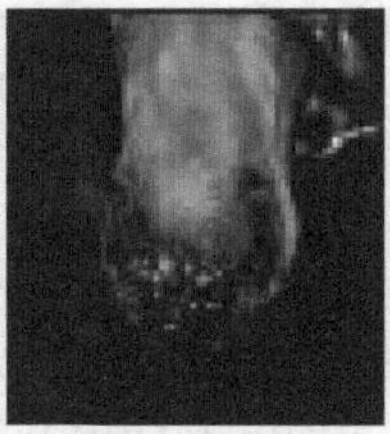

In horses, small, scattered papillomas develop on the nose, lips, eyelids, distal legs, penis, vulva, mammary glands, and inner surfaces of the pinnae, often secondary to mild abrasions. They can be a herd problem, especially when young horses are kept together, but regress in a few months, as a foal's immune system matures. When they develop in older horses, they often persist for >1 yr. So-called aural plaques are also thought to be a flat form of papilloma (verruca planum). Equine papillomas are disfiguring but benign. They should be distinguished from verrucous equine sarcoid.

In dogs, three clinical presentations of canine papilloma virus infection have been described. The first iscanine mucous membrane papillomatosis, which primarily affects young dogs. It is characterized by the presence of multiple warts on oral mucous membranes from lips to (occasionally) the esophagus and on the conjunctival mucous membranes and adjacent haired skin. When the oral cavity is severely affected, there is interference with mastication and swallowing. A viral etiology has been clearly established for these lesions. Azithromycin therapy has been shown to speed up regression in dogs. The second presentation is cutaneous papillomas, which are indistinguishable from the warts that develop on or around mucous membranes. However, they are more frequently solitary and develop on older dogs. Cocker Spaniels and Kerry Blue Terriers may be predisposed. A definitive viral etiology has not been established, and lesions may be confused with cutaneous tags. A syndrome characterized by papillomatosis of one or more footpads has also been described. Clinically, lesions appear as multiple, raised keratin horns. A viral etiology has been suggested but not proved. The third presentation iscutaneous inverted papillomas, which have more in common clinically with intracutaneous cornifying epitheliomas. In this disease of young, mature dogs, lesions most commonly develop on the ventral abdomen, where they appear as raised papulonodules with a keratotic center. Infrequently, viral papillomas in dogs may progress to invasive squamous cell carcinomas.

In cats, papilloma virus infection appears most commonly as a multicentric squamous cell carcinoma. The typical warty lesions associated with papilloma virus infection in most species are not present. Papillomas may affect the skin of goats, and infection on the teats has been reported to induce malignant transformation. In sheep, papillomas are rare and most commonly appear as fibropapillomas. In pigs, they are very rare and when present are identified as solitary or multiple lesions on the face or genitalia.

A cutaneous fibroma occurs in white-tailed, black-tailed, and mule deer, and in antelope, moose, and caribou. It is caused by a papilloma virus that resembles a bovine papilloma virus and is found only in the epithelium that covers the tumors.

Infectious papillomatosis is a self-limiting disease, although the duration of warts varies considerably. A variety of treatments have been advocated without agreement on efficacy. Surgical removal is recommended if the warts are sufficiently objectionable. However, because surgery in the early growing stage of warts may lead to recurrence and stimulation of growth, the warts should be removed when near their maximum size or when regressing. Affected animals may be isolated from susceptible ones, but with the long incubation period, many are likely to have been exposed before the problem is recognized.

Vaccines are of some value as a preventive but are of little value in treating cattle that already have lesions. Because wart viruses are mostly species-specific, there is no merit in using a vaccine derived from one species in another. An intralesional lymphocyte T-cell immunomodulator, which stimulates T cells and endogenous interleukin 2 levels, may benefit individual animals that develop multiple or persistent warts. In addition to azithromycin, topical imiquimod cream may also help the immune system to resolve these warts in dogs. When the disease is a herd problem, it can be controlled by vaccination with a suspension of ground wart tissue in which the virus has been killed with formalin. Autogenous vaccines may be more effective than those commercially available.

It may be necessary to begin vaccination in calves as early as 4–6 wk of age with a dose of ~0.4 mL intradermally given at two sites. The vaccination is repeated in 4–6 wk and at 1 yr of age. Immunity develops in a few weeks but is unrelated to whatever mechanism is involved in spontaneous regression. If the animal was exposed to the virus before vaccination, immunity may develop too late to prevent warts. A vaccination program must be in effect for ~3–6 mo before its preventive value will be evident. Vaccination should be continued for ≥1 yr after the last wart disappears, because the premises may still be contaminated. Stalls, stanchions, and other inert materials can be disinfected by fumigating with formaldehyde.

CONNECTIVE TISSUE TUMORS

BENIGN FIBROBLASTIC TUMORS

Collagenous nevi are benign, focal, developmental defects associated with increased deposition of dermal collagen. They are common in dogs, uncommon in cats, and rare in large animals. They generally are found in middle-aged or older animals, most frequently on the proximal and distal extremities, head, neck, and areas prone to trauma.

They are sessile to raised, dermal nodules, often with a papillated surface. Two forms are seen: one develops in the interfollicular dermis or subcutaneous fat that is not accompanied by adnexal involvement, and one incorporates adnexa and induces enlarged, often malformed follicles, sebaceous glands, and apocrine glands. This latter form has been called focal adnexal dysplasia. Excision of both forms is generally curative although, infrequently, expansive forms have been identified that may grow too large to be surgically removed.

Generalized nodular dermatofibrosis (dermatofibromas), recognized rarely in German Shepherds (believed to be an inherited, autosomal dominant trait) and even less commonly in other canine breeds, is a syndrome in which multiple collagenous nevi are associated with renal cystadenocarcinomas and, in females, multiple uterine leiomyomas. Skin lesions, first recognized when animals are 3–5 yr old, are characterized by the development of multiple collagenous nevi varying from barely palpable to large and nodular, generally on the limbs, feet, head, and trunk. They may be symmetrically distributed. Renal disease develops ~3–5 yr after the skin lesions are recognized. No known therapy can prevent development of the renal and uterine neoplasms.

Acrochordons (cutaneous tags, soft fibromas, fibrovascular papillomas) are distinctive, benign, cutaneous lesions of older dogs. These lesions are common, may be single or multiple, and can develop in any breed, although large breeds may be at increased risk. Most commonly, they appear as pedunculated exophytic growths, often covered by a verrucous epidermal surface. Treatment is optional, but a biopsy is recommended to confirm the diagnosis. Acrochordons are amenable to excision, electrosurgery, and cryosurgery, but dogs that develop one are prone to develop others over time.

Fibromas are discrete, generally cellular proliferations of dermal fibroblasts. Histologically, they resemble collagenous nevi or cutaneous tags. Fibromas occur in all domestic species but are primarily a tumor of aged dogs. Doberman Pinschers, Boxers (predisposed to developing multiple tumors), and Golden Retrievers are most at risk. The head and extremities are the most likely sites. Clinically, the lesions appear as discrete, generally raised, often hairless nodules originating in the dermis or subcutaneous fat. They palpate as either firm and rubbery (fibroma durum) or soft and fluctuant (fibroma molle). These lesions are benign, but complete excision is recommended if they change appearance or grow large.

SOFT-TISSUE SARCOMAS

This group of malignancies includes equine sarcoids, fibromatoses, fibrosarcomas, malignant fibrous histiocytomas, neurofibrosarcomas, leiomyosarcomas, rhabdomyosarcomas, and variants of liposarcomas, angiosarcomas, synovial cell sarcomas, mesotheliomas, and meningiomas. As a group, sarcomas are widely recognized, yet poorly characterized neoplasms. The confusion stems in part from the fact that spindle-cell sarcomas demonstrate much greater morphologic heterogeneity than carcinomas; often, features of one sarcoma are intermixed with features of another.

Consequently, it is widely accepted that the cell of origin of all soft-tissue sarcomas is a primitive mesenchymal cell that can differentiate in many different directions.

This makes it difficult to define histopathologic criteria necessary for making an unequivocal diagnosis of specific spindle-cell sarcomas. In addition, comparing neoplastic mesenchymal cells with the normal cell they most closely resemble does not imply origin from those cells.

Soft-tissue sarcoma involving foot pad, dog

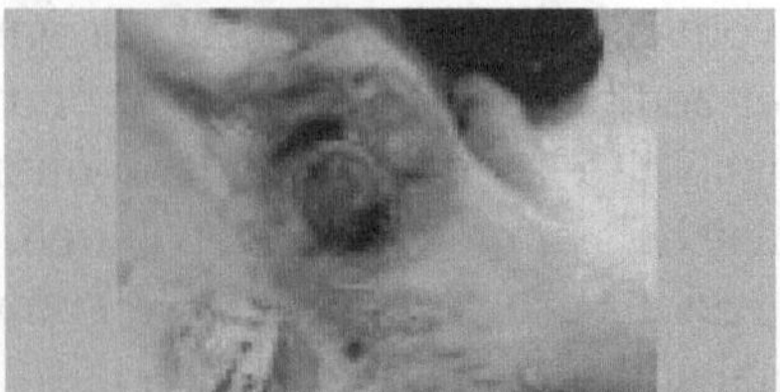

Soft-tissue sarcoma involving foot pad responding to treatment, dog

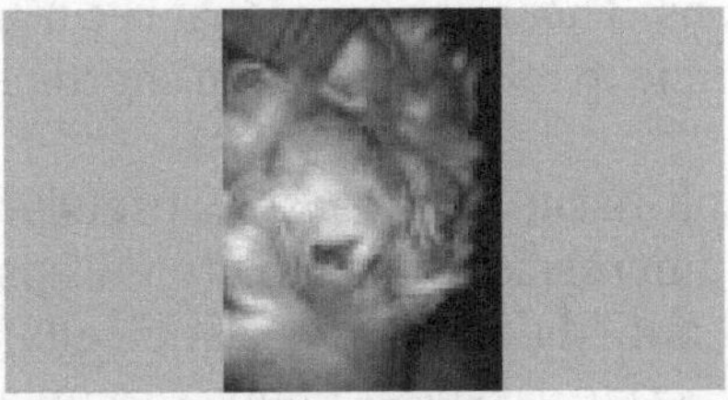

A second cause for the confusion stems from the difficulty in determining whether these are benign or malignant or what their biologic behavior will be in certain locations or breeds. Most spindle-cell sarcomas of domestic animals are locally infiltrative, difficult to excise, and yet seldom metastasize. Because, by definition, only malignant tumors have metastatic potential, these tumors should be considered benign; however, again by definition, benign neoplasms are not infiltrative, and those tumors should be considered malignant and treated aggressively from the start. In human pathology, infiltrative but nonmetastasizing mesenchymal spindle-cell tumors have been defined as sarcomas of intermediate malignancy.

Clinically, four general principles relate to spindle-cell sarcomas and soft-tissue sarcomas: The more superficial the location, the more likely the tumor is to be benign (deep tumors tend to be malignant). The larger the tumor, the more likely it is to be malignant. A rapidly growing tumor is more likely to be malignant than one that develops slowly. Benign tumors are relatively avascular, whereas most malignancies are hypervascular. The type of sarcoma, its size, location, stage and histologic grade (low, intermediate, high [or I, II, III]), which depends on the degree of differentiation, the number of mitotic figures per 10 high-power fields, and percent that is necrotic, will help guide treatment planning.

Fibrosarcoma of rostral maxilla, dog

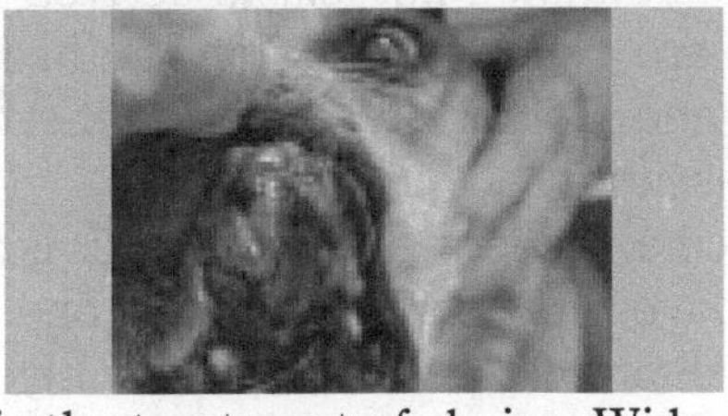

Surgical excision is the treatment of choice. Wide excision or amputation should be performed when anatomically feasible because spindle-cell sarcomas often infiltrate along fascial planes, making it difficult to determine from gross examination the peripheral margins of the tumor. The best, if not only, opportunity to completely remove a spindle-cell sarcoma is during the first surgical attempt. A presurgical biopsy should be performed and, if possible, tumor imaging with CT or MRI or ultrasound to provide a clear surgical plan that includes the intention of complete removal, with biopsy samples submitted for margin determinations. Those sarcomas that recur have a greater potential for metastasis, and the time to recurrence often shortens with each subsequent attempt at excision.

In addition, many soft-tissue tumors have a pseudocapsule, which on gross examination gives the impression of complete encapsulation; these tumors should not be "shelled out," because neoplastic cells are usually present in the pericapsular connective tissues. Many sarcomas are shaped like an octopus, with tentacles that extend deeply into the tumor bed. Except for equine sarcoids, cryosurgery is usually not used for these tumors because some types, most notably fibrosarcomas, are resistant to freezing. Spindle-cell sarcomas generally do not respond well to conventional doses of radiation; however, higher doses have been reported to control ~50% of them for 1 yr. Surgical debulking followed by intraoperative placement of carboplatinum-containing biodegradable beads, intracavitary chemotherapy with follow-up intratumor bed chemotherapy, and/or follow-up radiation therapy are options to enhance local control.

Chemotherapeutic protocols for sarcomas that include targeted therapy agents such as tyrosine kinase (T-K) inhibitors have become more accepted as a means of treatment. Most older protocols involve the use of adriamycin often in combination with other agents, including cyclophosphamide, vincristine, dacarbazine, and methotrexate. Some clinicians prefer to use a combination of carboplatin. T-K inhibitors, and metronomic chemotherapy, whereas others rotate carboplatin with adriamycin. Although chemotherapy may cause temporary adverse events, it can improve the length and overall quality of life; it is seldom curative.

Fibromatosis (aggressive fibromatosis, extra-abdominal desmoids, desmoid tumors, low-grade fibrosarcomas, nodular fasciitis) is a sclerosing and infiltrative proliferation of well-differentiated fibroblasts derived from aponeuroses and tendon sheaths. They are generally seen on the heads of dogs, especially Doberman Pinschers and Golden Retrievers, where they are commonly diagnosed as nodular fasciitis.

In veterinary medicine, the term nodular fasciitis is applied to two different diseases—one that behaves as a fibromatosis and one that commonly affects the periocular tissues. Fibromatoses are infrequently diagnosed in cats and horses. Grossly, fibromatoses are generally indistinguishable from infiltrative fibrosarcomas; however, they can be differentiated on histologic examination. Focal lymphoid nodules are scattered throughout the tissues. The fibromatoses are locally infiltrative, with essentially no metastatic potential. If feasible, wide, complete excision with local control techniques as described above is the treatment of choice at diagnosis. Recurrence is common, and radiation therapy may be of value for local control.

Low-grade fibrosarcoma in muzzle, dog

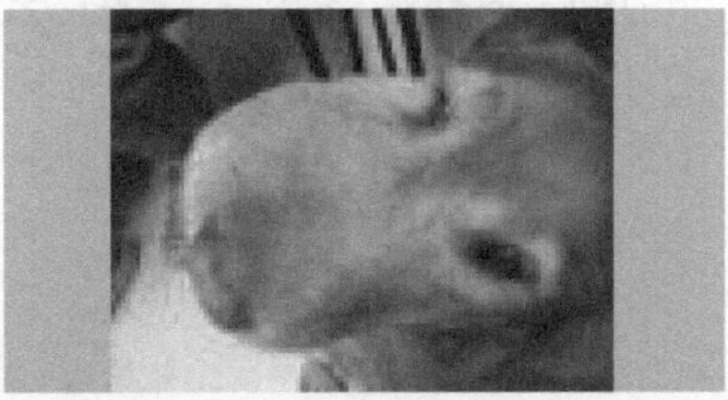

Fibrosarcoma, parakeet

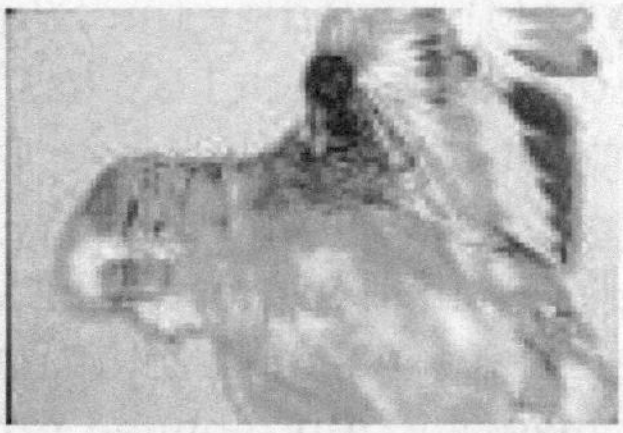

Fibrosarcomas are aggressive mesenchymal tumors in which fibroblasts are the predominant cell type. They are the most common soft-tissue tumors in cats and are also common in dogs but rare in other domestic animals. In dogs, these tumors are most common on the trunk and extremities. Gordon Setters, Irish Wolfhounds, Brittany Spaniels, Golden Retrievers, and Doberman Pinschers may be predisposed. Fibrosarcomas vary markedly in their appearance and size. Neoplasms arising in the dermis may appear nodular. Those arising in the subcutaneous fat or subjacent soft tissues may require palpation to identify. They appear as firm, fleshy lesions involving the dermis and subcutaneous fat and often invade musculature along fascial planes. When tumors are multiple, they are usually found within the same anatomic region. Fibrosarcomas with abundant interstitial proteoglycans (connective tissue mucins) are called myxosarcomas ormyxofibrosarcomas. Myxosarcomas remain poorly defined in veterinary medicine, and many of them could be characterized as variants of liposarcomas or malignant fibrous histiocytomas. Fibrosarcomas in dogs are invasive tumors; ~10% metastasize. Factors that affect whether a fibrosarcoma can be completely excised include the skill of the surgeon; rate of growth (as defined by the mitotic index and quantity of necrosis); degree of cellular atypia; and the tumor's infiltrative nature, size, and location (which may require imaging to define properly).

Vaccine-associated or injection site sarcoma, cat

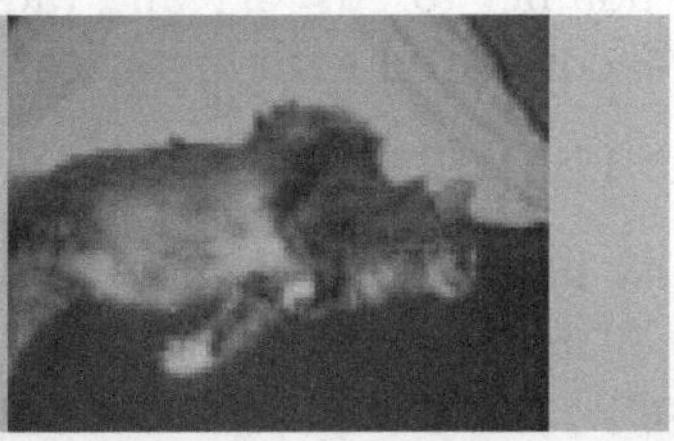

Fibrosarcoma, cat

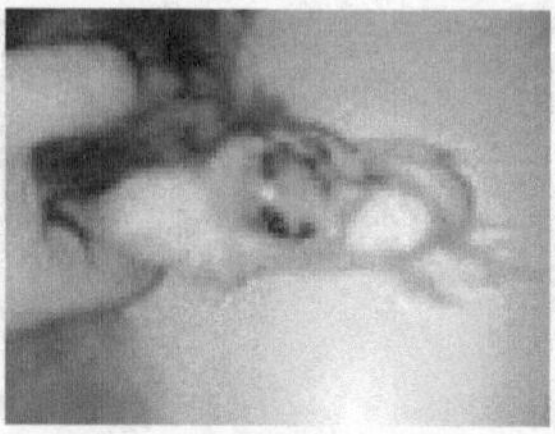

Three forms of fibrosarcoma are recognized in cats: a multicentric form in the young (generally <4 yr old) caused by the feline sarcoma virus (FSV); a solitary form in the young or old, in which FSV has not been implicated; and a fibrosarcoma that develops in the soft tissues where cats are commonly vaccinated. An association with rabies and feline leukemia virus vaccinations is better defined than with vaccinations for other viral or bacterial diseases. Aluminum hydroxide (commonly used in adjuvants) has been identified in vaccine-induced fibrosarcomas, and a prolonged proliferation of fibroblasts in response to the adjuvant may predispose these cells to undergo neoplastic transformation. These tumors appear as nodules or plaques between the shoulder blades, in the soft tissues of the proximal hindlimbs, or less commonly, over the lumbar areas. Although commonly classified as fibrosarcomas, vaccination-site sarcomas are extremely heterogeneous and may be appropriately called malignant fibrous histiocytomas (giant cell tumors), liposarcomas, osteosarcomas, or chondrosarcomas.

Wide and deep surgical excision is the treatment of choice for fibrosarcomas, but because the necessary margin extent is commonly underestimated, recurrence is common (>70% within 1 yr of the initial surgery). The rate of recurrence is >90% for vaccine-associated sarcomas. Even when surgical excision is clinically and histologically complete, recurrence is still the rule. Therefore, multimodal therapy combining presurgical imaging, aggressive surgery, intracavitary chemotherapy with carboplatin, followed by radiation therapy and IV carboplatin every 21 days for 4–6 treatments with tyrosine kinase (T-K) inhibitors as adjunctive therapy may yield the best results. Chemotherapy with carboplatin, doxorubicin and cyclophosphamide, or dacarbazine, along with T-K inhibitors, has been recommended for nonresectable tumors. Initial results using a biologic response modifier (used intratumorally before excision and followed by radiation therapy) appears promising. Further work suggests that its effectiveness as an adjunct to surgery and radiation may increase tumor-free intervals up to 20% compared with historical controls.

FIBROHISTIOCYTIC TUMORS

These pleomorphic, mesenchymal tumors composed of fibroblasts and histiocytic cells (often present as multinucleated giant cells) remain poorly defined in veterinary medicine. A lesion called canine fibrous histiocytoma(nodular granulomatous episclerokeratitis, nodular fasciitis, proliferative keratoconjunctivitis, conjunctival granuloma, Collie granuloma) is recognized at the episcleral junction and cornea primarily in young to middle-aged (2–4 yr old) Collies, but the histologic features are more suggestive of a granulomatous inflammatory response than a neoplasm. As might be expected for a noninfectious inflammatory process, these are generally responsive to sublesional injections of 10–40 mg of methylprednisolone.

Malignant fibrous histiocytomas (extraskeletal giant cell tumors, giant cell tumors of soft parts, dermatofibrosarcomas) are most frequently found in the skin and soft tissues of cats, occasionally found in horses and mules, and rarely in the skin of other domestic species, including dogs. In cats, malignant fibrous histiocytomas are most common on the distal extremities or ventral cervical regions of the aged but may also be diagnosed at vaccination sites. In horses and mules, these have been described as giant cell tumors of soft parts. Occurring in young adult to middle-aged Equidae, they are firm, nodular to diffuse swellings that are white on cut surface, with variable hemorrhage. Malignant fibrous histiocytomas are sarcomas of intermediate malignancy. They are locally invasive and tend to recur after attempts at complete excision but seldom metastasize. Radical excision is recommended.

PERIPHERAL NERVE SHEATH TUMORS

Amputation neuromas (traumatic neuromas) are non-neoplastic, disorganized proliferations of peripheral nerve parenchyma and stroma that form in response to amputation or traumatic injury. They are most commonly identified after tail docking in dogs or neurectomy in the distal extremities of horses. The most common clinical presentation is a young dog that continuously traumatizes its docked tail. In horses, such a lesion appears as a firm, often painful swelling at a neurectomy surgery site. Excision is curative. Neurofibromas and neurofibrosarcomas(perineuromas, neurilemmomas, nerve sheath tumors, hemangiopericytomas, neurothekomas, schwannomas) are spindle-cell tumors that arise from the connective tissue components of the peripheral nerve. They are believed to arise from Schwann cells, but they could also arise from mesenchymal cells, which produce the nonmyelinated connective tissues that surround the myelinated nerve fiber. In dogs, forms of this tumor can be virtually indistinguishable from hemangiopericytomas and may be the same tumor.

Peripheral nerve sheath tumor recurrence, left elbow, Great Dane mix

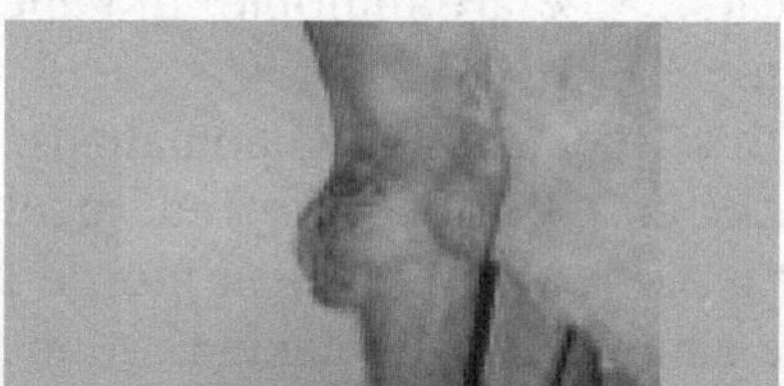

In dogs and cats, peripheral nerve sheath tumors of the skin are found in older animals. In cattle, they have a suspected genetic basis, may be multiple, can develop in both the young and old, and are generally an incidental finding at slaughter; they arise from the deep nerves of the thoracic wall and viscera, and cutaneous involvement is rare. Regardless of the species, these tumors appear as white, firm, nodules. Attachment to a peripheral nerve may occasionally be noted. Both benign and intermediate-grade malignant variants are recognized. Benign tumors are most common in cattle in which, because of their indolent nature, treatment is optional; also, additional tumors often develop spontaneously at other sites over time. In dogs, cats, and horses, most are locally infiltrative but do not metastasize. Complete excision is the treatment of choice. When margins are narrow or insufficient, followup radiation therapy, postoperative intralesional tumor bed chemotherapy (using the patient's serum mixed with the chemotherapy agent), electrochemotherapy, or systemic chemotherapy with carboplatin or metronomic chemotherapy may increase the tumor-free interval.

ADIPOSE TISSUE TUMORS

Lipoma, 12-yr-old, spayed female Aussie-Labrador mix

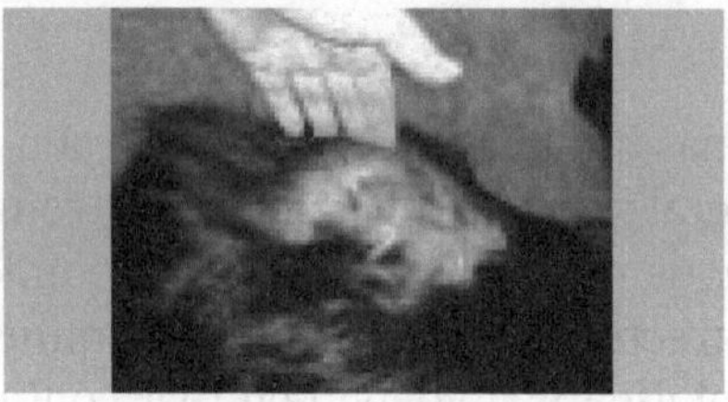

Lipomas are benign tumors of adipose tissue, perhaps more accurately characterized as hamartomas. They are common in dogs, occasionally identified in cats and horses, and rare in other domestic species. In dogs, they generally occur in older, obese females, most commonly on the trunk and proximal limbs. The breeds most at risk are Doberman Pinschers, Labrador Retrievers, Miniature Schnauzers, and mixed-breed dogs. Older, neutered male Siamese cats are predisposed, and tumors are most commonly found on the ventral abdomen. Obesity does not appear to be a factor in the development of lipomas in cats. Affected horses are generally <2 yr old. Lipomas typically appear as soft, occasionally pedunculated, discrete nodular masses, and most are freely movable. In dogs and cats, >5% are multiple. In general, these tumors float when placed in formalin.

A rare variant of this tumor, diffuse lipomatosis, has been identified in Dachshunds. Virtually the entire skin is affected, resulting in prominent folds on the neck and truncal skin. Many lipomas merge imperceptibly with the adjacent non-neoplastic adipose tissue, making it difficult to determine when the entire lesion is excised. Lipomas with an abundant connective tissue stroma (fibrolipomas), cartilaginous stroma (chondrolipomas), or a prominent vascular component (angiolipomas) are also recognized. Despite their benign nature, lipomas should not be ignored because they tend to enlarge over time, and their gross presentation may be indistinguishable from that of infiltrative lipomas or liposarcomas. Excision is curative. In dogs, dietary restriction to 70% of normal intake for several weeks before surgery may allow for better definition of the surgical margins of the tumor.

Infiltrative lipomas (intra- and intermuscular lipomas) are rare in dogs and even less common in cats and horses. In dogs, they are most common in middle-aged females, usually on the thorax and limbs. The breeds (dogs) most at risk are the same as those for lipomas. These tumors are poorly confined, soft, nodular to diffuse swellings that typically involve the subcutaneous fat and underlying muscle and connective tissue stroma. Infiltrative lipomas, which dissect along fascial planes and between skeletal muscle bundles, are considered sarcomas of intermediate malignancy. They rarely metastasize. Aggressive excision is recommended, and amputation may be necessary.

Liposarcomas are rare neoplasms in all domestic animals. Most are recognized in older male dogs in which they usually develop on the trunk and extremities; Shetland Sheepdogs and Beagles may be predisposed. In cats, feline leukemia virus infection has been infrequently associated with their development; whether this is a coincidence or such infections play a causative role remains unclear. Liposarcomas are nodular and soft to firm. They may exude a mucinous fluid when sectioned. Many have palpable, partially encapsulated areas, but these zones should not be construed as evidence of a benign tumor. Liposarcomas are malignant neoplasms that have a low metastatic potential but are frequently pseudoencapsulated. Wide excision is recommended. Recurrence is common, so followup radiation therapy is indicated in cases with insufficient margins.

VASCULAR TUMORS

Hemangiomas of the skin and soft tissues are benign proliferations that closely resemble blood vessels. Whether these are neoplasms, hamartomas, or vascular malformations remains undefined, and no clear criteria exist that allow for their separation. They are most commonly identified in dogs, occasionally in cats and horses, and rarely in cattle and pigs; they are an exceptional finding in other domestic animals. In dogs, they are tumors of adult dogs and most commonly develop on the trunk and extremities. Many canine breeds (including Gordon Setters; Boxers; and Airedale, Scottish, and Kerry Blue Terriers) are considered to be at risk. Cats most frequently develop hemangiomas when they are adults. Lesions are most common on the head, extremities, and abdomen. In horses, they are most common on the distal extremities of young (<1 yr old) animals. In cattle, they may be seen as congenital lesions or in older animals. Dairy cattle are predisposed to developing disseminated hemangiomas (angiomatosis) in the skin and internal organs. In pigs, these lesions generally develop in the scrotal or perineal skin of Yorkshire, Berkshire, and less commonly Chester White boars. In the first two breeds, the disease is believed to be genetically transmitted. Hemangiomas are single to multiple, circumscribed, often compressible, red to black nodules. The lining epidermis may be unaffected or ulcerated or papillated. Small, superficial hemangiomas that often appear as a "blood blister" are known as angiokeratomas. When erythrocytes are sparse or absent within vascular lumens, the term lymphangioma is applied. Hemangiomas are benign, but their tendency to ulcerate and grow quite large, along with the importance of confirming the diagnosis to make a prognosis, indicate removal. Excision is the treatment of choice; however, in large animals in which the lesions may be large and involve the distal extremities, this may be difficult. In these cases, cryosurgery or radiation therapy may be necessary. Except in dairy cattle with angiomatosis, development of additional tumors at new sites after complete excision is uncommon.

Hemangiopericytoma over the lateral right thorax, older Black Labrador

Hemangiopericytoma, right hindleg, mixed-breed dog

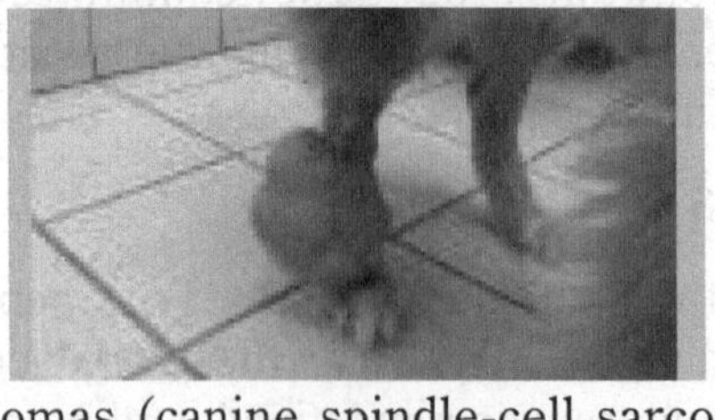

Hemangiopericytomas (canine spindle-cell sarcoma, canine malignant fibrous histiocytoma, canine neurofibrosarcoma, canine perineuroma) are common in dogs and rare in cats (if they occur at all). This tumor was initially named because it was thought to be derived from fibroblastic cells that surround small vessels; however, the appropriateness of the name remains a topic of debate.

These tumors develop most commonly on the distal extremities and thorax of older dogs. Females appear to be predisposed, and Siberian Huskies, mixed-breed dogs, Irish Setters, and German Shepherds are most at risk. Hemangiopericytomas typically present as firm, multilobulated, solitary lesions with irregular borders, most commonly in the subcutaneous fat but sometimes in the dermis. They are of intermediate malignancy and have limited metastatic potential. Complete excision with adequate margins is the treatment of choice but, because of the infiltrative nature of these tumors, excisions are incomplete or margins are narrow in most surgeries; therefore, ~30% recur. If the first excision of any sarcoma does not have adequate margins, ie, clean, wide surgical margins ("complete" may include narrow margins not considered adequate for local tumor control), followup surgery to remove the tumor bed and increase the margins is indicated. At surgery, intracavitary chemotherapy with carboplatin or 5-fluorouracil mixed with the patient's serum, and/or intraoperative radiation therapy with followup intratumor bed chemotherapy, and/or followup radiation therapy may increase the tumor-free interval.

Angiosarcomas, arguably the most aggressive of all soft-tissue tumors, are composed of cells that have many functional and morphologic features of normal endothelium. Although these tumors are often divided into hemangiosarcomas (of purported blood vessel origin) and lymphangiosarcomas (of lymphatic vessel origin), such a distinction is arbitrary. The term angioendothelioma is also used. These tumors generally arise spontaneously. In the nonpigmented skin of dogs with short, often white coats, chronic solar injury may induce transformation in the superficial vascular plexus, which initially appears as a hemangioma and then progresses to a malignant vascular tumor. The light-skinned breeds prone to actinically induced dermal angiosarcomas are Whippets, Italian Greyhounds, White Boxers, and Pit Bulls. Pathologists often diagnose these lesions as cutaneous hemangiosarcomas. Because histogenesis of dermal hemangiosarcoma is often related to that of chronic solar irritation, predisposed breeds should be protected from sun exposure. Dermal hemangiosarcomas have a moderately high potential for malignancy; therefore, wide surgical excision is the treatment of choice for tumors >0.5 cm. Cryosurgery or laser surgery can effectively control multifocal, small, early lesions. When metastasis occurs, it is by the hematogenous route to the lungs and spleen. Up to one-third of splenic hemangiosarcomas arise in the skin or peripheral tissue and end up in the spleen because of splenic filtering of cells in the bloodstream.

Angiosarcoma, hip, 13-yr-old Golden Retriever

Angiosarcomas of the skin and soft tissues are seen in all domestic animals but are most common in dogs, generally in adult or aged animals. In dogs, they most frequently develop on the trunk, hip, thigh, and distal extremities. In addition to the breeds prone to actinically induced angiosarcomas, Irish Wolfhounds, Vizslas, Golden Retrievers, and German Shepherds are also at risk. In cats, this tumor is seen most commonly in older, neutered males, on the extremities and trunk. Cats with skin, subcutaneous, or visceral involvement develop distant metastasis. Angiosarcomas can vary markedly in appearance. Most commonly, they appear as one or more erythematous nodules present anywhere in the skin or underlying soft tissues. Less frequently, they appear as a poorly defined bruise. All grow rapidly, often are associated with large zones of necrosis and thrombosis, and typically are red to black on cut section. Tumors often diagnosed as lymphangiosarcomas may have much less lumenal blood, and the vascular spaces are typically filled with serum. Characteristically, angiosarcomas create their own vascular space by dissecting through soft tissues. Distant metastasis, especially to the lungs and liver, is common. In other domestic animals, these tumors do not appear to behave as aggressively, and postexcisional recurrence rather than metastasis is more common.

Cutaneous hemangiosarcoma, dog

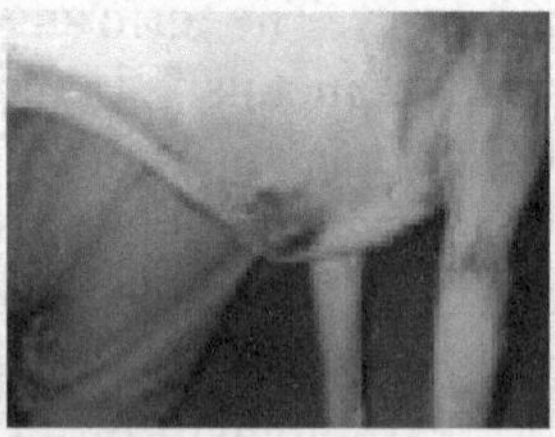

Solar-induced cutaneous hemangiosarcoma, older Bulldog

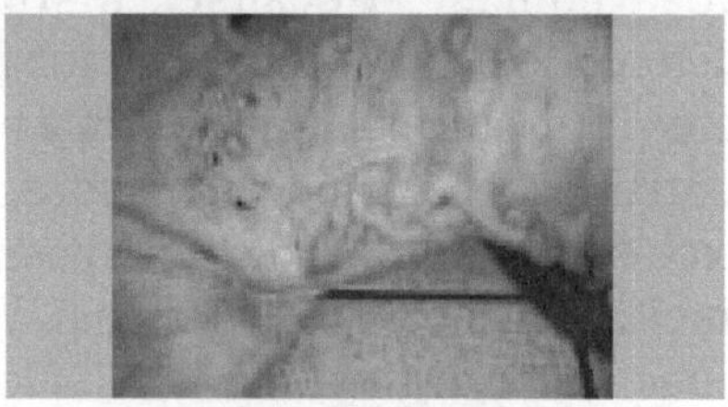

Solar-induced cutaneous hemangiosarcoma, Italian Greyhound

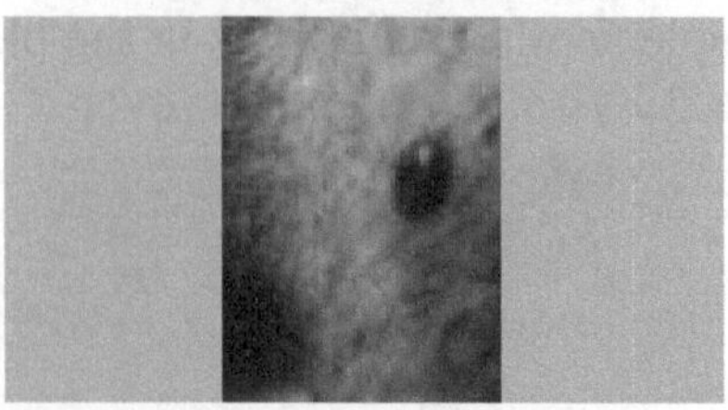

For all species, because of the aggressiveness of these nodules, wide excision with intraoperative placement of carboplatin beads or intracavitary carboplatin mixed with the patient's serum may be the treatment of choice. Solar-induced canine cutaneous hemangiosarcomas generally do not have an aggressive biologic behavior, although numerous lesions may continue to appear over a period of several years. Superficial lesions are easily destroyed with topical cryotherapy, which may control the disease for years; however, if lesions become large (>1 cm) or cystic, they are best controlled with excision. If multiple cystic lesions appear, the animal may benefit from a combination of local control and antiangiogenic therapy. Avoidance of further sun irradiation injury may reduce the development of new lesions; however, previously exposed skin may continue to develop lesions as a result of "field cancerization." Adjuvant chemotherapy consisting of vincristine, doxorubicin, and cyclophosphamide (the VAC protocol) has been reported to shrink angiosarcomas; carboplatin is also helpful. However, the effects of chemotherapy for systemic control or radiation therapy for local control and longterm survival remain to be defined. Electrochemotherapy delivering bleomycin and platinum-containing drugs into tumors has shown cytotoxic efficacy leading to endothelial cell death in 1 to 3 treatment sessions under heavy sedation. These agents may also be administered intraoperatively. The role of NSAIDs such as piroxicam, meloxicam, etc, is still not completely understood and may vary from drug to drug. Antiangiogenic or angiostatic compounds such as canine recombinant canstatin that attack the blood supply of tumors may control and prevent metastases; however, results of clinical trials remain elusive.

CUTANEOUS SMOOTH MUSCLE TUMORS

Because they either are not recognized or do not occur with any regularity in domestic animals, cutaneous smooth muscle tumors (leiomyomas or leiomyosarcomas) are diagnosed rarely. Those reported generally have been malignant and found in dogs and cats. Usually, they are firm cutaneous masses. Leiomyomas are small and tend to be limited to the dermis, whereas leiomyosarcomas are larger and most arise from (or extend into) the subcutaneous fat. The behavior of malignant smooth muscle tumors remains poorly defined. Complete excision is the treatment of choice for both leiomyomas and leiomyosarcomas. They have been responsive to vincristine chemotherapy given IV weekly for 6 wk then tapered; they may also respond to oral masitinib therapy and other tyrosine-kinase inhibitors.

LYMPHOCYTIC, HISTIOCYTIC, AND RELATED CUTANEOUS TUMORS

LYMPHOID TUMORS OF THE SKIN

Canine extramedullary plasmacytomas (atypical histiocytomas, cutaneous neuroendocrine tumors, reticulum cell sarcomas, cutaneous nodular amyloidosis) are relatively common cutaneous tumors. Although their derivation was long debated, neoplastic cells characteristically express cytoplasmic immunoglobulin and may produce primary amyloid, leaving little doubt as to their lymphoplasmacytic origin. These tumors of dogs and, rarely, cats are most frequently identified on the head (including ears, lips, and oral cavity) and extremities of mature to aged animals. Cocker Spaniels, Airedales, Scottish Terriers, and Standard Poodles are most at risk. The tumors are generally small (<5 cm) and sometimes pedunculated. Most of these tumors are locally confined, and complete but conservative surgical excision is the treatment of choice. Infrequently, extracutaneous plasmacytomas may be locally invasive or multiple (or both), especially when they occur in the oral cavity. Recurrence has also been correlated with the presence of amyloid. Treatment for these tumors targets local control using localized techniques described above. For recurrent, invasive tumors, more aggressive attempts at excision may be required. When tumors are multiple or when surgical excision is not feasible, radiation therapy appears to be the best secondary treatment. For tumors resistant to radiation and for those animals unable to receive radiation therapy, systemic treatment with chemotherapeutic agents, including melphalan, chlorambucil, cyclophosphamide, and glucocorticoids, have been recommended and have yielded longterm survival.

Cutaneous lymphosarcoma, dog

Cutaneous lymphosarcoma may occur as a disease in which the skin is the initial and primary site of involvement, or it may be secondary to systemic, internal disease. Cutaneous lymphosarcoma is uncommon but has been identified in all domestic species. In general, two distinct forms are recognized—an epitheliotropic form (in which there is infiltration by malignant lymphocytes into the epidermis and adnexa) and a nodular, nonepitheliotropic form. Both usually express surface and cytoplasmic antigens characteristic of T cells; this, along with the frequent identification of at least small foci of epitheliotropism in many cases of "nonepitheliotropic" forms in dogs and cats, suggest they may be different variants of the same tumor.

Epitheliotropic cutaneous lymphosarcoma (mycosis fungoides), 8-yr-old male Pit Bull

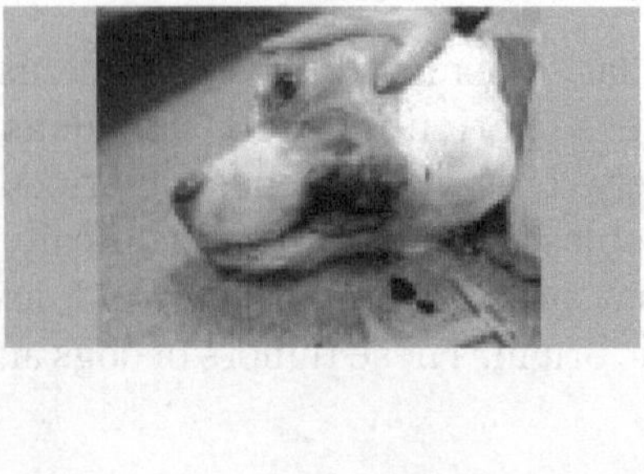

Mycosis fungoides, old toy Poodle mix

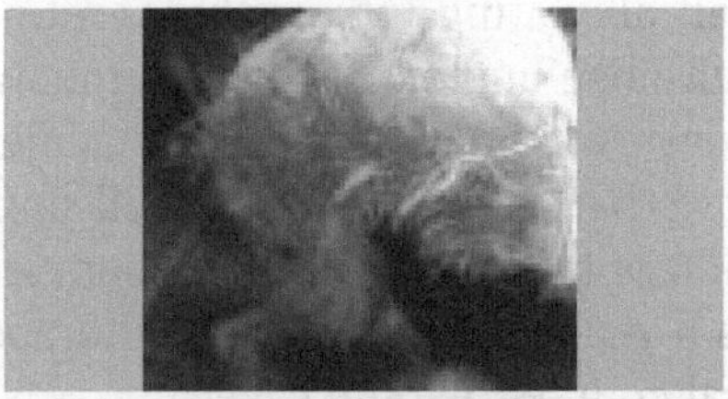

Mycosis fungoides lesion on prepuce, old Poodle mix

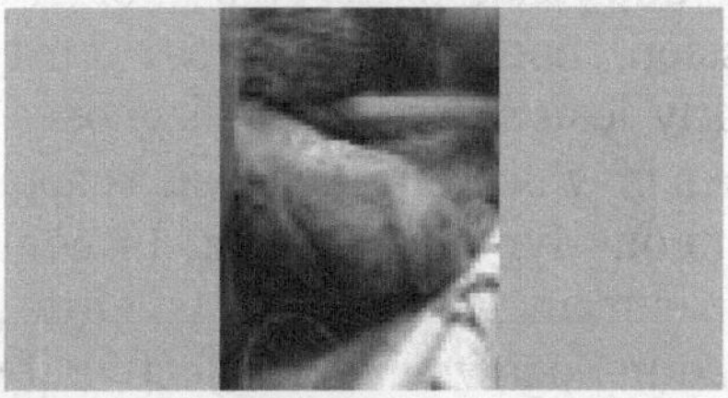

Epitheliotropic cutaneous lymphosarcoma (ECL, mycosis fungoides) is the most frequently recognized form of cutaneous lymphosarcoma in dogs and arguably cats. It is a disease of middle-aged and older dogs, and Poodles and Cocker Spaniels may be predisposed. Classically, the lesions progress from patch to plaque to tumor; however, one or any combination of these three primary lesions may be present.

For example, a form of ECL known as pagetoid reticulosis has minimal to no dermal involvement, and cutaneous lesions always appear as erythematous patches. Another common feature of the disease in dogs is the presence of areas of alopecia due to follicular atrophy caused by infiltration of neoplastic cells into the outer sheath and lumen of hair follicles. Although most cases are associated with diffuse cutaneous involvement, forms limited primarily to mucous membranes or the footpads have been identified. Because of the variable clinical appearance of this tumor, diagnosis based on clinical features can be very difficult, and early stages can be confused with allergic, autoimmune, endocrine, infectious, or seborrheic diseases. Most cases are limited to the cutis until late in the course of the disease. ECL with concurrent leukemia is known as Sézary syndrome.

In dogs, ECL is a slow to moderately progressive disease for which a number of therapies have been attempted. To date, all appear more effective in improving the clinical features of the disease than in prolonging an affected dog's life.

Methchlorethamine (nitrogen mustard) has been used in the past as a topical therapy, but because large areas of a dog's body may be affected (including the mucous membranes) and because of its sensitizing potential in people, it is infrequently used. The disease is often transiently responsive to steroids. Chemotherapeutic agents, such as combinations of adriamycin, chlorambucil, cyclophosphamide, doxorubicin, and vincristine, are variably effective. Lomustine (an alkylating nitrosourea), retinoids with and without glucocorticoids, and high-dose linoleic acid supplementation (safflower oil at 3 mL/kg, 3 consecutive days/wk) may occasionally achieve partial or complete remission. Studies show that a lymphocyte T-cell immunomodulator, which signals the differentiation of late stage CD_4 lymphocytes and causes apoptosis of malignant T cells, is a reasonable adjunctive therapeutic agent to use and may improve the clinical course in some cases.

ECL is rare in cats and tends to develop in older animals. Lesions often follow a defined progression, appearing initially as a crusty plaque that is variably pruritic. Biopsies of early lesions are often diagnosed as lymphocytic mural folliculitis. In many cases in which this diagnosis is applied, the lesions evolve into unequivocal cutaneous lymphosarcoma. In contrast to ECL in dogs, epitheliotropism is often extremely subtle in cats. Little is known about therapy or whether therapy used in dogs would be effective if used in cats; however, since the availability of a lymphocyte T-cell immunomodulator for use in cats with feline leukemia virus, it can be offered as a viable option.

Nonepitheliotropic cutaneous lymphosarcoma (NECL) is the most recognized form of cutaneous lymphosarcoma in all domestic animals but dogs and cats. In dogs, NECL is most common in middle-aged or older animals. Lesions are nodules or plaques that most commonly develop on the trunk. They generally are multiple, although solitary lesions may be noted, especially in cats. In many cases, NECL is grossly indistinguishable from the tumor stage of ECL.

A definitive diagnosis is important because NECL in dogs is generally more aggressive than ECL, and systemic involvement occurs commonly and early in the course of the disease. Various modes of therapy, including excision, chemotherapy, and less frequently radiotherapy, have been used as monotherapy and in combination. Excision is the choice when the disease is limited to a solitary tumor, and complete cures have occasionally been obtained. Excision or cryosurgery in more diffuse forms infrequently elicit longterm remissions. Chemotherapy or chemoimmunologic protocols used for other forms of canine lymphosarcoma should be considered as palliative. The average remission time is ~8 mo.

In cats, NECL is a rare disease of middle-aged or older animals. The role of feline leukemia virus remains undefined. The lesions are plaques or nodules that may be solitary or multiple, alopecic or haired, and ulcerated or lined by an intact epidermis. Feline NECL is aggressive; even when complete excision of a solitary nodule is attempted, recurrence is common. No definitive therapy is known; however, using a combination of lomustine, steroids, linolenic acid (oil of evening primrose) and injections of a lymphocyte T-cell immunomodulator may be of some value.

In horses, NECL (nodular lymphosarcoma, subcutaneous lymphosarcoma, lymphohistiocytic lymphosarcoma) may be recognized at any age but is most common in young and middle-aged animals. Firm, nonulcerated nodules are most common in the subcutaneous fat of the ventral body surface. Microscopically, two types of nodular lymphosarcoma are recognized in horses. The most common consists of a mixture of histiocytes and small, well-differentiated lymphocytes, occasionally with plasmacytoid features; the second consists of a monomorphic population of large atypical lymphocytes, with only occasional histiocytic cells.

Differentiation between these two forms is important because most cases of cutaneous lymphosarcoma in horses with a monomorphic pattern of cells have internal involvement, and the disease progresses rapidly. In contrast, the lymphohistiocytic form seldom is associated with internal involvement, and affected horses may live for years. As the lymphohistiocytic form progresses, the nodules tend to become more frequent on the ventral cervical regions. In many cases, euthanasia may be warranted when pharyngeal involvement induces dyspnea. Because of the expense of cytotoxic drugs, therapy is generally limited to glucocorticoids administered orally or intralesionally; remission, if induced, is usually short term.

In cattle, cutaneous lymphosarcoma is a disease of young animals (generally <4 yr old). It is one of the sporadic bovine leukosis syndromes that is not transmissible.

The term sporadic bovine leukosis is usually reserved for calf, cutaneous, and thymic types of lymphoma, which are defined by the young age of occurrence and the distribution of tumors. The cause or causes are not known. Only lymphomas caused by bovine leukemia virus infection should be termed leukosis or enzootic bovine leukosis. There may also be lymphosarcomatous conditions that do not fall into either the sporadic or enzootic bovine leukosis categories, ie, adult multicentric lymphoma with sporadic occurrence of unknown etiology. Cutaneous lymphosarcoma of young cattle is presumed not to be associated with bovine leukemia virus infection and presently has an unknown etiology. The lesions are typically nodular, involve the dermis or subcutaneous fat, and are often ulcerated. There is no known therapy.

CUTANEOUS MAST CELL TUMORS

Mast cell tumor, dog

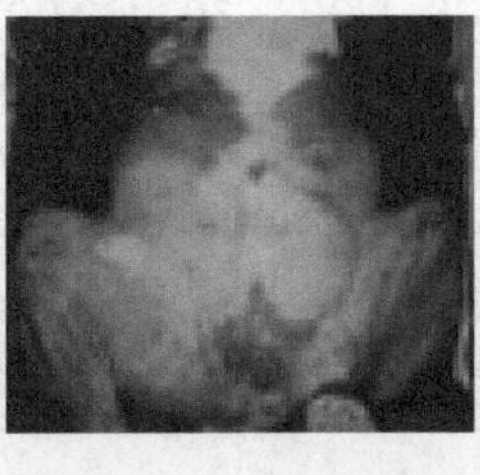

Mast cell tumor in hindfoot, dog

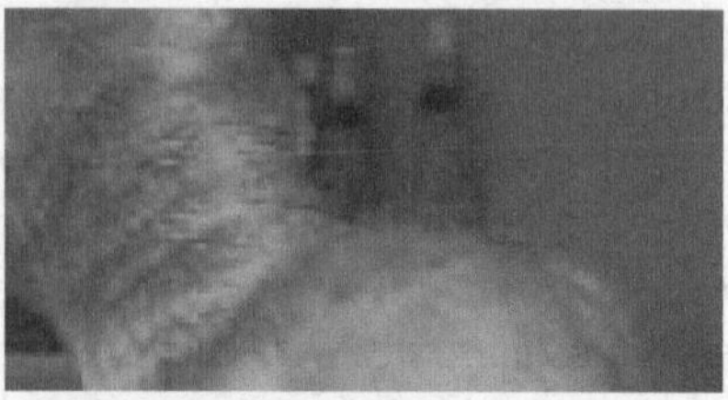

Mast cell tumor in the rostral upper lip, 10-yr-old English Bulldog

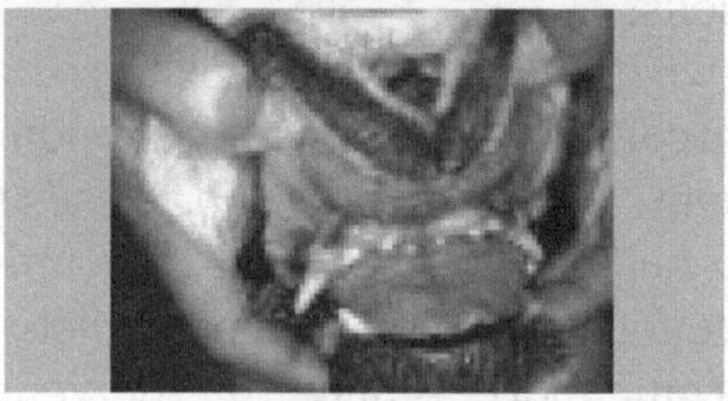

These tumors are the most frequently recognized malignant or potentially malignant neoplasms of dogs. In addition, visceral and leukemic forms can occur. A viral etiology has been speculated but remains controversial. Tumors may be seen in dogs of any age (average 8–10 yr). They may develop anywhere on the body surface as well as in internal organs, but the limbs (especially the posterior upper thigh), ventral abdomen, and thorax are the most common sites; ~10% are multicentric.

However, studies suggest that multiple tumors do not carry a worse prognosis when they are all adequately excised if grade 1 or low grade. Initial size at the time of surgery is highly predictive of outcome; tumors >3 cm correlate with decreased survival time. Tumors that are incompletely excised, grow rapidly, or arise from mucocutaneous junctions, muzzle, ventrum, prepuce, or subungual areas or in animals with clinical signs at diagnosis are associated with more aggressive biologic behavior. Many breeds appear to be predisposed, especially Boxers and Pugs (in which tumors are often multiple), Rhodesian Ridgebacks, and Boston Terriers.

The tumors vary markedly in size, and clinical appearance alone cannot establish a diagnosis. Most commonly, they appear as raised, nodular masses that may be soft to solid on palpation. Although they often seem encapsulated, mast cell tumors in dogs are seldom discrete. Rather, they consist of a highly cellular center surrounded peripherally by a halo of smaller numbers of mast cells that palpate as normal skin. Dogs can also develop clinical signs associated with the release of vasoactive products from the malignant mast cells. Most common is gastroduodenal ulceration that may be present in up to 25% of cases. Cytologic evaluation of Wright-stained, fine needle aspirates or impression smears can be used to establish the diagnosis of mast cell tumors in dogs. All skin tumors should be examined by fine needle aspiration cytology before excision to exclude mast cell tumor. If the surgeon is aware that the tumor is of mast cell origin, a surgical plan for wide and deep excision will yield the best results. All mast cell tumors should be submitted for biopsy to determine margins and grade, because cytology is not a substitute for histopathology—only the latter has been well correlated with prognosis. However, certain cytologic features can be used to identify high-grade tumors. Two systems of histopathologic grading have been defined, the Bostock system of 1973 and the Patnaik et al system of 1984. To avoid confusion, it is essential to know which system is being used. The commonly used Patnaik grading system has the following designations: grade 1 = low grade, grade II = intermediate grade, grade III = high grade. Up to 80% of canine mast cell tumors are classified as grade II and are further divided into subgroups of low grade II (mitotic index d"5) and high grade II (mitotic index >5) along with characteristics that may predict clinical behavior in this large category.

Although there is believed to be a benign variant of canine mast cell tumor, there is no clinical or microscopic means of identifying it. In addition, small mast cell tumors may remain quiescent for long periods before becoming aggressive. A subcutaneous variant of canine mast cell tumor, which does not have a primary infiltrative dermal involvement, occurred in ~10% of biopsied tissues in one study. This variant appeared most frequently in the hindlimb, with 66% having incomplete excision, most commonly at the deep margin. These tumors have intermediate histologic grade, a lower rate of recurrence (only 9%), and extended mean survival times, with only a 6% rate of metastases.

Up to 65% of incompletely excised mast cell tumors do not recur. This suggests that their biologic behavior is not always aggressive and that aggressive treatment may not always be warranted. Specialized stains can be used to distinguish tumor grades more adequately. Proliferation marker expression can help determine the likelihood that an incompletely removed mast cell tumor will recur or metastasize. The combination of Ki-67 >23/grid; proliferating cell nuclear antigen, Agnor count >54; mutations in exon 8 and exon 11 of c-KIT gene; and increased PCNA levels are prognostic for local recurrence and a poor prognosis.. Correlation of test results with survival times can be useful but is not always a reliable predictor of the outcome. There is agreement on cytologic features for high-grade mast cell tumors: seven or more mitotic figures/10 high-power fields (hpf), three multinucleated cells/10 hpf, three bizarre nuclei/10 hpf, and karyomegaly in 10% of the cells. These high-grade features confer a prognosis of a 4-mo survival time vs >2 yr for low-grade mast cell tumors. Because of the difficulty of subcategorizing canine mast cell tumors, all should be treated as at least potential malignancies. The Mast Cell Tumor Panel to determine prognosis and treatment guidance can be requested for grade II tumors at the Diagnostic Center for Population and Animal Health at Michigan State University.

Treatment depends on the clinical stage of the disease and the predicted aggressive biologic behavior. For Stage I tumors (a solitary tumor confined to the dermis without nodal involvement), the preferred treatment is complete excision with a wide margin; at least 3 cm of healthy tissue surrounding all palpable borders should be removed in an attempt to excise both the nodule and its surrounding halo of neoplastic cells. Intraoperative cytology (examination of impression smears at the excised tissue margins) can guide the surgeon, who should continue to remove tissue until the margins are adequate and free of mast cells. If histologic evaluation suggests that the tumor extends beyond the surgical margins, reexcision or tumor bed excision should be attempted. Because mast cells are sensitive to radiation, intraoperative radiation therapy is an option along with followup external beam radiation therapy, which may be curative if the remaining tumor is small or microscopic. Combination or multimodal treatments using radiation therapy (if affordable), with intralesional chemotherapy and or local hyperthermia or electrochemotherapy along with systemic chemotherapy including tyrosine-kinase inhibitors, may be more effective than radiation alone to control biologically and locally aggressive mast cell tumors.

Mast cell cancer in hindlimb, dog

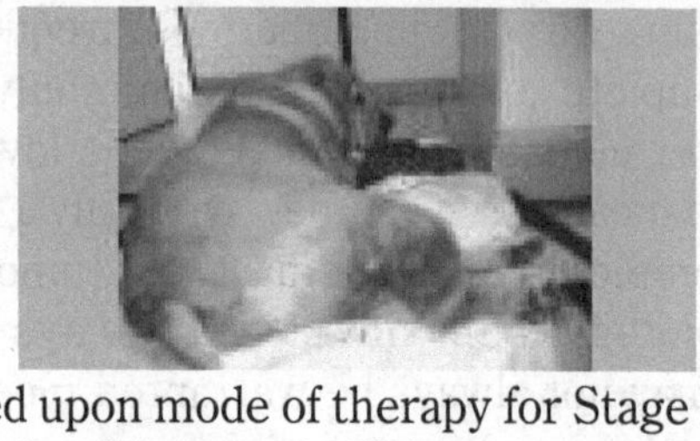

There is no agreed upon mode of therapy for Stage II-IV mast cell tumors; however, the advent of multikinase inhibitors represents a new option, which when used in combination with cytotoxic chemotherapy, requires dose reduction to avoid adverse events. Standard care options for Stage II tumors (a solitary tumor with regional lymph node involvement) include excision of the mass and the affected regional node (if feasible), prednisolone, and radiotherapy, used either singly or in combination. Triamcinolone or dexamethasone sodium phosphate, mixed with the patient's serum and injected evenly into the tissues of the tumor bed at the time of surgery or postoperatively as a followup series, may also help reduce recurrence.

Intraoperative radiation therapy and/or followup external beam therapy is still considered the highest standard of care but is often declined because of cost. Injections into the tumor bed after incomplete excision using hypotonic, deionized, or distilled water has been debated. Treatment of Stage III (multiple dermal tumors with or without lymph node involvement) or Stage IV (any tumor with distant metastasis or recurrence with metastasis) tumors is generally considered palliative. One recommended therapy is prednisolone (2 mg/kg, PO, for 5 days, followed by a maintenance dose of 0.5 mg/kg/day) or intralesional injections of triamcinolone (1 mg/cm diameter of tumor, every 2 wk). Treatment with H_1- and H_2-receptor antagonists for the peripheral and gastric effects of histamine, respectively, may be indicated for animals with systemic disease or clinical signs referable to histamine release. Chemotherapy with vinca alkaloids (vincristine, vinblastine), L-asparaginase, and cyclophosphamide has also been used with some effectiveness. Prednisone and vinblastine used as adjuvant chemotherapy to incomplete surgical resection conferred an apparent improvement over historical survival data using surgery alone, yielding a 57% 1- and 2-yr disease-free state and a 45% survival at 1 and 2 yr for dogs with Grade III tumors. In 19 dogs on a high dose of lomustine given every 21 days, 42% of mast cell tumors showed measurable responses, ranging from stable to partial with one complete response. Neutropenia appears 7 days after treatment, with neutrophil counts of 1,500 cells/ìL. Vincristine doses must be lowered if used in combination with multikinase inhibitors. Lomustine at metronomic doses given daily along with multikinase inhibitors may be helpful and may avoid the adverse events associated with high-dose lomustine.

Recurrent mast cell tumor, forelimb of 16-yr-old mixed-breed dog

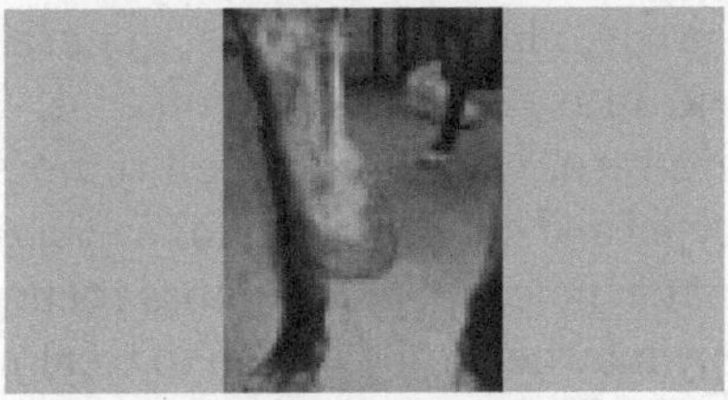

Novel small molecule multikinase inhibitors such as masitinib mesylate and toceranib phosphate are available to treat mast cell tumors and have been shown to be very helpful in management of difficult cases. They inhibit the activated or mutated proto-oncogene, c-KIT tyrosine kinase receptor, which is associated with the development of mast cell tumors. Tissue sampling to determine the biologic aggressiveness of mast cell tumors and to look for the presence of the mutated tyrosine kinase c-KIT receptor has been recommended before starting treatment, although evidence shows that many mast cell tumors respond regardless of their c-KIT status. A study of 202 client-owned dogs with or without prior treatment, having measurable cutaneous grade II or III mast cell tumors without nodal or visceral metastasis, found that masitinib (12.5 mg/kg/day, PO) was a relatively safe and beneficial treatment option.

Another clinical trial found that the receptor tyrosine kinase inhibitor, toceranib (3.25 mg/kg, PO, every 48 hr), results in inhibition of KIT phosphorylation in canine mast cell tumors and was of clinical benefit with continued use. Note that the toceranib dosage stated on the package insert may lead to unacceptable toxicity. Veterinary oncologists are the most informed sources for clinical application using these novel tyrosine kinase inhibitor drugs against mast cell tumors, which may also benefit other malignancies. In cats, cutaneous mast cell tumors are the second most common skin tumor; however, the disease is seen only occasionally in practice. In addition to cutaneous tumors, primary splenic, systemic, leukemic, and GI forms have been recognized. Two distinct variants of the cutaneous form occur—a mast cell type analogous to, but not identical with, cutaneous mast cell tumors in dogs, and a histiocytic type unique to cats. Feline cutaneous mast cell tumors may be either solitary or multiple. Primary splenic, systemic, recurrent, and multiple tumors (five or more) are associated with a guarded prognosis.

Cutaneous mast cell tumors, diabetic cat

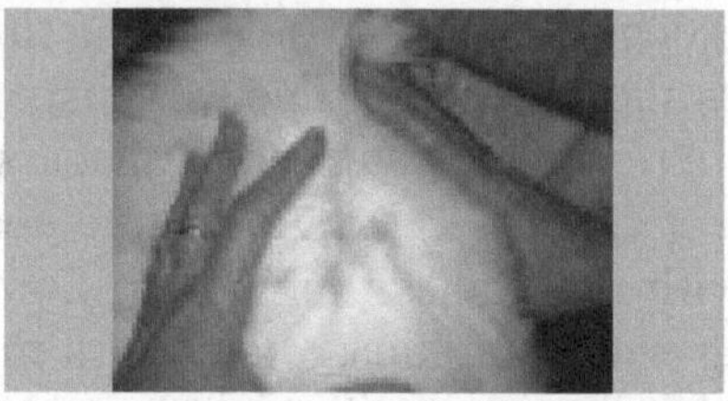

The mast cell type is most common. It is found primarily in cats >4 yr old and may develop anywhere on the body but most commonly on the head and neck. The tumors are single, alopecic nodules, generally 2–3 cm in diameter, that occasionally extend into the subcutaneous fat. Lymphoid nodules are common; eosinophils are rare. Unlike mast cell tumors in dogs, those in cats are generally benign; atypia and clinical behavior are poorly correlated. Surgical excision is the treatment of choice; 30% of tumors recur after surgery and some metastasize. Cryotherapy may be a good option to treat multiple recurrent small lesions to avoid anesthesia. Recurrent tumors may respond to chemotherapy, radiation therapy, and novel small molecule targeted therapy.

The histiocytic type of cutaneous mast cell tumor in cats is recognized primarily in Siamese cats <4 yr old. Lesions may develop anywhere on the body and appear as multiple (miliary), small (generally 0.5–1 cm in diameter), firm, subcutaneous papulonodules. Usually, the older the cat, the fewer the lesions. This variant may be difficult to distinguish morphologically from a granulomatous inflammatory response. Because some of these tumors are reported to resolve spontaneously, treatment may not be necessary.

In horses, mast cell tumors are uncommon and generally benign, although metastasis has been reported and should be considered. There has been debate as to whether mast cell tumors are actually a neoplastic process or an unusual inflammatory response; however, they are currently considered a neoplastic process caused by a gain of function mutation of the KIT proto-oncogene. Lesions may develop anywhere on the body but are most common on the head and legs. Typically, there is a single, solitary mass in the dermis or subcutaneous fat that rarely expands to involve the underlying musculature. Erythema and wheal formation, (Darrier sign) is not a clinical feature of equine mastocytoma.

Most affected horses are male, with an average age of 7 yr (range 1–18 yr). The tumor begins as a nodule composed of a generally monomorphic proliferation of mast cells. As the lesion evolves, the mast cells are limited to aggregates in a fibrous stroma that surrounds large foci of liquefactive necrosis containing numerous eosinophils. In the late stages, the necrotic foci undergo dystrophic mineralization, and mast cells may be very difficult to identify. Once mineralization occurs, the lesion is gritty on sectioning. The classification system used to grade mast cell tumors in dogs is unreliable in horses because of the variable histologic appearance of tumors in this species. Alopecia and ulceration are variable features.

A variant of cutaneous mast cell tumor is seen in newborn foals, in which the lesions may become generalized but regress over time, suggesting an equine equivalent of urticaria pigmentosa in people.

Conventional therapy for equine metastatic mast cell cancer is suboptimal. Excision is the treatment of choice; however, lesions have been reported to metastasize occasionally. Affordable protein-tyrosine kinase inhibitors or "small molecule" drugs such as masitinib mesylate and toceranib phosphate that selectively target mutated forms of the c-KIT tyrosine kinase receptors may become available in the near future.

In pigs and cattle, mast cell tumors are rare. In pigs, most appear as discrete, solitary, cutaneous nodules. Most are benign but disseminated, and leukemic variants do occur. In cattle, most are malignant and characterized by multiple cutaneous nodules often accompanied by systemic involvement; purely cutaneous forms have been recognized occasionally.

TUMORS WITH HISTIOCYTIC DIFFERENTIATION

Tumors with histiocytic differentiation comprise a group of poorly defined skin diseases all characterized by a proliferation of histiocytes (tissue macrophages) in the absence of any known stimulus.

Multifocal histiocytic nodules of the dermis and subcutaneous tissues, dog

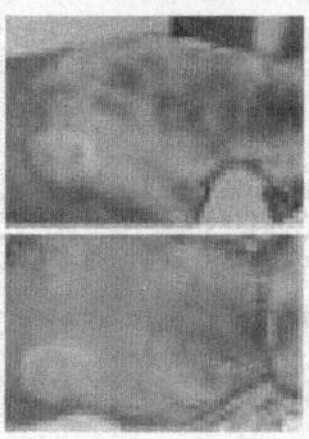

Cutaneous histiocytomas are common in dogs and rare in goats and cattle; it is debatable whether they are found in cats. Strong immunohistochemical evidence suggests that in dogs they are derived from Langerhans (intraepidermal antigen processing) cells. These tumors are typically seen in dogs <3.5 yr old but can occur at any age. English Bulldogs, Scottish Terriers, Greyhounds, Boxers, and Boston Terriers are most at risk. The head (including the pinnae) and limbs are the most common sites of involvement. These classic "button tumors" appear as solitary, smooth, pink, raised nodules that are generally covered by alopecic skin, or they may be ulcerated. They are freely movable. Although a common neoplasm, histiocytomas are not always easy to diagnose histologically and can be confused with granulomatous inflammation, mast cell tumors, plasmacytomas, and cutaneous lymphosarcomas. Canine histiocytomas should be considered benign, and most resolve spontaneously within 2–3 mo without treatment. Surgical excision is optional once the diagnosis is established (which can often be made via cytology).

In goats and cattle, histiocytomas are extremely rare and behave the same as those in dogs. Histiocytomas have also been reported in young cats; however, they most likely represent the histiocytic form of mast cell tumor in cats.

Cutaneous histiocytosis is associated with development of numerous plaques and nodules involving the dermis or subcutaneous fat. It is rare in dogs and can develop at any age but is most common in young adults. Chinese Shar Peis, Collies, Border Collies, Shetland Sheepdogs, Briards, Bernese Mountain Dogs, Golden Retrievers, and German Shepherds may be predisposed. Shar-Peis have excessive skin folding and severe cutaneous mucinosis due to an accumulation of dermal hyaluronan, a protein associated with angiogenesis and tumor cell motility. The nodules and plaques tend to wax and wane, and the extremities and trunk are involved most commonly. Lesions also occur on the face and nasal planum, and swelling of the nares may cause difficulty breathing. Lesions are nonpruritic, and larger lesions may ulcerate. Cutaneous histiocytosis seldom involves internal organs, but its diffuse nature and the unsightly appearance often force the pet owner to consider euthanasia. Various forms of therapy have been tried, including systemic glucocorticoids and a combination of glucocorticoids and chemotherapy and more recently in combination with masitinib and other tyrosine kinase inhibitors. Leflunomide without steroids has been used as well as azathioprine and cyclosporine. Response is variable; the lesions in some dogs respond rapidly and permanently, whereas in others, lesions are either transiently improved or unchanged.

The histiocytoses of Bernese Mountain Dogs are systemic, familial disorders of unknown etiology with two manifestations—a more indolent and generally cutaneous form known as systemic histiocytosis and a more aggressive form in which skin lesions are rare, known as malignant histiocytosis. Malignant histiocytosis has been infrequently identified in other canine breeds. In systemic histiocytosis, males (mean age at onset 4 yr) are affected more often than females. There are multiple cutaneous nodules, papules, and plaques involving the skin (especially of the scrotum), nasal mucosa, and eyelids. The lesions are poorly circumscribed and variably alopecic and may be ulcerated; they develop in waves and slowly regress, only to recur several months later.

The clinical disease tends to become more severe with each new wave of eruptions. Although the skin is the primary target organ, lesions may also develop in other organs, including lymph nodes, spleen, and bone marrow. The disease may be episodic in its clinical presentation, but it is progressive and eventually fatal.

Malignant histiocytosis in thorax, dog

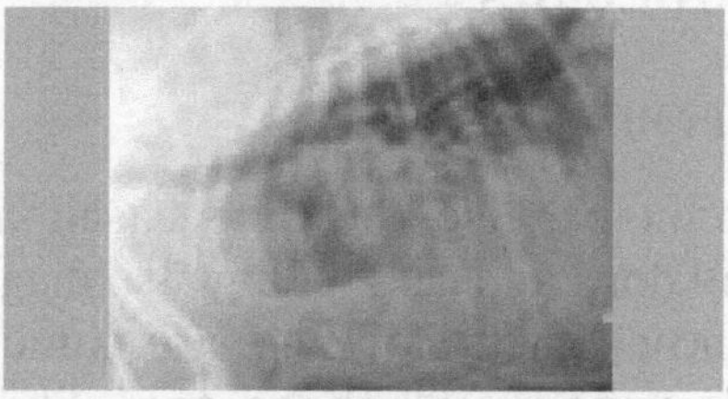

Malignant histiocytosis is seen in male Bernese Mountain Dogs (mean age at onset 7 yr) and, less frequently, in other canine breeds. The lungs, lymph nodes, and liver are the most common organs affected, and the disease tends to spare the skin. Grossly, the lesions appear as large, solitary, firm masses that may cause pleural effusion or efface large portions of affected internal organs. The disorder is rapidly progressive and does not wax and wane as does systemic histiocytosis. Few dogs survive >6 mo.

Various chemotherapeutic regimens mentioned above have been used to treat both forms; hopefully, masitinib and other tyrosine kinase inhibitors used in combination with vincristine or other combinations of antineoplastic drugs will be of benefit. Bovine thymosin fraction 5 may help induce remissions, especially in the systemic form. However, both forms of the disease are ultimately fatal.

EPITHELIAL AND MELANOCYTIC TUMORS OF THE SKIN

Skin tumors are relatively frequent, especially in dogs, horses, cattle and cats. There is a variable incidence, but especially a large diversity of tumor types in the different species of domestic animals. It is admitted that the classification of skin tumors is difficult due to the complex structure, as well as to the ectodermal and mesodermal origin of skin components, to which structural and physiological peculiarities of the skin in different animal species and breeds are added.

Regardless of the principles adopted by different authors, primary skin tumors involve the ectodermal and mesodermal origin of skin, along with epidermal adnexal structures.

Other studies have found the following incidence of skin tumors in cats, of a total number of 340 skin tumors: basal cell tumors 26.0%, mean age 10.3 years; mastocytomas, 21%, mean age 8.6 years; squamous cell carcinomas,15.5%, mean age 11.6 years; fibrosarcomas, 15%, mean age 10.2 years; the only tumor diagnosed under the age of 1 year was the mastocytoma.

Regarding the incidence of skin tumors in the horse, DIETZ and WIESNER (1982) mention: papillomas, fibromas, melanomas, carcinomas and lymphosarcomas.

In dogs, the main skin tumors seem to be mastocytomas with 16.1% and histiocytomas with 14.0%.

TUMORS OF THE EPIDERMIS

Basal cell tumor (basal cell epithelioma)

The basal cell carcinoma, also known as basalioma, is one of the most frequent skin tumors in dogs and cats, being also diagnosed in other animal species. Basal cells, in neoplastic disease, are similar to fetal germ cells, having a marked pluripotential character. This, as well as other considerations, under the conditions of neoplastic changes, requires concomitant or successive treatment of basal cell carcinomas and adnexal epidermal tumors (pilose follicles, sweat glands, sebaceous glands, ceruminous glands, perianal glands).

As a rule, basal cell carcinomas do not evolve concomitantly with adnexal epithelial tumors. The following risk factors are considered in the cytology of these tumors: genetic sensitivity as well as chemical and endocrine carcinogens. In humans, exposure to solar radiation is a risk factor, in animals this remains to be demonstrated.

The incidence of basaliomas is higher in dogs and cats. In dogs, statistics indicate a rate of 3–5–10% of all tumors, and in cats a percentage of up to 18%. In terms of age, in both dogs and cats, the adult age, 7 years in dogs and 9 years in cats, has a higher incidence. Males are more frequently affected by basaliomas, compared to females, and the dog breeds that seem to be more sensitive are Cocker Spaniel and Caniche. Persian and older cats have a predisposition for basaliomas, presenting ulcerated plaques at the level of the head, extremities and neck. These malignant tumors have continuity with the epidermis, evolving with local invasion, but without metastases. In dogs, basal cell carcinomas histologically show cornification, being termed basosquamous carcinomas that are found in old Saint Bernard, Scottish Terrier and Norwegian Elkhound dog breeds. They may be located in any body area, under the form of endo-exophytic nodules or plaques. They have local invasive growth, without metastases.

Locations and the macroscopic aspect are characterized by single or multiple formations, in both dogs and cats. In 1986, FEHRER and LIN reported a multicentric basal cell tumor in a cat. The most frequent location is in the head, neck and shoulder. In general, lesions are not extensive, being well circumscribed, up to 2.5 cm in diameter; a 10 cm tumor has been exceptionally reported in a dog.

Histologically, basalioma cells are characterized by oval prominent nuclei and a small amount of cytoplasm, they generally have small sizes and are uniform. Basalioma cells are very similar to epidermal basal cells. The mitotic index, in the case of this tumor, is high. Basal cells frequently contain abundant melanin amounts, especially in cats, which is why they should not be confused with pigmented tumors. The literature includes multiple classifications of epidermal tumors, all being based on the histological components of this skin segment.

Basal cell tumor, a benign neoplastic proliferation of cells that recapitulate the basal cell layer of the normal epidermis. The neoplastic cells are either small and round or polyhedral, with little cytoplasm and ovoid nuclei. Most tumors show little nuclear pleomorphism and variable mitotic activity. Central cystic degeneration of the tumor lobules may be found. Melanophages may be present in the stroma and within the center of the cyst. Melanin pigment can also be present within the cytoplasm of tumor cells. Basal cell tumor is differentiated from basal cell carcinoma by the lack of invasion and associated fibroplasias at the periphery of the benign tumors.

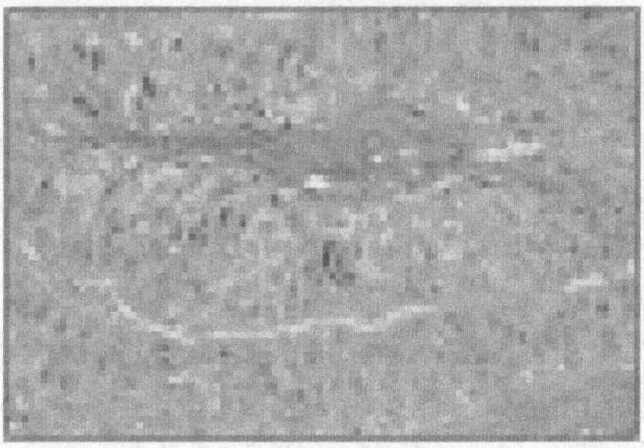

Fig. Basal cell tumor

Basal cell carcinoma, a low-grade malignant proliferation of cells that recapitulate the basal cell layer of the normal epidermis or the adnexa. Infiltrative basal cell carcinoma may have an association with the overlying epidermis. Cords of primitive basaloid cells extend into the dermis and the subcutis. The tumor cells have small, hyperchromatic nuclei with scant cytoplasm, and mitotic figures may be numerous. There is often an extensive proliferation of stromal fibroblasts.

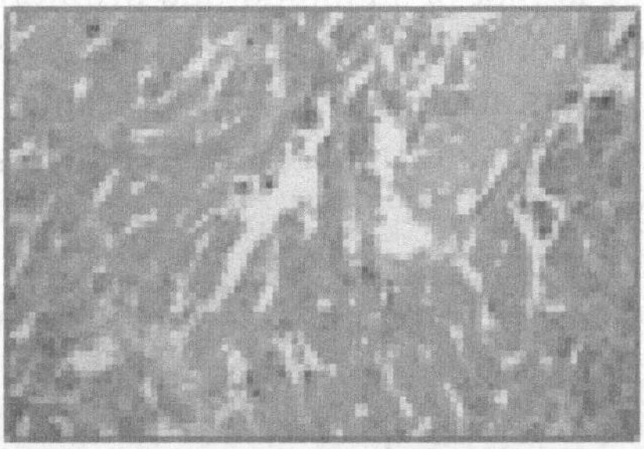

Fig. Basal cell carcinoma, infiltrative type.

Clear-cell basal cell carcinoma is an uncommon variant, the cells are large and polygonal and have clear or finely granular cytoplasm; the nuclei are ovoid and uniform, nucleoli are small, and the number of mitoses found may be quite variable.

The following 5 types are microscopically differentiated, depending on the cell arrangement:

1. Tumors disposed as ribbons or cords, having the appearance of a palisade with hyperchromatic ovoid cells;
2. Tumors formed by compact, trabecular cell groups - solid tumors;
3. Tumors with medusoid disposition;
4. Adenoid tumors;
5. Cystic tumors or basosquamous cell arrangement, frequent in cats, with a keratinized nucleus. In some cases, the same tumor can present a mixed arrangement.

The investigations performed by DITERS and WALSH (1984), on 124 cats with basal cell tumors, indicate a higher frequency of cystic forms compared to compact forms.

Basal cell tumors had a lower incidence in horses, the following tumors being histologically identified: adenoid tumors, solid tumors and medullary-like tumors, and a lower proportion with a cystic cord and space structure. Solid forms were intensely pigmented. In all the cases monitored, no postoperative recurrences were found, after 3 up to 8 years.

Cord-like cell tumors have the appearance of cell fascicles disposed as a palisade, with frequent mucinous degeneration of the stroma between the cords.

Solid tumors are formed by cells with variable sizes, the row of cells being frequently disposed at the periphery, nuclei are perpendicular to the stromal tissue delimiting the neoplasm.

Medusoid tumors appear as small nests formed by basal cells, which project into the surrounding stroma, under the form of fascicles, infiltrative growth aspect. Adenoid tumors display cells arranged as multiple islands, with round or ovoid spaces with homogeneous cells in the center, while at the periphery cells are perpendicular to the connective stroma, rich in fundamental mass. The general appearance is that of adenomatous proliferations.

Cystic tumors, more common in cats, may appear as a large single cyst or multiple smaller cysts. Cysts are formed by undifferentiated basal cells, without a limiting basal membrane, and squamous differentiations may frequently occur. Cysts have an eosinophilic content, probably as a result of the degeneration of basal cells from the solid mass,

These histological forms of basaliomas seem to suggest developmental stages of neoplasms. Basal cell tumors have a slow growth, they are usually encapsulated, they do not develop metastases and have no postoperative recurrences.

Squamous cell carcinoma

Squamous cell carcinoma is a highly malignant neoplasm, frequently associated with solar dermatosis, being the tumor of the cells of the malpighian layer from the epidermis.

It has a high incidence, being reported in all species of domestic animals, with a higher frequency in horses, dogs and cats, especially in adult and old animals. Although breed does not seem to be a risk factor, some authors report a higher sensitivity for Labrador and black Caniche dogs, with location in the digits, and for Dalmatian, Beagle, Whippets and white English Bull Terrier breeds, with location in the flank and the abdomen.

In horses, SCHWINK (1987) noted the presence of squamous carcinomas, with ocular location, in heavy breeds; BERRY et al. (1991) reported an exceptional location in a horse, at the level of the hoof, a squamous carcinoma, in the laminar corium, cells with a high mitotic index.

Squamous cell carcinoma in hens is extremely uncommon, consequently the report of the Japanese authors SUGIYAMA et al. (1991) deserves to be mentioned, according to which a 204 day old laying hen, with a weight of 1.55 kg, had a tumor in the scapular region. The authors described a microscopic image with the proliferation of epidermal cells, corium and subcutaneous connective tissue, with the presence of heterophils, lymphocytes and plasmocytes. Tumor cells had eosinophilic cytoplasm, the cell shapes were round and irregular, pale. The nuclei, round or oval, with 1–2 nucleoli. Mitoses were numerous, keratotic pearls eosinophilic and lamellar. Intercellular desmosomes were identified by electron microscopy.

A very peculiar location was reported by PRAJAPATI et al. in cattle, at the level of the horn core epithelium, under the form of squamous cell carcinoma. Keratotic pearls were microscopically present, in the center of which current alkaline phosphatase activity was found, as well as an increased succinate dehydrogenase activity, in the peripheral cells of keratotic pearls.

Actinic keratosis (solar keratosis), a localized proliferation of epidermal keratinocytes caused by chronic exposure to ultraviolet light. The affected cells have larger, more hyperchromatic nuclei and may exhibit dyskeratosis. There is epidermal hyperplasia and parakeratosis of the stratum corneum. Dermal elastin fiber degeneration may be found. These lesions may progress to squamous cell carcinoma.

Multicentric squamous cell carcinoma in situ (Bowen-like disease) is a malignant tumor of epidermal cells without invasion through the basal lamina zone. This type of squamous cell carcinoma is not associated with ultraviolet light exposure, but an association with papillomavirus infection has been noted. The epidermis and external root sheath in affected areas are thickened and are often sharply demarcated from normal epidermis. Atypical neoplastic squamous cells, often with large, hyperchromatic nuclei, are found primarily in the basal and spinous layer and extend down the external root sheath of hair follicles. Lesions frequently contain abundant melanin pigment. Hyperkeratosis, parakeratosis, and hyperpigmentation of the stratum corneum may be present.

Squamous cell carcinoma, a malignant tumor of epidermal cells with varying degrees of keratinocyte (squamous cell) differentiation. Most cases of squamous cell carcinoma consist of islands, cords, and trabeculae of invasive epithelial cells that almost always have an association with the overlying epidermis, in which there has been a breaching of the basal lamina zone. There is often the formation of keratin pearls (concentric lamellae of keratin within the tumor) by invasive neoplastic cells. The cells and nuclei are large, nuclei are hyperchromatic, and chromatin often appears clumped. Nucleoli vary in size and may be prominent. Whereas those tumors that are well differentiated from keratin pearls, poorly differentiated tumors only show keratinization of individual cells.

Two uncommon histologic variants of squamous cell carcinoma occur:

- - *acantholytic squamous cell carcinoma,* in which there has been dyshesion and degeneration of neoplastic cells resulting in cyst formation with a single peripheral layer of neoplastic cells, producing a pseudoglandular pattern;
- - *spindle cell squamous cell carcinoma,* in which neoplastic cells are fusiform; cytokeratin stains are often positive and help identify these rare tumors as squamous cell carcinoma.

Basosquamous carcinoma, a low-grade malignancy composed of basal cells with foci of squamous or follicular infundibular differentiation, may have an association with the epidermis. Dyskeratotic cells may be seen among the basaloid cells. The keratinized cells display atypia, and occasionally numerous mitotic figures are observed. Melanin may be present in the peripheral basaloid cells. MADEWELL and THEILEN (1987) synthesized the different locations by body regions and species, with moderate/high incidence of squamous cell carcinoma:

Location	Species
eyes and periorbital tissue	cattle and horses
lips, nostrils and nasal septum	cats, dogs and horses
ears	cats, sheep
penis and prepuce	horses
vulva, anus, perineum	cattle, horses
digits	dogs
abdomen	dogs
horn base	cattle
mammary gland	cats, dogs, goats
skin, generalized	chickens

Ocular squamous cell carcinoma in cattle has a variable incidence in different geographical areas, but it has been diagnosed in all countries where cattle are raised. The risk factors incriminated are: genetic predisposition, UV radiation, nonpigmented skin in the orbital area, also including irritations produced by insects, chemical substances and even some viruses. The Hereford and Holstein breeds have a higher sensitivity to squamous cell carcinoma. In India, neoplasms have been diagnosed in buffaloes and sheep. Bovine herpesvirus-5 antigens have been identified in the cytoplasm of neoplastic cells. Bovine papillomavirus has been detected in bovine ocular lesions such as: conjunctival plaques, conjunctival papilloma, eyelid papilloma and cutaneous horn.

Ocular squamous cell carcinomas are preceded by precancerous lesions of the cornea and sclera and/or eyelids.

Ocular lesions progress gradually, starting with precancerous plaques located on the cornea and the sclerocorneal limbus. Conjunctival papillomas develop subsequently. The carcinoma *in situ* develops directly from the plaque or the papilloma. In a more advanced stage, the carcinoma invades the cornea and sclera.

According to the cited authors, histological aspects, starting with benign lesions, towards malignant lesions, have the following succession: acanthosis with ulcerative foci; keratosis, located at the limits of the mucocutaneous junction; papillomas located in the periocular skin, with keratotic lesion appearance; carcinoma *in situ,* difficult to diagnose due to the resemblance to inflammatory ulcers; infiltrative squamous cell carcinoma with diffuse proliferative, frequently hemorrhagic character, and with necrotic foci, superficial bacterial infections.

Metastases frequently occur in regional lymph nodes, but it can occasionally appear in the lungs and other visceral organs. A special case was that of a cow with a suspicion of tuberculous mastitis, with retromammary lymph node hypertrophy. Microscopy showed squamous cell carcinoma metastasis in the retromammary lymph nodes, with tumor lesions on the third eyelid, without affecting ocular structures and without metastases in other organs or lymph nodes[9].

Ocular squamous cell carcinoma in horses has a lower incidence compared to cattle. Advanced age, the absence of periocular pigmentation and UV exposure are risk factors.

In a study performed on 41 horses with ocular squamous cell carcinoma, SCHWINK (1987) found a higher incidence in heavy breeds (46.3%), which had recurrent postoperative neoplasms in a proportion of 42.4%, and 6% developed metastases. Similar incidences have been reported in USA, South Africa, as well as in other countries. In 15 up to 20% of cases, carcinomas have a bilateral evolution; the age group with the highest incidence is over 9 years.

Ocular squamous cell carcinomas in horses require differential diagnosis from other lesions with similar locations (eyelid sarcoid, ocular lymphoma, hemangiosarcoma, mastocytoma, bacterial or parasitic inflammation). Squamous cell carcinoma in other species has been diagnosed in sheep, dogs and cats. Incidence is much lower compared to other species. In sheep, location is ocular and periocular, risk factors such as UV exposure and probably papillomavirus infection being incriminated.

In *dogs,* this neoplasm is uncommon, of a total number of 202 ocular neoplasms only 2% were squamous cell carcinomas; ocular papillomatous proliferations had a benign evolution, without being progressive. Squamous cell carcinoma induced by exposure to solar radiation is most frequently reported in dogs. The cited authors used with promising results etretinate treatment, in early squamous cell carcinomas or in preneoplastic lesion carcinomas. In the case of preneoplastic lesions, evolution was stationary or progressive. In the case of multifocal lesions, treatment did not influence the disease evolution. The authors administered 1 mg/kg etretinate, two times a day, for at least 90 days.

In *cats,* this carcinoma is located in the skin and conjunctival mucosa of eyelids. A higher sensitivity is found in white cats, in which squamous cell carcinoma can develop simultaneously in eyelids, ears, nostrils and lips. Ocular neoplasms in the early stages are circumscribed, precisely located, facilitating surgical intervention. Metastases occur in regional lymph nodes. LIU et al. (1974) diagnosed 8 squamous cell carcinomas in cats, of which 6 were located in the head area, in the gums, causing osteolytic or osteoblastic periosteal lesions in the mandible. In two other subjects, location was in the interdigital skin, inducing osteolytic lesions in digital bones, with metastases in the regional lymph nodes, lungs and eyelids. The malignancy of squamous cell carcinomas in the cat manifests in a similar way to those of the dog, including the invasion of adjacent bones.

Squamous cell carcinoma located in the lips, ears and nostrils is more frequently found in dogs, cats and horses, and less frequently in other species. Squamous cell carcinoma especially occurs in old animals, with non-pigmented skin and UV exposure.

Cutaneous squamous cell carcinoma in the cat is usually located on the external side of the auricle or/and on the margin of the ear tips, lips and nostrils. In California, an incidence of 26.9 cases/100 000 cats is estimated, at the age of over 5 years, in white subjects. The risk of squamous cell carcinoma in white cats is 13.4-fold higher compared to colored cats.

BAER and HELTON (1993) studied 12 cats with the mean age of 12 years, with multicentric squamous cell carcinomas. The affected skin was pigmented, and the neoplasms were located in the trunk, limbs, feet, head and neck, without having been exposed to solar radiation. Multiple skin lesions were 0.5–3.0 cm in diameter, irregular, slightly bulging, having the appearance of plaques or papillae, partially with alopecia. Following microscopic examination, the authors established differences between the multicentric squamous cell carcinoma *in situ* of the cat and precancerous Bowen dermatosis in humans. ASHLEY (1978) showed that human Bowen dermatosis is an intradermal type of cancer rather than a precancerous condition. The result of the study on multicentric squamous carcinoma in cats demonstrates that this is a neoplasm morphologically similar to the bowenoid squamous cell carcinoma *in situ* present in humans. In spite of this, the cat disease is not identical to human Bowen disease and should be referred to as multicentric squamous cell carcinoma *in situ.* The neoplasm is limited to the epidermal epithelium and the follicular infundibulum.

Multicentric squamous cell carcinoma *in situ* in cats differs from Bowen disease in humans, by the multicentric aspect in 75% of the examined cats. In contrast with the predominantly solitary lesions found in human Bowen disease, multiple development centers are typical in humans with solar keratosis; however, the lesions of the cats included in the study did not appear in poorly pigmented or non-pigmented skin areas, and the majority of cats were minimally exposed to solar radiation.

In dogs, the locations in the lips, ears and nostrils are similar to those mentioned in cats, being associated with the lack of pigmentation of these skin areas. Malignant evolution with deep infiltrations and ulcerations is associated with metastases in regional lymph nodes.

Subungual keratoacanthoma (nailbed keratoacanthoma), a benign neoplasm, is symmetrical and unencapsulated with irregular scalloped borders and a central core of keratin; it is composed of sheets of keratinocytes that have abundant pale eosinophilic cytoplasm, distinct cell borders, and large vesicular nuclei. A variable number of apoptotic keratinocytes may be found; this tumor may cause lysis of the phalangeal bone.

Squamous cell carcinoma located in the epithelium of the horns and nails is found in cattle, a particularly high incidence being reported in zebu cattle, in India. The authors report an extremely high frequency of squamous cell carcinomas at the base of the horns, mentioning 10% of the animals killed in slaughter houses, exclusively castrated males, around the age of 5 years. The possibility of actinocancer and/or repeated traumas associated with castration is mentioned. Epidemiological studies show the implication of the hormonal factor (castrated or cryptorchid males), repeated traumas, genetic predisposition and UV rays. The neoplasm in its progressive evolution causes osteolysis in the bone plug, invades the frontal sinus, and can extend to the nasal cavity, frontal bones, pituitary fossa and optical fossa (PACHAURI and PATHAC, 1969).

The location in the nail epithelium has been relatively frequently reported in dogs, with chronic inflammations, repeated traumas and non-pigmentation of the area. The cited authors mention the Poodel, Schnauzer and Labrador breeds with a higher incidence of squamous cell carcinoma in the nail epithelium. Following a study in 213 dogs with cutaneous squamous cell carcinoma, the same authors found 128 cases (60%) located in the nail epithelium, at a mean age of 9 years, without sex-dependent differences.

The location in the nail epithelium manifests with local invasion, the invasion of phalanges, of the metacarpal or metatarsal bone, and metastases occur in lymph nodes, lungs and other visceral organs.

Tumors of the hoof wall in horses are rare; BERRY et al. describe a squamous cell carcinoma of laminar cells, with a high mitotic index and local invasion.

Surgical treatment, in the case of dogs, with the excision of the digit in the early stages, has good results.

Squamous cell carcinoma of the prepuce, penis, vulva, anus and perineal region. Squamous cell carcinoma of the penis and prepuce is more common in old horses and stallions, as well as in other animal species. Tumors are present in unpigmented skin areas and develop progressively as premalignant inflammatory processes that evolve into infiltrative and ulcerative squamous cell carcinoma. The accumulation of smegma in the prepuce is associated with the initiation of carcinogenesis.

In cows and mares, squamous cell carcinoma is frequently located in the vulva. In sheep, LAGADIC (1980) reports an enzootic of vulvar squamous cell carcinoma. The author mentions a perineal carcinoma enzootic in Angora goats from South Africa, studied by THOMAS (1929), while the same type of perineal neoplasm is reported in India by DAMODORAN and PARTHASARTH (1972). In most cases, in sheep and goats, lesions are located at the mucocutaneous junction or in the perineum, with depigmented skin. Lesions are of precancerous type such as keratosis, cutaneous horn or papillomas, and subsequent ulcerations occur, as well as secondary infections or other lesions. YERUHAM et al. (1993) described neoplastic lesions with the same characteristics in two goats, with perineal location, emphasizing that skin was pigmented and the possibility of UV effects more reduced. BABA (1995) reported a case of squamous cell carcinoma in sheep, located in the left thoracic side, caused by the application of hexachlorocyclohexane on a wound and exposure to solar radiation. The neoplasm developed slowly, but became generalized after several months by metastases in the majority of lymph nodes and in the lungs.

Vulvar cancers, in Ayrshire cattle, from Kenya, under the form of epizootics, diagnosed as squamous cell carcinomas, were reported by BURDIN. About 3% of the livestock were affected, reaching 30%, and the neoplasm started with precancerous lesions, such as papilloma, hyperacanthosis, and hyperkeratosis.

Squamous cell carcinoma in the ventral abdominal region in the dog has epidemiologically proved to be an actinotumor. Tumors develop in skin areas with little or no pigmentation, following exposure to sunlight. The carcinoma develops as a multicentric neoplasm that is ulcerated, forms crusts and eschars. Microscopically, precancerous lesions appear, as well as epithelial hyperplasia, keratosis, and finally invasive carcinoma *in situ* and metastatic carcinoma, sometimes with scirrhoid reaction.

The morphology of squamous cell carcinoma is characterized by papillary growths of variable sizes, usually with a cauliflower-like appea-rance, firm and rough on palpation, of white-gray color, unclearly delimited, ulcerated and with red striae. At other times, the tumor develops as an ulceration covered by crusts, gradually becoming crateriform. In the case of the tip of the ear, in cats, the lesion starts as an erythema of the ear tip and margins, with progressive hair loss, desquamation, formation of crusts and ulcerations, with necroses of the ear structures (skin and cartilage).

Histologically, the neoplasm has a characteristic aspect, which facilitates a positive diagnosis. Hyperkeratotic, parakeratotic, ulcerative and acanthotic lesions are predominant. Neoplastic cells are easily recognized, having the characteristics of spinous cells, a polyhedral shape, vesicular nuclei and extremely prominent nucleoli. Cells are arranged irregularly or under the form of cords. Carcinomas with numerous large keratotic pearls and obvious cell differentiation with low mitotic index are considered to have a lower malignancy grade and slow evolution. Keratotic pearls are concentric arrangements of parakeratin layers and/or hyperkeratosis. The high malignancy forms are composed of undifferentiated cells, large, round, atypical cells with eosinophilic cytoplasm, sometimes with multinucleate cells and a high mitotic index.

The periphery of neoplastic cell islands is always delimited by basal membrane and the presence of basal cells.

Laboratory studies have attempted to establish some criteria that could assess the malignancy grade, prognosis and optimal treatment for different evolution stages, forms and locations of squamous cell carcinoma.

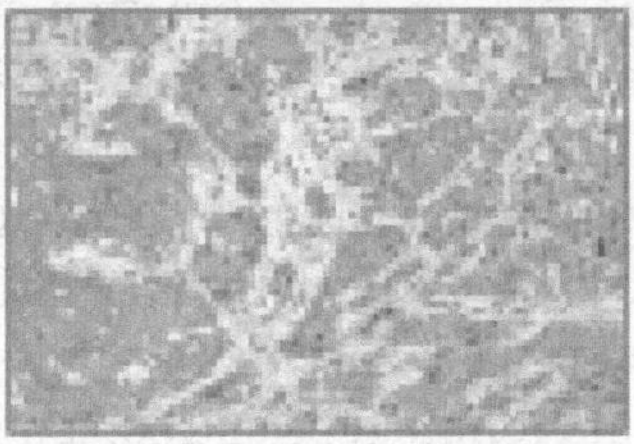

Fig. Squamous cell carcinoma. Islands and cords of malignant squamous cells surrounded by desmoplasia.

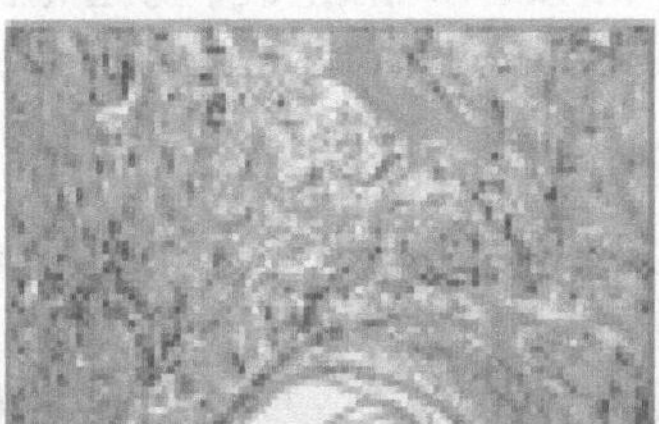

Fig. Squamous cell carcinoma, poorly differentiated.

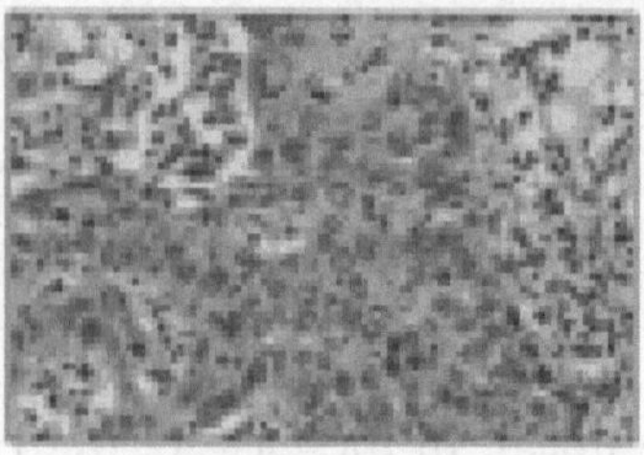

Fig. Squamous cell carcinoma, poorly differentiated.

BARRY and SHARKEY (1986) studied the possibility to establish a histological scale for the evaluation of squamous cell carcinoma malignancy. The authors used morphometric methods, assessing the keratinization degree and architectural aspects of the tumor. In this sense, the volume of neoplastic cells (basal cells, spinous cells and keratinized cells) was estimated, data being statistically processed. It should be mentioned that investigations were performed in 6 squamous cell carcinomas from affected individuals and from transplantations of these tumors in nude mice. The conclusion is that histological differentiation can be highly reliable and that the method is sensitive enough to be applied in the solution of practical problems of histological importance.

BABA and PRICĂ (1988) performed morphometric studies on squamous cell carcinomas in cattle. In all cases the nucleus/cytoplasm ratio was in favor of the nucleus; the mitotic index was two-fold higher in poorly keratinized or non-keratinized carcinomas, compared to intensely keratinized carcinomas; in the case of intensely keratinized carcinomas, mitoses were predominantly in the basal layer; nucleoli had larger sizes and were less numerous in carcinomas with a higher mitotic index, while nucleoli from carcinomas with a low mitotic index were smaller and more numerous. The conclusion is that one single parameter (nucleoli/nucleus, etc.) in the estimation of squamous cell carcinoma malignancy is not sufficient, and the morphometric assessment of several parameters is required in order to appreciate the malignancy grade.

Studies of a completely different nature, regarding the reaction of organisms with carcinomas, were performed by JUN and JOHNSON (1979 a, b). The authors found that in sheep with carcinomas, the blastogenic response of peripheral lymphocytes to phytomitogens and squamous cell carcinoma extracts decreases significantly with the increase in tumor maturity. The tumor avoids its own rejection by the direct or indirect formation of serum blastogenic inhibitor that causes the immunosuppression of T cell activity.

Papilloma and fibropapilloma

Papilloma and fibropapilloma are benign tumors, frequently found in cattle, horses and dogs, and exceptionally in cats; their etiology is recognized as being of viral nature, the papillomavirus, *Papovaviridae* family. Papillomaviruses produce two types of lesions: squamous cell carcinoma and fibropapilloma, while intermediate lesions can also be found. In the etiology of these tumors, in addition to papillomavirus infection, some co-carcinogenic factors are also incriminated, such as: genetic factors; *immune* depression; hormonal factors, sunlight (UV) exposure of pigmented skin; chemical synthesis substances, etc.

The incidence of these benign tumors is not generally influenced by sex and race; in cattle they may appear under the age of 2 years, and in horses more frequently between 1 and 2 years of age. Angora goats seem to be more sensitive, and in the Saanen breed papillomatosis is more frequent during the lactation period.

Papilloma and fibropapilloma in cattle are usually located in the head, neck, shoulders, neck folds, limbs and udder[89]. As it has been mentioned, viral etiology has been demonstrated, to which risk factors have been added. Tumor formations are multiple, they can be generalized in almost all body areas, having a typical papilloma appearance, of variable sizes, from 1–2 cm to large structures, which are rough, dense, cauliflower-like. Tumors are exophytic, sometimes with a large attachment base or pediculate.

The typical papilloma has a size of 1–2 cm, a filiform wart aspect, with exophytic growth, epidermal hyperplasia through the expansion of the spinous layer against a dermal structure. The cells of the spinous layer undergo hydropic ballooning of the cytoplasm that contains large eosinophilic keratohyaline granules; nuclei are vesicular, containing viral particles, easily detectable by electron microscopy, and viral antigens detectable by immunofluorescence.

Fibropapilloma: characteristic lesions consist of hyperkeratosis, acanthosis and growth in depth under the form of epithelial growth in a massive dermal proliferation. Fibroblast cell proliferation is massive under the form of perpendicular fascicles. Frequently, epidermal proliferation is minimal, with slight acanthosis.

Clinical and experimental studies have demonstrated that the development of one or the other tumor form, papilloma or fibropapilloma, depends on the anatomical location and histological structure, on the virus type, as well as on virus infection or inoculation, which can be superficial or deeper. In this way, the proliferative tissue reaction can be superficial (epidermal) or deep (dermal).

Genital fibropapilloma is found in adult cattle, in females on the vulvar mucosa and in males on the penis. The virus is transmitted by breeding. Macroscopically, it starts with small formations that gradually reach 2–5 cm in diameter; since they are well vascularized, tumors bleed easily, and superinfections occur. Microscopic structure is characteristic of fibropapillomas with the deep penetration of the epithelium into the conjunctival vascular stroma.

Mammary gland papilloma appears with a multiple aspect, in adult cows, with two microscopic forms: fibropapilloma or papilloma and squamous cell papilloma.

Digestive tract papillomatosis is of viral nature, although feeding on fern is also incriminated, sometimes in association, or the transformation of benign forms into malignant ones, of carcinoma type. In an extensive study, from an area with enzootic hematuria, the disease was undoubtedly caused by the consumption of fern *(Pteridium acquilinum),* and digestive papillomatosis was sporadically present.

Digestive papillomas are located in the oral cavity, tongue, esophagus and rumen, in both young and adult cattle.

Fibropapilloma in the horse is of viral nature, and macro- and microscopic aspects are characteristic, having a pediculate or sessile appearance. It is commonly located in the lips, nose and legs, but it can also be found in the oral cavity, ocular area and genital system, at over 3 years of age. Proliferations usually are 2 to 10 cm in diameter, especially in solitary forms with ocular and genital location.

The literature reports skin papilloma cases in 1–5 day old foals, which suggests the possibility of papillomaviruses infection during intrauterine life.

Macroscopically, lesions are exophytic, circumscribed, horny masses and a cauliflower-like surface. Microscopically, the tissue reaction is that of squamous epithelium.

Canine papillomatosis, in the form of multiple skin proliferations, is most probably non-infectious. In multiple skin papillomatosis in dogs, canine papillomas can have the following locations: oral location, skin location, usually on the limbs, the posterior abdominal area and occasionally on the body; ocular locations, in the cornea and conjunctiva.

The oral location with typical lesions has been reported in dogs, coyotes and wolves. In internal genital organs, multiple papillomas have been described in males, after the administration of synthetic estrogens. Multiple oral papillomas are more common in young subjects, being located on the mucosa of the lips, gums, tongue and epiglottis. Papillomas develop relatively rapidly, as pediculate forms or in the depth of the mucosa, but sometimes they regress without completely disappearing. Eyelid papillomas may appear as juvenile papillomatosis or as a solitary lesion in old dogs.

Histologically, the surface of the oral papilloma is covered with stratified squamous epithelium, with acanthosis and hyperkeratosis. Spinous cells show obvious tonofibrils and cytoplasmic inclusion material in the cells of the granular layer. In the vesicular nucleus of spinous cells, intranuclear inclusions characteristic of papillomaviruses are identified by electron microscopy and immunofluorescence. *The skin location* usually appears as a solitary tumor in old dogs, the cases of multiple cutaneous papillomatosis being more rare. Early lesions have a smooth surface, becoming rough in time, with deep crevasses. The literature does not mention the possibility of experimental transmission, although there is evidence that cutaneous papilloma in the dog is of viral nature.

The ocular location has been reported in both puppies and adult dogs. In tumor cells, viral particles have been evidenced, the lesion has been experimentally reproduced.

Papillomatosis in sheep and goats develops with locations in different body areas. In sheep, primary tumors occur in the head and ears, as papillomas and fibropapillomas, squamous cells become hypertrophied, forming horny skin plaques. In goats, neoplasms appear both in pigmented and non-pigmented skin areas. They can be located in the head, neck, trunk and mammary gland, usually a multicentric form. In this species, viral etiology could not be demonstrated, but the action of solar rays seems to favor and/or induce proliferation. Clinical and epidemiological data demonstrate the evolution of papillomas and fibropapillomas into squamous cell carcinoma.

Characteristics of papillomavirus infections in mammals (Theilen, 1983; Jubb, Kennedy and Palmer, 1993)

Virus	Carrier	Location	Histology
Human papillomaviruses	Human	Different	Papilloma
Equine papillomaviruses	Horse	Skin (lip)	Papilloma
Canine papillomaviruses	Dog	Oral mucosa and skin	Papilloma
Rabbit papillomaviruses (Shope)	Rabbit	Skin	Papilloma
Rabbit oral papillomaviruses	Rabbit	Oral mucosa	Papilloma
Mastomys natalensis papillomaviruses	Rodents	Skin	Papilloma
Chaffinch papillomaviruses	Birds	Skin	Papilloma
Bovine papillomaviruses			
type 1	Cattle	Skin	Fibropapilloma
type 2	Cattle	Skin	Fibropapilloma
type 3	Cattle	Skin	Papilloma
type 4	Cattle	Digestive tract	Papilloma
type 5	Cattle	Udder	Papilloma
type 6	Cattle	Skin	Papilloma
Sheep papillomaviruses	Sheep	Skin	Fibropapilloma
European elk papillomavirus	Elk	Skin	Fibropapilloma
Deer fibromavirus	Deer	Skin	Fibroma

DIAGNOSIS in the case of papilloma or fibropapilloma is relatively easy to make, considering the somewhat specific locations for the different animals; in dogs, oral papillomas require differential diagnosis from melanotic nodules, by histological examination.

TREATMENT, for all species, will be surgical or electrosurgical, usually with good results. The possibility of the spontaneous disappearance of papillomas is considered. In dogs, the oral form involves the risk of the extension of the proliferation to the pharynx and larynx; consequently, electrosurgical excision is required. Vaccinations have very good results and autovaccination is recommended especially for cattle, dogs and horses. The vaccine is prepared from one part triturated tumor tissue in nine parts physiological serum, which is filtered and stored at 4°C. Inoculations are performed intradermally, in dogs 0.5–1 ml once a week, three treatments; in horses and cattle, intradermally as well, 1–5 ml once a week, three treatments being sufficient.

Autohemotherapy, in cattle with cutaneous papillomas, has promising results, the success rate being 70% in cows treated with 30 ml blood harvested and immediately inoculated subcutaneously, 3–4 injections at 7–10 day intervals.

Prognosis is favorable for skin infections in cattle, horses and dogs. In the case of goats, sheep and rabbits, papillomas or fibropapillomas may evolve into squamous cell carcinomas.

Infundibular keratinizing acanthoma (keratoacanthoma, intracutaneous cornifying epithelioma

This tumor is also known as scaly papilloma or intracutaneous cornifying epithelioma and has only been observed in dogs.

Incidence is higher around the age of 5 years, males being more frequently affected than females, and the breeds selected for hunting seem to be more sensitive, developing a generalized form.

The tumor consists in epidermal invagination, the keratoacanthoma has the appearance of a nodule, cyst, 1–3 cm in diameter, being located in the back, neck, thorax, more rarely in the limbs, head and tail. It usually occurs as a single tumor, but in some breeds it can also be multiple or in succession.

Microscopy shows a dermal cavity filled with keratin, which has an orifice towards the skin surface. The wall is formed by a thick layer of stratified and well differentiated scaly epithelium. Keratin appears as a concentric lamellar mass, with a keratotic pearl aspect. Some spinous cells are arranged as epithelial cords or islands in the keratin mass. The tumor formation exerts a compression force on the adjacent connective tissue, conferring an encapsulated appearance.

Microscopic diagnosis can be difficult due to certain similarities to squamous cell carcinoma (keratotic pearls).

SEBACEOUS GLAND TUMORS

Sebaceous gland tumors may be classified depending on histological structure, on cells participating in proliferation. In the majority of cases, the evolution of tumors is benign. The following proliferative forms of sebaceous glands have been identified: hyperplasia; adenoma; epithelioma; and carcinoma.

Sebaceous gland hyperplasia appears as multiple superficial multilobular nodules, several mm (0.1–0.5 cm) in diameter, having a yellow color and easily ulcerating. In cross section, they appear as hyperplastic and hyperkeratotic epidermal nodules. Histologically, they are characterized by symmetric hyperplasia of the lobules, the gland is significantly enlarged, the canal is enlarged and obliterated, lined with keratinized cells. Sebaceous cells are mature, and basal cells are poorly evidenced.

Sebaceous gland hyperplasia is common in old dogs, especially males.

Sebaceous adenoma is a benign tumor, common in dogs, which is morphologically difficult to differentiate from hyperplasia. The lobular proliferation of the gland is important but less symmetric, hyperplastic basal cells are evidenced, while the gland duct is more difficult to detect. As WILCOCK (1993) notes, sebaceous adenoma is difficult to differentiate from sebaceous gland hyperplasia.

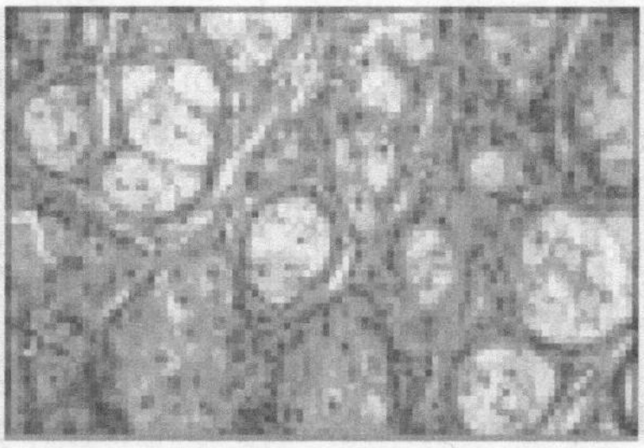

Fig. Sebaceous adenoma. Lobules composed primarly of sebocytes.

In dogs, in which adenomas are frequent, nodules have different aspects depending on the affected skin area. Frequently, they are found in the trunk, head and neck as multiple nodules, 0.5–3 cm in diameter, moderately compact, non-capsulated, with a white or black fatty content, with muddy appearance.

Sebaceous ductal adenoma is a benign tumor characterized by a preponderance of ducts, and fewer sebocytes and basaloid cells. There is extensive ductal differentiation, but the ducts are haphazardly arranged and interspersed with basaloid reserve cells and lesser numbers of mature sebocytes. There is no well-organized pattern of ducts and associated glands, as there is with sebaceous hyperplasia.

Sebaceous epithelioma is a structure with the proliferation of basal cells, and in many respects is similar to basalioma. It is frequently located on the eyelid margin, originating from the *meibomian* gland. The proliferation of undifferentiated basal cells is nodular, which can represent a differentiation criterion. Lipid vacuoles appear in the cell cytoplasm. In the center of the gland there are groups of mature sebaceous cells, some of which are disintegrated, forming amorphous material cysts. Many sebaceous epitheliomas are pigmented due to melanocytes infiltrated among basal cells. The epithelioma may be bordered by lymphocyte and plasmocyte infiltrate.

The epithelioma has a slow evolution, it remains well circumscribed, but it can relapse after incomplete removal.

Sebaceous carcinoma is uncommon in all species, being occasionally found in dogs, and rarely in cats. In a study performed on 32 sebaceous carcinomas in dogs, STRAFUS (1976) mentioned the age of 8–12 years, with locations in the head, abdomen and fore limbs. Tumors were single or multiple, frequently having a 2 cm diameter.

Structure is characterized by infiltrative growth, cells are little delimited, pleomorphic, with large variations in shape and size, the cytoplasm is eosinophilic, with fine vacuoles, and nuclei are large and vesicular. The characteristics of malignancy are obvious by a high mitotic index, anisocytosis, anisokaryosis and hyperchromasia.

Metastases in the regional lymph nodes and lungs have been reported.

Meibomian adenoma, a benign tumor arising from the meibomian (tarsal) glands located on the inner aspect of the eyelid. The overlying epidermis is often hyperplastic and may be papillomatous. Secondary inflammation is common. A severe granulomatous response, with multinucleated giant cells, occurs when the tumor ruptures and releases the sebaceous secretion into the surrounding stroma. There is a tendency for these tumors to be highly melanized.

Meibomian ductal adenoma, a benign tumor arising from the meibomian glands with a preponderance of ducts, and few sebocytes and basaloid cells; similar to sebaceous ductal adenoma.

Meibomian epithelioma, a tumor of low-grade malignancy arising from the meibomian glands with a preponderance of basaloid cells, and few sebocytes and ducts; similar to sebaceous epithelioma.

Meibomian carcinoma, a malignant tumor; its histologic features are similar to those described for sebaceous carcinoma.

APOCRINE AND MODIFIED APOCRINE GLAND TUMORS

Apocrine glands are present and more developed in the paw pads of dogs and cats. Sweat gland tumors are more frequent than sebaceous gland tumors, being described in dogs and cats, and more rarely in other species. In 1954, COTCHIN mentioned a 2.4% incidence of sweat gland tumors, of all skin tumors. The age of the affected dogs was over 6 years, the following types being mentioned: cysts, hyperplasia, cystic hyperplasia, adenoma, adenocarcinoma and mixed tumors.

The more common locations are in the head, neck, back and flanks, but the highest incidence is reported in paws.

Cystic dilation can be caused by dilation with or without epithelial hyperplasia. These formations appear as well delimited nodules, 1–4 cm or even more in diameter, and in cross section they have a waxy content, of gray or light brown color.

Epithelial hyperplasia is histologically found, with aspects suggesting physiological states of secretory activity; aspects can be catalogued as cystic adenomas, with papillary proliferations.

Apocrine adenoma may evolve as cystic adenoma and papillary adenoma.

Cysts can be single or multiple, lined by epithelium with well differentiated prismatic cells, with obvious secretory activity. The microscopic aspect is reticular, with large cystic spaces, delimited by narrow septa, lined by unistratified epithelium. The cystic content causes compression atrophy in the epithelium, and in content-free formations the epithelium is cylindrical. The tumor has benign features, being well circumscribed, with 1–4 cm diameter or more.

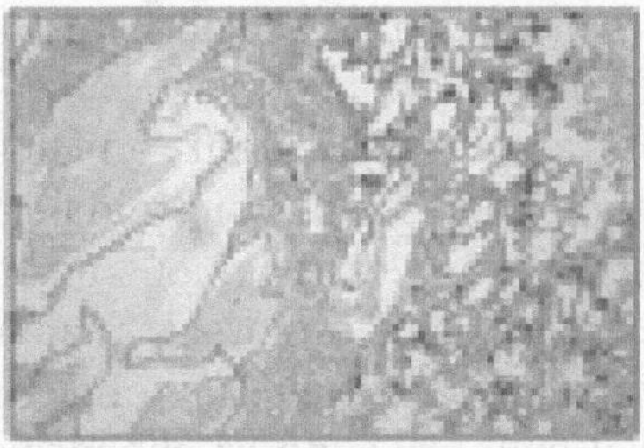

Fig. Apocrine multiple cysts.

Papillary adenoma is formed by cuboidal or cylindrical cells, proliferated in the lumen of the nodule, which is bordered by fibrous tissue, sometimes with monocyte infiltrate. Simple or ramified papilliferous formations proliferate in the cyst lumen. Epithelial cells that cover papilliferous proliferations have secretory activity, which can exert compressions on the epithelium. Myoepithelial cells have been found between epithelial cells and the basal membrane.

Microscopically, sweat gland carcinomas may develop as papillary and tubular types. WEISS and FRESE (1974) describe 4 types: papillary, tubular, solid and signet-ring cell carcinomas.

Papillary carcinoma may be similar to benign forms, but the shape and size of the nuclei, the mitotic index and cellular anaplasia should be considered. The tumor is situated under the epidermis and penetrates deeply in the subcutaneous connective tissue. It is poorly delimited, ramified, with many interpenetrating papilliferous formations. The epithelium that covers the tumor usually consists of 2–3 rows of cubical or cylindrical cells, which have oval nuclei with weak basophilic staining and finely granulated, eosinophilic cytoplasm. In some areas, the epithelium is multistratified. Epithelial cells grow towards the lumen, with a bud or palisade appearance, without having a connective skeleton. Numerous cells present cytoplasmic vesicles, giving the cell the aspect of a signet ring. Mitoses are present.

Tubular carcinoma is the most common malignant neoplasm, with aggressive infiltrative character in the dermis, connective tissue and even muscles. It develops under cystic, solid and scirrhous forms.

According to the cited authors, *solid carcinoma* is the most frequent. Metastases are common in regional lymph nodes, but may also occur in visceral organs. Tumor cells grow infiltratively, arranged as lobules or as round or irregular cords, and they can be disseminated or grouped tightly. These formations are delimited from the rest of the connective tissue by a basal membrane, frequently with ramified canals that communicate directly with the skin surface. In subcutaneous tissue, the growth of canals is complete. Tumor cells are undifferentiated and polymorphic, they have a polyhedral form and their cellular limits are hardly detectable or invisible. Nuclei have variable sizes, poor basophilic staining, one or several nucleoli. The cytoplasm is finely granular and poorly eosinophilic. Mitoses are numerous.

Sweat gland tumors are less frequent, and may be either benign or malignant. The proliferation of epithelial cells and periglandular myoepithelial cells is structurally remarked, and cartilaginous or bone nodules can appear, aspects also found in mammary gland tumors. In malignant forms, pleomorphic cells grow infiltratively, being arranged as solid, compact cords, with the presence of myoepithelial cells. Hyaline or chondroid nodules appear in the stroma. Anal sac gland carcinoma (carcinoma of the apocrine glands of the anal sac), a simple malignant tumor arising in the wall of the anal sac with differentiation to apocrine secretory epithelium, is more uncommon, but is a highly malignant tumor in old female dogs, while in males it is rare. The solid type consists of sheets of tumor cells subdivided by thin bands of fibrous tissue. Cells have round-to-oval, euchromatic-to-hyperchromatic nuclei with a prominent nucleolus, little pleomorphism, and a small amount of eosinophilic cytoplasm. In the rosette type, tumor cells with peripherally located nuclei and eosinophilic cytoplasm are radially arranged around a small central focus of eosinophilic secretion. The tubular type has accumulations of eosinophilic secretion within tubular lumina, and neoplastic cells often have abundant eosinophilic cytoplasm, hyperchromatic nuclei, and variable mitotic activity. The tumor is invasive and may evoke a strong desmoplastic response.

Histologically, it has the structure of a tubular to solid carcinoma. It usually produces metastases in regional lymph nodes and viscera. Location and especially massive growth as a visible compact mass does not pose diagnostic problems. Eccrine adenoma, a simple benign tumor with differentiation to eccrine secretory epithelium, is a rare tumor in domestic animals. The tumor cells have very pale eosinophilic cytoplasm with basally located nuclei. Little nuclear pleomorphism or mitotic activity is evident.

Eccrine carcinoma, a malignant tumor with differentiation to eccrine secretory epithelium, is a very uncommon tumor. In some cases, the neoplastic cells have pale eosinophilic cytoplasm and small hyperchromatic nuclei, suggesting they are differentiating to eccrine glandular epithelium. This tumor must be differentiated from metastatic carcinomas, particularly in the cat.

METASTATIC TUMORS TO THE SKIN

Metastatic spread of tumors to the skin is uncommon and occurs via lymphatics, the hematogenous route, or implantation metastasis following surgery. Immunohistochemistry and the appropriate history of a primary tumor at a distant site are helpful in establishing the diagnosis of a metastatic tumor.

Pulmonary carcinoma with digital metastases is the most common tumor metastatic to the skin in the cat. There is involvement of the dermis, subcutaneous tissue, and third phalanx by neoplastic epithelium arranged in solid islands, nests, and glandular structures. Short papillary proliferations may extend into the glandular lumina. Focal squamous differentiation may be found, but the majority of the cells have abundant eosinophilic cytoplasm with basally located ovoid, euchromatic nuclei; cilia may be present on neoplastic cells.

Cutaneous metastasis of mammary carcinoma is most often seen in the dog, involving the inguinal area, due to direct invasion of dermal lymphatics or by retrograde metastasis of tumor cells from the external inguinal lymph node to the dermis of the inner thigh. Differentiation of primary apocrine and mammary carcinoma is very difficult; it is therefore essential that the presence of a primary mammary tumor be established.

Implantation metastasis secondary to prior surgery for prostatic carcinoma, transitional cell carcinoma, and colonic carcinoma may be found in the skin, or there may be retrograde lymphatic metastasis to the inguinal skin (103).

CUTANEOUS CYSTS

Epidermal cysts appear as single or multiple circumscribed formations, generally 1 cm in diameter, located in the dermis or subcutaneously, more commonly in dogs and less frequently in other species (cats, cattle, horses). Young dogs and especially the middle age group are affected. In dogs, location is in the head, neck and the sacral region, the cyst having a semifluid brown content or a stratified cornified content. Microscopy shows the stratified keratinization of the lumen, sometimes the content is calcified, the cyst periphery is formed by stratified horny epithelium, subject to content compression atrophy, with the presence of hairs. The content may be infected or, more rarely, the cyst may turn into squamous cell carcinoma.

Surgical treatment, by total removal, results in rapid healing without complications.

Dermoid cysts are similar to epidermal cysts, except that they include in their composition adnexal skin structures. Cysts are filled with keratin and sebaceous material, cholesterol crystals and coiled hairs. Lesions have been more frequently reported in the midline of the body in a breed of Rhodesian dogs.

Follicular cysts are most frequently found in animals, and they develop by the accumulation of follicular or glandular products following the congenital or acquired obliteration of the excretory or follicular duct orifice. The cyst wall is lined by cylindrical basal cells or cubic squamous cells and an abrupt keratinized area. The cyst contains horny or lamellar cells, hairs and cholesterol crystals.

Around the cyst, sebaceous or sweat glands or atrophied pilose follicles are frequently found. Inflammatory complications are common, with specific reaction in the surrounding area.

Infundibular cyst (epidermoid cyst, epidermal cyst, epidermal inclusion cyst), a simple cyst lined by stratified squamous epithelium. All four layers of the normal epidermis or infundibulum, including a granular cell layer, are present. Lesions are solitary or multiple, but usually unilocular. The basal, spinous, and granular cell layers may all be well developed. The cornified cells that occupy the cyst are loosely packed (basket-weave orthokeratosis).

Dilated pore, a dilated infundibular cyst with hyperplastic epithelium forming the deepest portions of the cyst wall. The lining epithelium has a prominent granular cell layer that produces compact laminated keratin that can protrude through the pore, forming a cutaneous horn.

Isthmus cyst, a simple cyst lined by cells that resemble the middle segment of the anagen follicle and the lower segment of the catagen follicle. The cyst is lined by stratified squamous keratinizing epithelium, but lacks a granular cell layer. The keratinocytes above the basal layer have abundant pale eosinophilic cytoplasm, and intercellular bridges are difficult to identify. The cyst contents are less eosinophilic and less compact or laminar than those of an infundibular cyst.

Panfollicular (trichoepitheliomatous) cyst, a cyst in which two or all three layers of follicular epithelium are identified. The cyst is lined by stratified epithelium with foci of changes as described above for the infundibular cyst and isthmus cyst; small basophilic cells that show abrupt keratinization to shadow cells. Small foci of cells with intracytoplasmic trichohyalin granules may be found between zones of infundibular and matrical keratinization. These cysts may progress to trichoepithelioma.

Dermoid cyst (dermoid sinus), a congenital dermal and/or subcutaneous cyst lined by epidermis with mature dermal and appendageal structures in the cyst wall. The cyst may communicate with the overlying epidermis via a small pore, and is often found on the midline of young animals. It is lined by stratified squamous keratinizing epithelium with hair and keratin in the cyst lumen and hair follicles and sebaceous glands within the dermis that surrounds the cyst.

Sebaceous duct cyst, an intradermal cyst lined by a thin squamous epithelium (sebaceous duct) and surrounded by hyperplastic sebaceous glands; it consists of basaloid cells and squamous cells. The inner lining of the cyst is brightly eosinophilic and corrugated. Around the cyst are hyperplastic sebaceous glands. Apocrine cyst(s) (apocrine cystomatosis), an intradermal cyst lined by apocrine secretory epithelium, with clear secretion. The cyst is lined by a single layer of apocrine secretory epithelium that may show decapitation secretion.

Ciliated cyst, a simple cyst lined by ciliated epithelial cells, by cuboidal-to-columnar epithelial cells.

Subungual epithelial inclusion cyst, a simple cyst lined by a stratified squamous epithelium that is present within the bone of the third phalanx.

HAMARTOMAS

The term hamartoma (a mass of disorganized but mature specialized cells or tissue indigenous to the particular site) was chosen to describe these lesions, rather than the term nevus, in order to avoid any possible confusion with the melanocytic cells that make up the pigment lesions, referred to as nevi in humans.

Epidermal hamartoma (pigmented epidermal nevus), a (presumed) congenital proliferation of benign epidermal cells; the lesions, which are discrete plaques, show a variable degree of hyperkeratosis, acanthosis, and papillomatosis. There is hyperpigmentation of the lower epidermis, but no accompanying melanocyte proliferation.

Follicular hamartoma, a (presumed) congenital proliferation of hair follicles, consists of aggregates of enlarged primary follicles surrounded by a fibrous root sheath and a variable amount of dermal collagen with normal or modestly hyperplastic sebaceous glands. The lesion may be focal and nodular or extensive and plaque-like.

Sebaceous hamartoma, a (presumed) congenital proliferation of sebaceous glands, present in the mid to deep dermis. There is excessive proliferation of mature sebaceous glands within the dermis, some of which have an association with normal or hypoplastic hair follicles. Dermal collagen bundles tend to envelope the sebaceous glands.

Apocrine hamartoma, a (presumed) congenital proliferation of apocrine glands, within the dermis and/or within the panniculus adiposus. The overlying epidermis is often hyperplastic.

Fibroadnexal hamartoma (adnexal nevus, focal adnexal dysplasia, folliculosebaceous hamartoma), a focal proliferation of the pilosebaceous units associated with increased dermal fibrous tissue. It is likely that some lesions are congenital and others may not be true hamartomas, but rather occur following trauma to the area; composed of folliculosebaceous units lying in an abundant dermal collagenous matrix. These foccliculosebaceous units are arranged in a haphazard and disorganized fashion. The lesions are often secondarily inflamed, with focal aggregates of neutrophils and pyogranulomatous foci within the lesion.

TUMORLIKE LESIONS

Squamous papilloma, a non-neoplastic papillated mass composed of epidermis and supporting dermal stroma. The histology is similar to the viral-induced lesion. The epidermis shows normal differentiation, but enlarged keratohyalin granules, koilocytes, keratinocyte viral cytopathic effect, and intranuclear inclusions are not present. Pressure point comedones, multiple simple cysts lined by the stratified squamous epithelium of the follicular infundibulum with accumulations of hair and keratin within the cyst lumens. The cyst is lined by follicular infundibular epithelium with a prominent granular cell layer, and has hairs and loose keratin within the markedly dilated lumen. The cysts may be multiple; disruption of the epithelial wall evokes a severe granulomatous inflammatory response.

Cutaneous horn, a circumscribed exophytic lesion composed of dense, compact keratin. On the surface, the compact stratum corneum, which is primarily orthokeratotic but may include foci of parakeratosis, forms columns.

Warty dyskeratoma, a focal or multifocal proliferation of the epidermis that exhibits marked intralesional dyskeratosis and acantholysis; the lesions are endophytic, well demarcated from the adjacent epidermis.

Fibroepithelial "polyp" (cutaneous tag, skin tag, acrochordon), focal or multifocal exophytic lesion composed of epidermis and dermis. The core of the lesion consists of redundant dermal fibrous tissue covered by a hyperplastic epidermis with few, if any, adnexal structures.

Fibropruritic nodule (acral lick granuloma), a focal or multifocal hyperplastic dermal nodule(s) associated with chronic self-trauma and inflammation. The dermis often has marked fibrosis; the normal dermal collagen is replaced by thick, coarse bundles of collagen; and there is often loss of adnexa. A hypersensitivity pattern of inflammation is present in the adjacent dermis.

MELANOMA AND MELANOCYTOMA

Pigmentary tumors have been known since the antiquity, being called by CELSIUS "melas", due to their black color. In 1806, LAENNEC introduced the term "melanosis" for the pigmentary skin disorder, also making a first classification: encapsulated melanosis, infiltrative melanosis and diffuse melanosis.

In 1837, CARSWELL proposed the term melanoma for malignant pigmentary tumors, and in 1864 VIRCHOW classified melanomas into three types, depending on their macroscopic structure: simple melanoma, characterized by tumor cell agglomerations situated in the connective tissue; melanocarcinoma, characterized by alveolar or cord-like proliferations; melanosarcoma, characterized by a fusocellular structure. VIRCHOW described the structure of mixed melanic and amelanotic tumors.

In 1969, CLARK et al. published a classification of pigmentary tumors, estimating the invasion level of melanoma depending on the degree of involvement of anatomo-physiological structures of the skin. One year later, in 1970, BRESLOW proposed the assessment of the malignancy grade by measuring the maximal tumor invasion thickness. The melanocytic system is formed by melanotic cells of epithelioid and dendritic types, disseminated among the cells of the epidermal basal layer. These are precursors of melanoblasts that migrate from embryonic neural crests up to the skin; the nervous (neuroectodermal) origin of the melanocytic system is unanimously accepted.

During fetal development, melanoblasts migrate and are located at the dermoepidermal junction of the skin, at the root of and around the pilose follicle. In time, some melanocytes remain in the dermis, where they form the *blue naevus,* others are found in different structures of the eye, ear, meninx, adrenal glands, vascular intima and endocardium. Nevus is defined as a birthmark.

Dermal melanoblasts and cutaneous melanocytes possess in their cytoplasm *premelanosomes* and *melanosomes,* organelles specialized for the elaboration of melanic polymer or that can synthesize it. Melanin synthesis is initiated in the premelanosome, starting with tyrosine transformed into dopaquinone under the action of the tyrosinase enzyme. Dopaquinone is turned into dopachrome and finally results in a brown-yellowish or blackish compound, which is melanin. In premelanosomes, tyrosinase activity is present from the beginning and during melanin synthesis, while in melanosomes, since melanization is complete, tyrosine activity disappears.

In the epidermis, the mean number of normal melanocytes is of 1 to 40 basal cells. Melanin from melanosomes is directed towards the dendritic extensions, from where it is transferred to keratocytes situated in the proximity. A similar procedure is found in the pilose follicle. Two types of melanin have been identified in humans, one that protects skin against UV radiation - black melanin or *eumelanin,* and red melanin, *pheomelanin,* which is responsible for the photoaggravation of UV aggression.

The correlation between the exposure to ultraviolet rays and the development of melanomas that exists in humans has not been demonstrated in animals.

By examining the melanic cells from the thalamus of newborn animals (goat, rabbit and dog), BESTETTI et al. (1980) identified three types of cells: *type 1,* round, without extensions, or with short extensions and uniform granules; *type 2,* oval, with long and ramified dendritic extensions, with uniformly sized pigment granules; *type 3,* round cells, without extensions, with pigment under the form of irregular spherules, of yellow-brown to black color. Following electron microscopy, the authors concluded that type 1 and 2 cells were melanocytes, and type 3 cells were melanophages.

Researches at three organization levels of melanic pigmentation have established the presence of *melanin* at molecular level, of *melanoproteins* at macromolecular level, and of *melanosomes* at subcellular level.

Melanogenesis only occurs within melanosomes, which are secretory granules delimited by a membrane. Melanosomes may be transferred to neighboring cells, such as keratinocytes from mammalian skin or hair. The genetic information for this regulation is stored in the nuclear genome, whose expression is controlled by the intra- and extracellular environment. As premelanosomes become biosynthetically active, they mature into melanosomes by fusing with vesicles derived from the trans-Golgi network and from the plasmalemma level, including and incorporating different compounds and membrane components from within and the surface of the cell. In conclusion, melanosomes are final regulators of melanin biosynthesis.

The investigations performed by SLOMINSKI et al. (1988) have demonstrated that alpha-tyrosine and alpha-dihydroxyphenylalanine (L-DOPA) are affected by the activity of their own mechanism, and they can act as autoregulators. Tyrosine (a metabolic rate limiting enzyme, with an essential effect on melanogenesis) acts at the level of melanosomes.

The authors have demonstrated that the regulation of melanogenesis is specific and extremely complex, including melanosome synthesis, the increase in tyrosinase activity and changes in the subcellular distribution of tyrosine and acid phosphatase. Ultrastructural changes suggest that alpha-DOPA has mainly caused an increase in tyrosinase activity, while alpha-tyrosine has stimulated melanosome synthesis, determining an increase in tyrosinase activity, and concomitantly changing the subcellular distribution of the enzyme.

SLOMINSKI et al. (1988) consider their study, along with that of MCEWEN and PARSON (1987) as the first studies that confirm the regulating effect of melanin precursors on tyrosinase *in vivo* and on the formation of melanosomes in particular.

These studies show that the precursors of the melanogenetic mechanisms can be positive regulators of the subcellular melanogenetic apparatus in living cells. At the same time, regulation involves alpha-tyrosine that serves as a substrate and possible regulator of the catecholamine biosynthesis mechanism.

Melanogenic tumor lesions in humans and production animals (according to GOURREAU et al., 1995)

In certain cases, melanocytes may multiply focally, being at the origin of tumors formed by cells called *"Unna naevi"*, hence the name *naevus* or *pigmentary naevus* or *naevocellular naevus* given to these tumor proliferations. The evolution of these tumors may be benign or malignant.

Regarding terminology, which should also be used in the case of melanic tumors in animals, we mention that the term *melanoma* is used for malignant forms and *melanocytoma* for benign forms. This terminology is also used in human oncology.

For the purpose of uniform terminology in human and veterinary oncology, we present the corresponding terms used for the various forms of melanogenic tumors, according to GOURREAU et al. (1995).

Location	Stem cells	Tumor character	Humans	Animals
Intradermal junctional	–	Benign	Lentigo	Lentigo
	+	Benign	Ephelide	–
		Benign	Cell *naevus*= pigmentary or naevocellular *naevus*	Junctional melanocytoma (naevus)
		Malign	Atypical junctional *naevus* melanoma (epithelioid naevocarcinoma)	– Epithelioid melanoma
Dermal and/or hypodermal		Benign	Dermal *naevus:* cellular (round, small melanocytes, extended fields) – fibromatous (blue *naevus*), fusiform melanocytes	Cell melanocytoma (dermal *naevus)* Dermal fibromatous melanocytoma
		Malign	Melanoma: – fusiform cells	Fusiform cell melanoma Dendritic and spiral cell melanoma
Mixed, junctional and dermal	+	Benign	Mixed cell *naevus*	Mixed melanocytoma *(naevus)*
		Malign	Invasive mixed malignant melanoma	Metastasizing mixed melanoma (epithelioid and fusiform cells)

The histological classification of melanogenic tumors, made by WEISS and FRESE in 1974 and accepted by WHO, should be reviewed and adapted to the new terminology used in human oncology, compulsory in veterinary oncology. For this purpose, we propose the following histological classification:

- Melanocytomas:
  - Intradermal, lentiginous;
  - Junctional naevus;
  - Dermal and/or hypodermal, cellular type, or dermal naevus, fibromatous type;
  - Mixed, junctional and dermal, mixed naevus.
- Melanomas:
  - Junctional, epithelioid cells;
  - Dermal and/or hypodermal, fusiform cells, and dendritic and spiral cells;
  - Mixed, junctional and dermal, epithelioid and fusiform cells.

Using CLARK's classification, LARSEN (1978) analyzed 60 primary skin melanomas, with the following result: 49 superficial melanomas, 6 nodular melanomas and 5 unclarified melanomas. The classification adopted by WHO takes into consideration both the invasive grade of the melanoma, the Clark level, and the Breslow scale.

In the case of invasive level I, malignant melanocytes are only located in the epidermis, exceeding the basal membrane, the tumor being termed junctional melanoma or melanoma *in situ.*

Invasive level II is histologically characterized by the invasion of the superficial dermis, the basal membrane being exceeded over limited, discontinuous portions. Melanocytes are present in the superficial papillary dermis.

Clark level	Breslow scale	Prognosis
I. junctional melanoma or melanoma *in situ*	under 0.75 mm	Favorable
II. discontinuous melanoma in the superficial papillary dermis	0.76–1.25 mm	Reserved
III. massive continuous melanoma in the papillary dermis	1.26–2.25 mm	Unfavorable
IV. melanoma in the reticular dermis	2.26–3.00 mm	Unfavorable
V. melanoma in the hypodermis	over 3.00 mm	Unfavorable

Invasive level III presents the invasion of the superficial dermis over a large area, and connections with blood and lymphatic vessels are established.

Invasive level IV will present a massive infiltration of the middle and deep reticular dermis.

Invasive level V is characterized by massive diffuse infiltration of all skin layers, including the hypodermis, with the disorganization of normal structures.

In comparative pathology, a histological classification of all tumors of domestic animals was initiated by WHO. The histological classification of benign and malignant melanomas in animals was made by WEISS and FRESE, in 1974, and accepted by WHO. The authors proposed the following histological classification of melanomas:

A. Benign melanoma:
   - Junctional activity melanoma;
   - Dermal melanoma;
   - Cellular;
   - Fibromatous.

B. Malignant melanoma:
   - Epithelioid;
   - Fusocellular;
   - Mixed, with epithelioid and fusiform cells;
   - Dendritic and verticillate.

SMITH et al., 2002 (107) use the term melanoma for all malignant melanocytic tumors, whereas melanocytoma refers to benign forms.

The histological characteristics of these tumor forms have a particular practical importance. In this sense, we will emphasize the aspects and peculiarities of each type of benign and malignant melanoma.

MORPHOLOGICAL CHARACTERISTICS

It should be mentioned that histological examination is the only one that can establish the malignant or benign nature of a melanic tumor, and that a certainty diagnosis is difficult to make, being subject to errors. Morphological data should be corroborated with history data regarding the time of appearance of the first melanic lesions, the evolution in time, the lymph node aspect, multiple tumor formations, the age and color of the animal, etc. Primary melanotic tumors always originate in the skin. PET (pigmented extraepidermal tissue) mice have all body tissues pigmented, the embryonic musculature containing melanocytes. Although neoplastic extradermal tissue melanocytes have not been found in PET mice, it could be supposed that under certain conditions, extradermal melanocytes might become cancerous.

The presence of melanic tumor cells is histologically found in the epidermis, dermis and hypodermis. Tumor cells are grouped into clusters or columns, and sometimes they can invade tissues up to the underlying muscles. Tumor melanocytes multiplied in the dermis cause the disorganization of the basal layer, they compress the granular layer and induce the thickening of the horny layer, with the appearance of necrotic foci surrounded by inflammatory reaction. In advanced forms, the epidermis is disintegrated and ulcers occur.

Tumor melanocytes in the dermis are frequently grouped around vessels or nerves, into a collagen mass with fibroblasts and histiocytes. The intravascular presence of melanocytes is a sign of high malignancy. The melanin amount from melanocytes varies widely and does not represent a malignancy criterion.

Melanic tumor cells have either a fusiform appearance, similar to that of fibrosarcomas, or an epithelioid shape, mimicking undifferentiated carcinomas. Even in highly malignant melanomas, mitoses are less numerous, but giant cells can be noted.

Junctional melanic tumors are histologically characterized by an alteration of melanocytes or naevus cells, which are solitary or under the form of nests, with their obvious passage from the epidermis or pilose follicles to the dermis. Tumor cells have an extremely varied aspect and they are frequently recognized by their pigment content and their cluster or nest arrangement.

In the epidermis, melanocytes show a dark colored, dense nucleus and a clear cytoplasm. In the superficial area of the dermis they are cuboid or oval, similar to epithelioid cells. In the lower area of the dermis, cells are elongated, with a nucleus similar to fibroblasts. In general, the pigment content tends to diminish from the surface towards deeper areas. Junctional melanocytes are both a dermal and epidermal component, in an extremely variable proportion.

In some melanic tumors, almost all cells are situated at the dermoepidermal junction, while in others, their presence at the level of the junction is extremely scarce, almost all tumor cells being located in the dermis.

Dermal melanocytes, unlike Junctional ones, do not show changes and seem to be derived from the dermis. The majority of melanic dermal tumors can be classified as fibrous and cellular tumors, but some or them may also have a mixed structure. In the fibrous type, melanocytes are flat or dendritic, being grouped in irregular bundles mixed with fibroblasts, with an interlaced aspect. Cells are usually disposed parallel to the epidermis.

Compared to tumor cells, melanophages are present in a varied number, they are larger and contain rough melanin granules.

In dermal melanocytomas of cellular type, dense cell islands are found, with round or fusiform shapes. Larger islands are frequently subdivided into alveolar cell groups.

Melanomas, known by the old nomenclature as malignant melanomas, may develop from normal melanocytes from the epidermis and the buccal epithelium, as well as from dermal or junctional elements of melanocytes. In dermoepidermal junctional melanomas, a considerable junctional activity is noted, with a descending flow of anaplastic melanocytes form the epidermis to the dermis. Tumor cells frequently invade the upper epidermis, so that the epidermis disintegrates, inducing ulcerations. Reaching the dermis, tumor cells can manifest significant pleomorphism, having cuboid or fusiform shapes. Cells tend to form alveolar structures or irregularly ramified cords.

Melanomas, especially amelanotic ones, are formed by fusiform cells, resembling fibrosarcomas, while tumors formed by predominantly epithelial cells are similar to undifferentiated carcinomas. Mitoses in giant cell tumors, with a bizarre aspect, are not associated with a high malignancy grade.

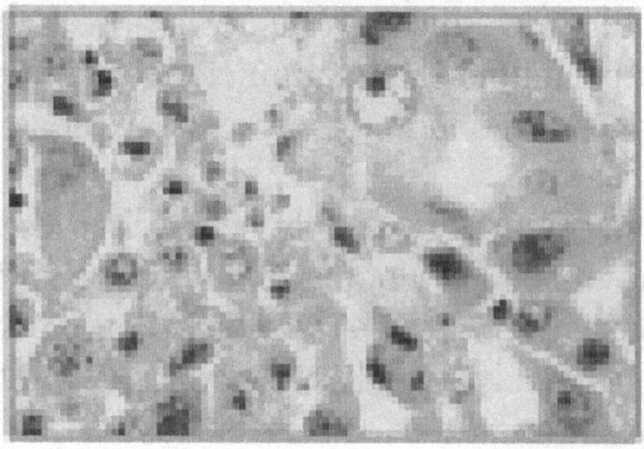

Fig. Malignant melanoma, pleomorphic and giant cell type.

Malignant melanocytes contain variable melanin amounts, which can be evidenced by ammonium silver nitrate or, in fresh tumor tissue, the DOPA reaction may be performed. The melanin amount is not associated with the malignancy grade.

The diagnosis of a malignant tumor, a melanoma, compared to the benign form, the melanocytoma, is difficult, all the more so when prediction is involved. The difficulty consists in the differentiation of a junctional melanoma and the early recognition of the malignant transformation of the junctional melanocytoma. Another difficult problem is represented by diagnosis and the recognition of the malignant transformation of the dermal melanocytoma. The presence of mitoses and necrotic foci is indicative of malignancy. At least in the dog, fibrous dermal melanocytoma, *blue naevus,* has a malignant potential.

Melanomas can metastasize by blood and lymphatic route, regional lymph nodes usually being the first affected. Metastases frequently occur in the lungs.

The intraepidermal lentiginous melanocytoma is characterized by the presence of melanocytes that are located strictly intraepidermally, having the aspect of a pigment spot, without malignancy characteristics. However, malignant developments have been noted in humans.

The junctional naevus melanocytoma histologically presents melanocytes grouped in 3–20 cell nests, having a polygonal or round shape. Melanocytes are located in the epidermis, without exceeding the basal membrane.

The dermal and/or hypodermal melanocytoma of cellular or dermal *naevus* type has no junctional activity, a variable pigment content, and cells have a disorderly arrangement, being both melanocytes and melanophages. Mitoses are absent or completely sporadic. Tumor formations located in the chorion do not exceed the size of a nut.

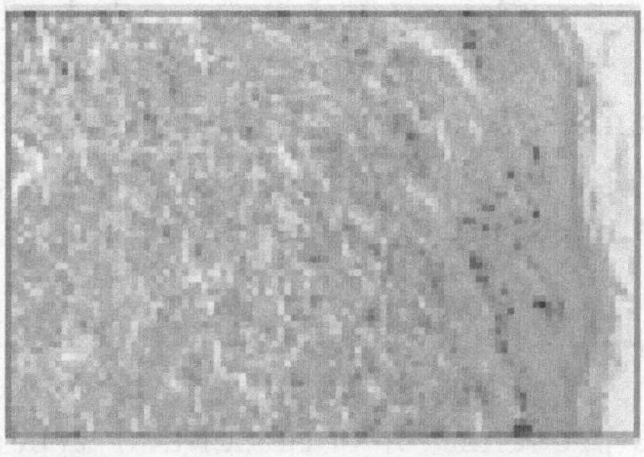

Fig. Dermal melanocytoma, spindle cell type.

Fig. Dermal melanocytoma, spindle cell type.

The mixed, junctional and dermal melanoma, of mixed *naevus* type has the histological characteristics of the previous types, being considered as a benign tumor predisposed to malignization.

The epithelioid junctional melanoma is formed by cells similar to those of the carcinoma. Cells are grouped densely, resembling the medulla of the carcinoma or they can be arranged as lobules or trabeculae. The cytoplasm of melanocytes contains less pigment compared to other melanoma types, and nuclei are larger.

The dermal and/or hypodermal melanoma with fusiform cells is formed by typically bipolar fusiform cells, with variable sizes. Nuclei are elongated, with scarce chromatin, and nucleoli are obvious. The number of mitoses is greater than in other melanomas. Cells give the tumor a sarcomatous or fibromatous aspect, against a background of reticulin fibers.

Fig. Melanoma, intraepithelial nests with atypia and pleomorphism.

The dermal and/or hypodermal melanoma with dendritic and spiral cells is formed by cells with a dense, disorderly or band-like arrangement, with an extremely high melanin content. Cellular details can only be seen after discoloring. Cells have polyhedral or round shapes, being present in numerous small dendritic cells. Mitoses are not numerous, but necrotic foci are frequently found.

The metastasizing mixed, junctional and dermal melanoma, with epithelioid and fusiform cells is frequently found on the buccal mucosa or skin. Microscopically, the presence of both cellular types is found, with the predominance of one type.

BRESLOW SCALE	>0.75 mm	0.76-1.25 mm	1.26-2.25 mm	2.26-3.00 mm	>3 mm
CLARK LEVEL	I *in situ* junctional melano-cytoma	II Disconti-nuous dermal melano-cytoma	III massive continuous melano-cytoma	IV reticular dermal melanoma	V hypoderma l melanoma

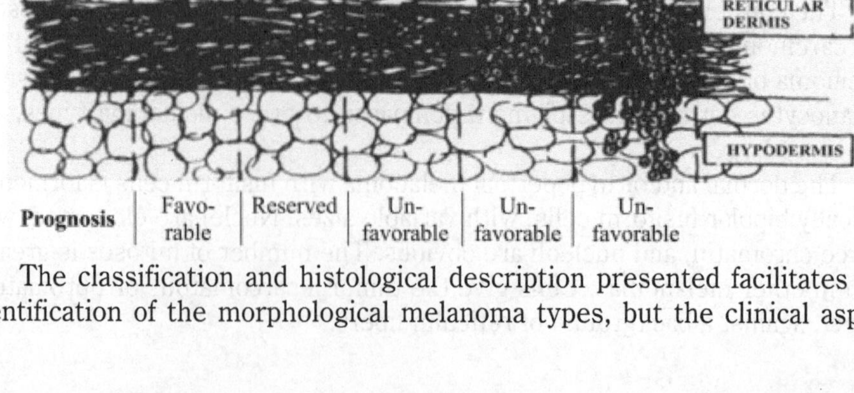

Prognosis	Favo-rable	Reserved	Un-favorable	Un-favorable	Un-favorable

The classification and histological description presented facilitates the identification of the morphological melanoma types, but the clinical aspect remains deficient, in particular prediction and prognosis. We consider valuable the use for diagnostic purposes and in veterinary oncology of the invasion level proposed by CLARK et al. (1969) and of the BRESLOW scale (1970), for the assessment of the malignancy grade according to the maximal thickness of the tumor invasion. For this purpose, we propose a scheme for the evaluation of the malignancy grade, adapted from CLARK (1969) and BRESLOW (1970):

Classification and assessment of the malignancy grade of melanic tumors, adapted from CLARK (1969) and BRESLOW (1970)

Usually, histological diagnosis does not pose special problems; rapid growth, metastases will be considered for malignant forms, especially in the lung and lymph nodes. Melanomas are frequently bordered by inflammatory reaction.

Amelanotic melanomas, as well as malignant forms with benign cells, pose diagnostic difficulties.

The following histological criteria are considered:

- Presence of dispersed and ungrouped melanic cells in the deep epidermal area and melanocyte pleomorphism;
- Sporadic presence of tumor cells in the superficial epidermis;
- Presence of inflammatory infiltrate, in the absence of traumas or infection and the presence of melanocytes in the lumen of lymphatic or blood vessels;
- Presence of abnormal mitoses and cytonuclear atypias;

Differential diagnosis is made with other more or less pigmented malignant skin tumors, such as epidermoid carcinomas and fibrosarcomas, but also with skin melanosis, brown or black tissue pigmentation.

ETIOLOGY

In animals, as well as in humans, the problem of the etiology of melanoma still remains at the stage of theories and suppositions, with controversies and certainties that are subsequently contradicted partially or totally. A multifactorial etiology may be admitted, in which there are favoring determining factors or risk factors.

Regardless of whether the neoplasm is initiated by chemical carcinogens or ultraviolet (UV) light, melanomas in all species share a similar biology in that they frequently recur and are predisposed to metastases to regional lymph nodes. Animal melanomas are common and provide a useful model for a deadly human disease. Initiation of as many as 65% of cutaneous melanomas in humans and in most of the melanomas diagnosed in Angora goats is thought to occur secondary to mutations generated by both UVA and UVB solar radiation.

In humans initiation within benign, the most important category of precursor lesion is the dysplastic nevus; malignant transformation of benign lesions is very uncommon in animals. Promoters can include chronic trauma, chemical exposure, burns, hormones, infections, drugs, and other causes for reactive hyperplasia.

THE HEREDITARY FACTOR

The hereditary factor seems to be at least for some animal species (horses and swine) scientifically documented.

A phenomenon occurs in Lipizzaner horses that seems to be genetically programmed. At birth, foals of this breed have black hair and skin. With age, depigmentation of hair occurs, which becomes white-silver, and the skin remains totally or partially depigmented. When we refer to the color of the horse, we consider the color of hair, not of skin. Skin is usually black, but it can also have depigmented, pink or white spots, that are visible on the glabrous skin (mouth, perineum, etc.). The preferential areas for the development of melanomas will always be pigmented.

GEBHART and NIEBAUER (1977) and DESSER et al. (1980) maintain that depigmentation as well as the development of tumors in Lipizzaner horses are genetically determined. The color of hair is inherited according to Mendel's laws, the order of dominance being: white, bay, bluish-gray, brown, black and chestnut, but colors do not form an allelomorphic series. It has been noted that some lines of bluish-gray horses are more susceptible to develop the melanotic disease, with a more severe form, at a younger age (under 10 years), especially when consanguineous breeding has been practiced. SEARLE states that the "G" factor is responsible for the discoloring of animals, and the pleomorphic effect of this factor could also be responsible for the form of melanic tumors. The execution of this genetic program must be related to an activity or block of the hormonal, enzymatic, immunological or other regulating system. JIMBOW et al. establish in Leghorn hens the genetic programming of death and the disappearance of pigment cells, which results in the completely white color of their feathers. This makes GEBHART and NIEBAUER (1977) affirm that something similar occurs in the case of Lipizzaner horses, in which melanocytes can no longer be identified histologically (DOPA reaction) or electron microscopically in depigmented spots. The place of these cells is taken over by cells whose significance is not yet known. The authors do not exclude the possibility that melanocyte migration deficiencies might occur in unpigmented spots.

The intervention of the genetic factor in the development and evolution of melanic tumors could be observed in Sinclair pigs. By evaluating the results obtained following an extensive study of a population of 349 pigs, HOOK et al. found in 1979, an increased incidence (54%) of melanomas in the newborns whose parents had both melanomas. By breeding a female without melanomas with a male with such tumors or a female with melanomas and a male without melanomas, the authors reported a decrease of up to 22% and 21%, respectively, in the incidence of melanomas in newborns. The pigs of parents without melanomas only had the lesion in 2% of the cases. These results suggest the existence of a transmissible factor from parents to offspring, in the appearance of melanic neoplasms. The incidence of melanomas in Sinclair pigs can be influenced by selection, which brings new information on the etiopathogenesis of melanic tumors in this species. New work hypotheses can be advanced that may contribute to the elucidation of the mystery that still surrounds the black neoplasm.

PET (pigmented extraepidermal tissue) mice manifest the pigmentation of all body tissues, melanocytes being found in the embryonic musculature. Although extradermal tissue melanocytes of PET mice have not been found under a cancerous form, it may be supposed that under certain circumstances, extradermal melanocytes can become cancerous.

Studies of the molecular base of carcinogenesis in melanomas have been oriented in three directions: activated oncogenes; cytogenetic analysis of chromosomal abnormalities; analysis of familial forms, with the research of the gene that can favor the appearance of uncontrolled melanocyte proliferation.

Oncogenes are capable of inducing malignant transformation features, after they are introduced in a melanocyte culture. Results suggest that in melanomas, the frequency of the activation of ras genes, detected by transformation experiments, is around 10%. Experiments performed with murine sarcoma viruses have shown that viral oncogenes induce malignant transformations in melanocytes.

Chromosomal alterations, such as translocations, deletions or variations in the number of chromosomes, have been noted in malignant melanocytes. Chromosomal abnormalities are found during the "vertical" growth of melanomas and especially in metastatic lesions. Researches have demonstrated that the "melanoma sensitivity" gene is related to the Rhesus blood group, situated on the short arm of chromosome 1. If an autosomal dominant gene is involved in familial melanoma and dysplastic naevus syndrome, its nature and location remain to be determined.

Genetically or environmentally initiated DNA instability facilitates subsequent neoplastic transformation. Human melanomas with p53 expression are common, 20–40% among all cases, and may have a different pathogenesis compared to those without p53. A canine case of multicentric melanocytoma was demonstrated to lack p53; p53 is not involved in equine melanoma tumorigenesis; most melanoma cases in humans lack detectable p53 mutations.

Expression of the apoptosis suppressor gene bcl-2 was detected in 16 of 18 human cutaneous melanomas examined by MORALES-DUCRET et al. 1995 (cite) but the bcl-2 gene product was also found in resting melanocytes in normal skin. Atypical bcl-2 overexpression has been detected in human uveal melanoma.

Proto-oncogene mutation to oncogenes also favors proliferation and tumor development. Oncogenes detected in in vitro and in vivo melanoma studies include c-myc, c-erb-B-2, c-yes, c-kit and ras.

In the case of malignant melanomas, 5 to 10% can be considered familial. Familial melanoma frequently develops from a preexisting lesion, the dysplastic naevus, also present in other animal species. This lesion may correspond to an early stage of cancerization, a proliferation caused by chromosomal instability or a genetic predisposition of melanocytes to proliferation or mutations under the effect of physical stimuli, such as ultraviolet light.

The hereditary component, as it has been demonstrated in some pathological cases, needs the intervention of an ecological factor for the clinical expression of the disease. The existence of the hereditary pathological potential allows the understanding of particular individual reactions to the aggressive action of ecological factors.

THE HORMONAL FACTOR

In the first place, sex hormones in animals, contrary to observations in humans, do not seem to have a determining or triggering role in the onset of melanomas. The authors who have described melanomas in horses, swine and dogs do not remark differences regarding frequency, evolution and prognosis between males and females.

Hypophyseal hormones, as well as pineal gland melatonin, have not been proved to play a role in the triggering and development of melanomas.

Irritating factors, of chemical nature, have been successfully used in the genesis of some neoplasms. In hamsters, skin melanomas have been induced by repeated paintings with carcinogenic substances.

Low intensity, long duration traumas, such as harness sores, have been incriminated in the etiology of skin melanomas, but the involvement of this factor has not been proved with certainty. It rather seems that skin melanomas become, as a result of their development, easily irritable sites, which subsequently ulcerate. The possibility that traumas subsequent to the development of tumors might favor the dissemination and onset of secondary neoplasms is not excluded.

Congenital neural crest abnormalities have been made responsible for the production of polymorphic dysplasias. LEVENE (1980) points out that melanomas in horses are located in certain body areas, such as the area of the head, neck, parotid glands and rhomboid muscles. The neural crest is the origin of pigment cells, autonomous and sensory ganglia of the peripheral nervous system, with supporting cells of the APUD cell system, the head and neck mesenchyma and the skull. The author maintains that following neural crest abnormalities, melanocytic ectopias would be initiated, which would subsequently become proliferation sites of melanic neoplasms. The intervention of the diffuse endocrine system in the development of melanocytes, sometimes with uncontrolled proliferations, may be supported.

The melanin production imbalance has been incriminated in the genesis of large melanin deposits by macrophagy or histiocytic infiltration. There is a similarity to lipid storage diseases or Gaucher's disease, some forms of cellular myeloblastoma, xanthoma and hypomelanic reticulosis. Practical observation and the great number of cases made JAGER affirm as early as 1909 that melanosarcomatosis in bluish-gray horses could be the manifestation of a metabolic abnormality. Current knowledge allows us, if this supposition is accepted, to consider the possibility of an inborn error of melanocytic cell metabolism. The role of the hereditary factor in the induction of melanotic neoplasms is rediscussed.

The blue naevus in animals is nothing else than a deposit of melanic pigment seen through the transparency of the dermis and epidermis. Comparative studies with what is known of humans have been performed by LEVENE (1980), who mentions the larger sizes and more intense pigmentation of the blue naevus in animals. The blue naevus has been observed more frequently in Terrier dogs and especially in fibrotic forms. Fibrotic pigmentary tumors are more frequently found in dogs and cats.

Horses that change their hair color, becoming white or gray, have a tendency to develop melanotic spots as blue naevi, which can subsequently become cancerous.

The brief review of some possible etiological factors of melanomas in animals includes no reference to the existing theories that are accepted in the etiology of cancer disease in general and that can also be adopted for melanomas.

Melanic neoplasms, like other diseases, seem to be generated by a complex of factors, on a favorable terrain. In this case, risk factors may be involved, which include both ecological conditions and the terrain, the organism with its genetic patrimony, more or less favorable to the onset and evolution of a neoplasm.

EPIDEMIOLOGICAL DATA ON MELANIC TUMORS IN ANIMALS

Epidemiological data reveal the presence of melanic tumors in all vertebrate species (mammals, birds, fish). Melanic tumors are commonly found in dogs, horses and some breeds of pigs. Incidence is lower in cattle and goats and extremely rare in cats.

In dogs, according to various bibliographic sources, melanic tumors represent between 4 and 7% of all tumors and between 9 and 20% of all skin tumors. The highest incidence is in dogs aged between 7–14 years and more frequently in Airdale Terrier, Cocker Spaniel, Scottish Terrier, Boxer and Boston Terrier breeds. Incidence is in general increased in dogs with pigmented skin. It seems that in males, incidence is higher, compared to females. Tumors are usually solitary, located on the skin or the buccal mucosa. Approximately 90% of oral tumors are malignant, while skin tumors are benign.

In a study over 40 years (1953–1974), by analyzing 3388 tumors from dogs, BASTIANELLO (1983) established the following locations: mesenchymal tissue, 33.7%; skin and skin adnexae 10.8%; female genital tract 10.2%; lymphohematopoietic tissue 8.9% and male genital tract 5.8%. The mastocytoma was the most common tumor type, with a frequency of 12.7% of all tumors, followed in decreasing order by lymphosarcoma, melanoma, squamous cell carcinoma, basalioma, hemangiosarcoma and histiocytoma. The main skin tumors were: basaliomas, squamous cell carcinomas, perianal gland tumors and melanomas. Over 80% of the tumors of the female genital tract were breast neoplasms, while the main tumors of the male genital tract were located in the testis, of which Sertoli cell neoplasms were the most frequent. The majority of digestive tract neoplasms were found in the oral cavity, where most of them were inflammatory epulis and melanomas. Osteosarcomas, neurofibromas and thyroid carcinomas were the most frequent neoplasms of the skeleton, nervous system and endocrine system, respectively, and pulmonary adenocarcinomas, melanomas and cholangiocarcinomas were the tumors of the lung, eye and liver.

The same author found in a total number of 243 neoplasms from cats the following frequency of locations: skin, lymphoid tissue, digestive tract and genital tract, representing 76.6% of all tumors. Squamous cell carcinoma represented 26.7% of the 243 neoplasms analyzed, being the most frequent type of skin tumor, followed by basalioma, mastocytoma and melanoma. The digestive tract was involved with 13.5% of all tumors, with the following types: squamous cell carcinoma, lymphosarcoma and intestinal adenocarcinoma. Mammary gland tumors represented 9.5% of all tumors, and of these 61% were carcinomas, especially skin carcinomas, followed by cutaneous and ocular melanoma, and osteoma, hepatocellular carcinoma and hemangiocarcinoma.

In cats, melanic tumors are extremely rare, representing less than 1 % of all tumors and approximately 2% of all skin tumors. In this species, the following order of tumor frequency is noted: squamous cell carcinoma, lymphosarcoma and breast carcinoma.

In horses, these tumors represent 6–15% of skin tumors. A relation between the incidence of these tumors and the gray color is found, although horses of other colors are occasionally affected. There is an increased tendency to the development of melanic tumors in gray horses with age; approximately 80% of gray horses over 15 years old develop skin melanic growths that are clinically detectable.

In pigs, melanic tumors have been found in 3–5% of the killed pigs, and in some breeds incidence exceeds 20%. Melanocytic tumors occur congenitally or at a very young age and almost exclusively in the Duro-Jersey breed, in dwarf Sinclair and Harmel pigs.

In cattle, melanic tumors represent less than 2% of all tumors, with the mention that the literature reports a high incidence, between 17 and 24% of all skin neoplasms.

HORSE MELANOMA (EQUINE MELANOTIC DISEASE, EMD)

The first report of a horse melanoma belongs to GOHIER, in an extensive veterinary surgery study, published in Lyon in 1813. The natural history of EMD. A white stallion, of a rare beauty, was bred with a great number of mares, the majority of foals inheriting the color of their father. Over the next years, black perianal formations appeared in this stallion. In time, nodular formations extended, including the scrotum and prepuce. Nodules gradually increased their volume, becoming confluent, their development accelerated and the stallion soon died. The author underscored the fact that the disease was only transmitted to white offspring, while dark color horses and their offspring were not affected. GOHIER gave an extensive and detailed description of clinical signs and lesions found on necropsy. The attempt to induce the disease experimentally, by subcutaneous inoculations of tumor triturates in bay and light bay horses, in mules and dogs, was not successful. The author showed that the disease was not contagious, since it was not transmitted by copulation when some of the partners had melanomas on their genital organs. The idea of hereditary transmission was advanced, and the castration of white or gray stallions was proposed.

CLINICAL PICTURE IN EMD

Primary locations are almost constant in black skin, with a smooth aspect and scarce fine hairs. The most affected body regions are: the inferior side of the tail, the lips, the anus, the vulva, the prepuce, the perineum. Few cases with locations at the level of the ear concha or in the external auditory canal are reported. Almost without exception, in more advanced cases, external lymph nodes are involved.

Equine melanocytic neoplasms have traditionally been grouped according to one of the three growth patterns. Some grow slowly over many years without metastasizing, whereas others grow slowly initially, with a subsequent increase in the rate of growth after a few years. A third subset grows rapidly and is malignant from the beginning. VALENTINE, 1995 (cite 107) has suggested that there are actually at least four possible syndromes of equine melanoma, three of which have the potential for metastatic behavior. Two of these three categories, dermal melanoma and dermal melanomatosis, are histologically very similar and can only be classified based on clinical features. Dermal melanomas are usually solitary, discrete lesions that are surgically excisable occurring in a wide age range of gray horses. Dermal melanomatosis denotes the presence of many lesions, often coalescing and usually occurring in typical locations, such as the genital or perineal region, of gray horses older than 15 years. These are not surgically curable and are much more likely to metastasize internally. The third category refers to anaplastic melanoma in aged, non-gray horses. Although rare, it is the most aggressive, leading to death within months of diagnosis. The fourth category is the melanocytic nevi, which are benign, pigmented lesions predominantly occurring in horses less than 6 years old.

EMD seems to be found almost exclusively in white or bluish-gray horses. In 1918, MOREL examined 400 000 horses and found melanotic sarcomas in two bay subjects and one brown subject. The incidence of the disease in horses that change their hair color from black to white or bluish-gray with age reaches 100%. EMD usually appears during adulthood or old age, but cases are also reported at the age of 2–4–5 years.

Up to 15% of all equine skin tumors are melanocytic. More than 90% of these tumors are benign at initial presentation, but approximately two thirds are thought to progress to malignancy and are capable of widespread metastasis. Arabian, Lipizzaner and Percheron breeds seem to have a predilection for cutaneous melanoma.

Tumor formations with a nodular aspect are well circumscribed, adherent to deep tissues and less to skin. The diameter of nodules varies from several millimeters to several centimeters. Nodules are hard, they are not painful, and in time the skin that covers them becomes ulcerated, exposing a black mass and a black oily liquid.

In rare cases, general disorders are reported, manifesting by dyspnea, ascites, digestive disorders, cachexia, and sometimes limping may appear due to muscular locations.

Secondary tumors appear in the subcutaneous connective tissue of various body regions. Secondary neoplasms are frequently found at the level of the parotid gland, which is enlarged, becomes bosselated and hard on palpation. Clinical evolution is fluctuating, usually tumor nodules develop slowly or remain stationary for years. Sometimes, the increase in volume and the appearance of tumor formations occurs in a short time, causing metastases in internal organs, and death occurs by exhaustion and cachexia. Cases with galloping development are rare, but always result in visceral metastases, sometimes only in the abdominal cavity, at other times with generalization in all organs and the serous membranes of the thoracic and abdominal cavities. Following the removal of tumor nodules, the evolution of the disease has not been influenced.

PATHOMORPHOLOGICAL PICTURE OF EMD

On necropsy, in the preferential areas of primary tumors, in the subcutaneous connective tissue, nodules of various sizes appear, sometimes with a rosary appearance, interconnected by fine black striations, which are the lymphatic vessels. Nodules have a black color, they are hard, so that they are very difficult to dissect. Tumor nodules are encapsulated and bands of fibrous connective tissue can be identified in sections, which traverse the brown blackish mass. Tumor nodules of various sizes (0.5–5–8–10 cm in diameter) are seen in the cavities, which are disseminated over the serous membranes and the organ capsules. Sometimes, huge black tumor formations (3–5 kg), with a bosselated surface appear in the abdominal cavity (on the epiploon and the mesentery). Tumor nodules or large neoplastic masses are found in the pulmonary parenchyma, the myocardium, the hepatic and renal parenchymas. The ovaries and the uterine serous membrane are covered with grape-like melanomas. Cases have been reported in which no organ has been excepted, even (muscular, adipose, connective) tissues being invaded by black neoplasms.

Locations in certain muscle groups, such as the rhomboid muscles at the level of their fascia of insertion onto the capsule, have made PEYRONNY consider this area as the site of development of primary tumors. Well structured or diffuse melanotic nodes have been identified in various muscle groups, on fascias and in the interstitial connective tissue. Cases are reported in which the whole dorsal musculature has been transformed into a huge melanotic mass. Locations at vertebral level are more rare, and paraplegias may clinically occur.

Intraocular melanomas are altogether exceptional in horses. Only 4 cases were mentioned until 1979. MURPHY and YOUNG (1979) reported a case of intraocular melanoma in a light color horse. Fusiform and dendritic cells invaded the crystalline lens capsule and the vitreous body, the superficial and peripapillary retina. Histologically, the tumor had a malignant character. DAVIDSON et al. (1991) described an uveal melanoma, with secondary keratitis, cataract and glaucoma. Finally, the authors concluded that the classification of ocular melanomas cannot be extrapolated from humans to animals. In a synthesis study, FOLLY et al. (1991) reported the identification of 4 congenital melanocytomas and 11 acquired melanocytomas, up to the age of 1 year and 3 cases before the age of 2 years. Epithelioid, mixed and fusiform cells were dominant in the histological structure, with large euchromatic nuclei, some of them having a marked cellular pleomorphism. Mitotic activity was generally low. In conclusion, the authors estimated that tumors were benign, having the characteristics of human benign melanocytic naevus.

KIRCHER et al. (1974) carried out a classification and a study of incidence of ocular and adnexal tumors in animals.

The *histological study* of melanomas requires the discoloration of the sections, since overloading with melanic pigment makes impossible their examination. Early changes in EMD occur at the level of pilose follicles. Melanic pigment gradually accumulates around the pilose bulb, and extends to the sweat glands, without the participation of the pigmented basal layer of the epidermis or of the outer hair sheath. Cells have a melanic content, they are either fusiform or dendritic, about 30 microns in diameter. With the development of the tumor nodule, polyhedral cells appear on a reticulin and collagen background. According to LEVENE (1971), during this developmental phase there is a considerable histological similarity to human blue naevus. In older evolution forms, numerous multinucleate cells are noted, with abundant cytoplasm, numerous nuclei (sometimes more than 10 nuclei), usually situated at the periphery of the cell or in the cytoplasmic mass without a precise grouping. Melanocytes appear as dendritic cells with a large, clear nucleus and cytoplasm with melanin granules. Melanophages are large reticuloendothelial cells.

Regional lymph nodes will present pigment cells that have reached this site by afferent lymphatic pathways, inducing tumor embolism at sinus level. Lymphoendothelial cells, dendritic cells and multinucleate cells are found, whose cytoplasm contains melanic pigment. The proportion of these cellular types can vary from one case to another and even from one lymph node to another.

Metastasizing can be achieved by blood vessel route, with the presence of emboli and neoplastic obliterations in the veins, as well as by lymphatic vessel route, when afferent regional lymph nodes will be affected.

In parenchymatous organs, liver, spleen, lung, kidney, digestive tract, genital tract, etc., melanic proliferations have a nodular character, with the described histological structure. However, melanic pigment loading of reticuloendothelial cells, histiocytes and even cells of the white series may be found. The macroscopic aspect of circumscribed nodular tumor is contradicted by the microscopic image, of distant infiltrations with melanocytes and melanophages.

According to BABA et al. (1981), the *electron microscopic examination* evidences dendritic melanocytes and melanophages. Dendritic melanocytes had ovoid or slightly elongated shapes, with short conical cytoplasmic extensions. Both the cytoplasm and the extensions contained melanosomes. The nucleus was eccentric, slightly flattened and elongated; in melanophores, melanosomes were uniformly dispersed, the endoplasmic reticulum was well developed, mitochondria were large and numerous. In melanophages, melanosomes were numerous, under the form of compact masses, delimited by a membranous structure.

Inside these huge lysosomal formations, images of melanosome disaggregation were observed. The endoplasmic reticulum was poorly developed, and mitochondria less numerous. With 7000 and 15000 times magnifications, the cytoplasmic membrane had a trilamellar structure. Some melanosomes had a blurred contour, fine granular structure, and light gray color. Mature melanosomes were intensely black, in cross sections elementary granules were disposed concentrically, sometimes with a lamellar aspect. In longitudinal sections, structure was lamellar, with lamellae parallel to the long axis of the melanosome. The densification of the melanosome started from the axial area, extending to the periphery.

By using the technique of centrifugation of tumor cells from equine melanomas, DURAND et al. (1972) studied the cell organelles from these neoplasms electron microscopically. The ultrastructural study allowed to evidence all stages of melanin synthesis. From ribosomes, the information reaches the Golgi vesicles, where protein is synthesized. The premelanosome, which at first is fibrillar, becomes coiled, striated, mature, ready to receive melanin that is formed under the action of tyrosinase. Melanin binds to the premelanosome, which is a supporting protein structure, the assembly constituting the melanosome. This organelle, surrounded by a membrane, is situated in the dendritic extension of the melanocyte. The cited authors conclude that the structure of the horse melanosome would be very similar to that of human melanosome, with a crystalline, fibrillar structure, arranged under the form of a screw.

Melanophages contain an extremely high amount of melanosomes, grouped into large vesicles, phagosomes, with degeneration images as a result of the action of proteolytic enzymes.

Malignant melanocytes have melanosomes with atypical forms, such as: modalities of binding melanin to the protein, abnormal structures or even the absence of melanin (damaged, folded proteins, rosette-like or clear center melanosomes). Dysplastic states are identical to those found in human melanomas. By the centrifugation technique and subsequently, by electron microscopic examination, the authors found melanosomal structure abnormalities similar to those of human malignant melanoma. Melanosomes at different developmental stages, irregular protein (rosette or ring-shaped melanosomes) were identified; some melanosomes had an almost normal protein support, but with abnormal melanin hooking, having a gray nuance or a non-homogeneous, mixed aspect, immature melanosomes being frequently noted.

Laboratory researches have attempted to establish some correlations between EMD and the levels of some plasma compounds. DESSER et al. (1980) determined the following constants from the blood and plasma of Lipizzaner horses: polyamines, putrescine, cadaverine, spermidine, diaminopropanol and histamine. The blood level of polyamines diminished with age and was independent from tumor growth. Aliphatic polyamines increase in a non-determinable manner with slow tumor growth, with a tendency to proliferation and weak destruction.

MAYR et al. (1979) conducted researches on total leukocytes, the leukocyte formula, the proportion of T and B cells, the proportion of proliferative cells (autoradiography) from peripheral blood, comparing the values in horses with and without melanomas. The authors found that animals with melanomas had an increased number of lymphocytes that incorporate ^{3}H-thymidine. This could be explained by either the heterogeneity of blood lymphocyte population, or the appearance of new properties in the lymphocytes of animals with melanomas.

By further observations, VOGEL et al. found in 1980 that in 4 hour cultures, the methyl-^{3}H-thymidine incorporation rate was by 40.3% lower in horses with melanomas compared to healthy horses. No significant age-dependent influence was found. The authors concluded that animals with melanomas probably possess a factor that inhibits proliferation and becomes inefficient if lymphocyte culturing lasts for a longer time. A parallel is made with the presence of soluble immune complexes in the blood serum of patients with cancer or inflammations and diseases appearing as a result of low ^{3}H-thymidine incorporation, and the consequence could be a decreased cell division rate or cell inhibition during the division phase, which determines a prolonged cell cycle.

Melanotic proliferations in horses occur with hair whitening, when melanocyte activity from the pilose bulb slows down. Tumor growths develop in certain preferential sites, having a local character, but in the majority of cases there is a tendency to invasion with the involvement of regional lymph nodes. Evolution is slow, and both melanocytes and melanophages can be structurally identified. In the case of increased malignancy, cell divisions become a prominent characteristic, concomitantly with a decrease in melanogenesis. As LEVENE (1971) underscored, although there are no comparable etiological factors between EMD and malignant blue naevus, histological similarities and evolution allow to establish parallels and even a similarity between the two neoplasms. Both are manifestations of neoplastic growth potentials of melanocytes.

Primary and secondary tumors that grow slowly, for example scirrhous mammary carcinoma that can imperceptibly grow over 10–20 years, are comparable to EMD, being dermoplastic. In the rapid evolution of EMD, melanin deposits, instead of being uniformly colored, have a color that varies in intensity, and amelanotic neoplastic cells appear. In this case, numerous cells with an embryonic character are present, with numerous mitotic figures, vesicular nuclei and multinucleated giant cells.

When consanguinity was practiced between dark color horses with melanomas, EMD had an acute malignant evolution. The suggestion of using carcinogenic substances on the skin of different body regions in dark color horses and white or bluish-gray horses seems promising for the verification of some hypotheses regarding the etiopathogenesis of EMD and even of melanotic neoplasms, in general.

DOG MELANOMA (CANINE MELANOTIC DISEASE, CMD)

The incidence of melanomas seems to be higher in dogs with intensely pigmented skin. Tumor formations in CMD are usually cutaneous, being located on the oral mucosa and intraocularly. Cutaneous locations at the level of distal limb extremities and on the scrotum usually have a malignant evolution. According to BASTIANELLO (1983), the melanoma occupies the third place in terms of incidence, after the mastocytoma and lymphosarcoma.

COTCHIN, 1955, and GOLDSCHMIDT and SHOFER, 1992 (cite) established: melanoma is relatively common in dogs, accounting for 3% of all neoplasms and up to 7%; the most frequently affected sites are the oral cavity (56%), lip (23%), skin (11%), and digit (8%), with other sites, including the eye, accounting for only 2%.

Cutaneous melanoma occurs more commonly in dogs with heavily pigmented skin, with the Miniature Schnauzer, Standard, Schnauzer, and Scottish Terrier at increased risk of developing neoplasms. Most dogs with melanoma are more than 10 years old with a range of 1–7 years; an increased frequency has been described in male dogs.

By examining 63 dogs with oral melanomas (32 males, 28 females and 3 castrated females), HARVEY et al. (1981) established the following locations: 38 gingival, 12 labial, 8 palatal and 5 lingual. Following clinical, radiological and necroptic examination, the authors mentioned that in 45 of the 63 dogs (71 %), the cause of death was melanoma. All melanotic tumors of the oral cavity are considered as malignant, whether they present metastases or not. This is so because they have the histological characteristics of malignant tumors, and evolution is also malignant, with metastases, generalization and death. Melanomas in dogs, in any location, have an unfavorable prognosis. Even in early surgical interventions, this neoplasm produces recurrences, lymph node, pulmonary metastases, and death. Surgery is the main treatment method of dogs with melanoma. In conclusion, no correlation can be established between the prognosis of dogs with melanoma located in the oral cavity and the following parameters: the microscopic aspect of the tumor, the mitotic index and the degree of pigmentation, on the one hand, and early surgery, sex of the dogs, intraoral location, volume and type of surgery (removal by scalpel using the cryosurgical method), on the other hand. Survival in the case of 7 untreated dogs was 65 days compared to 242 days for dogs submitted to surgery.

Regardless of location, the following histological types can be diagnosed in CMD: junctional melanoma; cellular and fibromatous dermal melanocytoma; epithelioid melanoma; type A and B fusocellular melanoma; mixed melanoma and dendritic melanoma. In humans, osteogenic melanomas have been described, in dogs one case has been reported, with gingival location, with a microscopic nest or lobule arrangement of obviously anaplastic cells, containing melanin, and in the stroma, abundant osteoid.

Eyelid and conjunctival melanocytomas do not differ from skin melanocytomas.

Intraocular melanoma in dogs originates either in the iris or in the ciliary body, unlike in humans, in whom tumor growth starts in the choroid. At the level of the ocular globe, melanomas can have intraocular or epibulbar locations. Intraocular melanocytomas in dogs are exceptional. Intraocular melanoma is histologically characterized by the proliferation of fusiform, epithelioid, dendritic and mixed cells.

Mixed melanoma seems to be the most frequent. RYAN and DITERS (1984) found the following incidence of ocular melanomas in dogs: melanomas at the level of the sclerocorneal limbus 22.5%; fusocellular intraocular melanomas, type A 30%; fusocellular intraocular melanomas, type B 22.5%; mixed intraocular melanomas 17.5%.

Epidemiological researches have revealed a higher incidence of ocular melanomas with distant metastases in the German Shepherd breed, especially at an advanced age. Anterior uveal melanoma is the most frequent primary intraocular tumor in dogs. These primary tumors induce metastases in the kidney and the lung. Amelanotic melanomas with fusiform cells have been diagnosed.

A simplified classification of uveal melanomas in dogs is proposed - benign forms called melanocytes, and malignant forms called melanomas. In general, uveal neoplasms are white-gray or black. They have a diffuse growth pattern, occupying most of the inside of the ocular globe. Polygonal cells are large, with abundant and pale cytoplasm, and fusiform cells are arranged in spirals or are interlaced, being usually stratified. Hematogenic metastases are frequent.

The iridociliary location in dogs appears as a primary malignant tumor. Macroscopically, cells have various shapes, from small nodules to masses that fill the ocular globe. Melanocytomas are well differentiated, with cuboid or columnar cells, with rare mitoses. Melanomas are formed by pleomorphic polygonal or fusiform, columnar cells, with frequent mitoses. Hematogenic metastases are rare and are usually pulmonary.

The clinical and histopathological study of 14 dogs with intraocular melanomas, performed by BUSSANICH et al. (1987) revealed intense pigmentation and extensive necrosis in the tumor mass. The authors established a higher incidence in dogs aged between 6 and 15 years, with a mean age of 10 years, mentioning that there were 9 females and 5 males, the left eye being involved in 8 cases, and the right eye in 6 cases.

Clinically, the inflammation of the anterior segment of the ocular globe was found in 10 dogs. In one dog, necropsy revealed metastases in the lung, liver and kidney.

Tumors were located in the ciliary body, in the iris, frequently filling and distorting the anterior segment of the ocular globe.

Histologically, 3 types of tumor cells were identified: round, fusiform and oval.

Round cells. The authors identified, in all 14 cases, round cells, which sometimes were dominant. Round cells presented numerous melanin granules, uniformly dispersed in the cytoplasm. Nuclei were pyknotic and slightly eccentric, and mitoses were practically absent. Electron microscopy showed small, irregular nuclei, with rough chromatin, and premelanosomes and mature melanosomes were present in the cytoplasm.

Fusiform cells formed a small tumor mass, nuclei were fusiform, nucleoli were difficult to detect, and the cytoplasm was abundant and less pigmented. Electron microscopy showed small, elongated nuclei, with rough chromatin, but with few melanosomes.

Ovoid cells presented the widest variation in number. Thus, in some tumors they were dominant, in others, they were virtually absent. Ovoid cells were well circumscribed in an obvious cytoplasmic membrane. Nuclei were centrally arranged, with a marginal chromatic mass, vesicular nuclei, and abundant cytoplasm with few melanin granules. Electron microscopically, ovoid cells had the largest, prominent nuclei, with fine, uniformly dispersed chromatin. In the cytoplasm, the rough endoplasmic reticulum and free ribosomes were abundant, but melanin granules were scarce.

Based on the analyzed material, the cited authors propose a classification of melanic tumors depending on the identification of the three types of cells: tumors with predominantly round cells, tumors with predominantly fusiform cells, and tumors with predominantly oval cells.

By analyzing the clinical evolution and morphological peculiarities, a similarity is found between melanic tumors with predominantly round cells in dogs and human melanocytomas. Regardless of the histological aspect, canine melanotic tumors are less aggressive than human tumors.

Limboscleral melanomas appear as slightly bulging, highly pigmented structures, situated between the sclera and the limbus. The duplication of the conjunctiva and corneal erosions are evident. Two cellular types have been histologically differentiated: small fusiform cells, with margins that are difficult to differentiate, and large, round, pigmented cells. Neoplastic cells have been constantly identified at the level of the sclerocorneal junction.

Type A fusocellular melanomas were clinically expressed by hydrophthalmia, pigmented tumor mass, identified following enucleation as anterior uveal melanoma. Histologically, the dominant population was formed by large, round cells, with fine pigment granules arranged uniformly in the cytoplasm, nuclei were round and vesicular, chromatin and nucleoli prominent. These pigmentary cells were numerous at the periphery of the tumor mass. Fusiform cells had oval, fusiform nuclei, cytoplasmic borders were poorly delimited. Mitotic images were rare.

Type B fusocellular melanomas clinically manifested by massive tumor growths, corneal edemas and hydrophthalmia. In all studied cases, the anterior tract was affected, and in some cases the tumor extended through the sclera towards the musculature of the ocular globe. Histologically, type B tumors were similar to type A tumors, except for larger and more pleomorphic type B fusiform cells, with round to oval nuclei and obvious nucleoli. Cytoplasmic borders were poorly delimited. The mitotic index was low. Round, large, pigmented cells were less numerous than fusiform cells. Type B melanomas were less pigmented than type A melanomas.

Epithelioid melanomas had no melanic pigment. The neoplasm included the iris, the ciliary body and the choroid. It sometimes extended to the muscles of the ocular globe, and in one case, hepatic, adrenal, pulmonary, myocardial and ganglionic metastases were found. The tumor extended along the optical nerve, up to the optic chiasma. Microscopically, the tumor was formed by large, round cells, the cytoplasm was abundant and eosinophilic. Nuclei were large and oval, nucleoli were evident and in numerous cells, they were multiple. Multinucleate cells were present, and the mitotic index was high.

Mixed cell melanomas were characterized by the constant involvement of the choroid. In one case, the anterior eye chamber was filled with a pigmented tumor mass. This case showed hepatic, splenic, pulmonary, renal and ganglionic metastases. Mixed melanomas were formed by round, large pigmented cells, epithelioid cells and fusiform cells. The most numerous were fusiform cells, and the less numerous large round cells. Epithelioid cells were grouped or dispersed in the tumor mass. The mitotic index was situated between that of pure epithelioid melanomas and fusiform melanomas.

Following the study performed, RYAN and DITERS (1984) estimated that no strict correlation could be established between the development and frequency of ocular melanomas in dogs, on the one hand, and sex, breed and age, on the other hand. However, the authors found the following: females were affected in 61% of cases; the German Shepherd breed had a three-fold incidence, and the Boxer breed a two-fold incidence compared to a control group; the German Shepherd breed was more frequently affected by type A fusocellular melanoma, located at the level of the sclerocorneal limbus; in the Boxer breed, melanoma was of mixed type.

The metastatic dissemination of ocular melanomas in dogs takes place by hematogenic route, and locally, dissemination occurs along vascular and nerve branches. The risk of metastasizing is highest in mixed and epithelioid melanomas. Type A and B fusocellular melanomas are less aggressive, so that prognosis will be more favorable compared to mixed and epithelioid melanomas.

The treatment applied in the case of intraocular melanomas is the enucleation of the ocular globe.

Epibulbar melanomas in dogs are generally located at the level of the sclerocorneal limbus. The cases presented by MARTIN (1981) showed that epibulbar melanomas in dogs originated exclusively in the sclerocorneal limbus; intraocular extension, located on the surface of the ciliary body, was noted in one case. In man, corneal invasion develops between the epithelium and the

Bowman's membrane, and it tends to remain superficial, rarely penetrating the ocular globe. The invasion of the cornea in the dog was deeper or approximately up to the mid-stroma, probably involving a deeper origin of the tumor.

Epibulbar melanomas have a favorable prognosis if surgery is performed early. Lamellar keratectomy is preferred to the enucleation of the ocular globe, in melanomas of the sclerocorneal limbus. The author noted that in younger animals, aged between 2 and 4 years, the development of the tumor was more rapid, compared to 11 year-old subjects, in which slow evolution, with stationary states manifested. It is recommended that the graft should be used through all the thickness of the cornea and sclera, in order to avoid the dissection and dilaceration of the neoplasm. Based on the reported cases, the cited author considers prognosis favorable not only for survival, but also for the maintenance of eye function in younger animals, while old animals will only be monitored for signs of neoplastic development.

Feline melanoma is uncommon, accounting for less than 1 % of all feline oral neoplasms, and approximately 0.5% of feline skin tumors. The ocular and cutaneous forms are generally more common than the melanoma of the oral cavity. The most common cutaneous sites are the head, tail, distal extremities, and lumbar area. The prognosis is generally poor because of recurrence and regional metastasis in up to half the cases.

PIG MELANOMA (SWINE MELANOTIC DISEASE, SMD)

Melanic pigmentations have been described in the cerebellum of Hampshire pigs. The presence of melanocytes has been mentioned in the pia mater, the arachnoid, in and around vessel walls, in the presented cases melanosis being of genetic nature. The literature before 1964–1966 considered SMD as completely sporadic and studies were limited to the presentation of isolated cases. The appearance of new pig breeds, their cross-breeding and the creation by selection of the laboratory dwarf pig has led to a sudden increase in the SMD incidence and has implicitly broadened the range of this field.

In 1965, STRAFUSS et al. described the first cases of cutaneous melanomas in dwarf Sinclair swine. Subsequently, the dwarf Sinclair pig became the object of extensive researches, macro- and microscopic studies, selective and genetic reproduction studies, intended to clarify SMD. In Romania, BABA et al. (1981) reported a case of malignant cutaneous melanoma in a Bazna pig. In this case, melanocytes were numerous in the papillary dermis, with the dislocation of the basal epidermal layer; in the peritumoral hypodermis there was a lymphohistiocytic reaction. Dendritic melanocytes were arranged as nests, surrounded by fusiform cells and with numerous melanic granules in the cytoplasm.

Potentially malignant melanomas were described, by NOMURA et al. (1974), in Duroc x Hampshire crossbred pigs.

The studies performed on Sinclair swine aimed to macroscopically examine a great number of subjects, to perform optical microscopic and electron microscopic examinations. The authors mentioned that the benign exophytic tumor was formed by dendritic melanocytes, similar to those of human naevus. Melanophages were easy to detect ultrastructurally. In pulmonary melanomas, melanocytes that contained atypical forms of melanosomes and mitochondria were identified. There were lysosomes without melanosomes, while some phagolysosomes showed different stages of melanosome degradation. The authors mentioned that benign exophytic forms were similar to type B KLUG melanoma. Malignant tumors were similar to type A melanoma, with numerous nuclear changes, incomplete and atypical melanization.

MILLIKAN et al. (1973; 1974), as well as MANNING et al. (1974), identified several developmental stages of cutaneous lesions: smooth spots, fine nodules, red tumor and regressing tumor, which could coexist in the same individual.

SMD is more frequent in pig breeds with pigmented skin (Duroc, Jersey, Bazna, etc.) and their crossbreeds. The presence of exophytic melanomas at birth is surprising, which demonstrates their histogenesis during intrauterine life. The involution and complete regression of pigmentary cutaneous lesions has been noted even in intrauterine life.

HOOK et al. (1979) demonstrated the influence of selective reproduction on melanomas in Sinclair pigs. The incidence of melanomas was highest in offspring whose both parents had tumors. In offspring with only one parent affected, incidence was higher compared to descendants of normal parents and lower compared to the first category. The cited authors remarked the absence of the influence of a strictly maternal factor or of a sexual component. Results demonstrated that only 24.5% of cutaneous melanomas were present at birth and that 75.5% developed later. This observation does not exclude the possibility that melanomas might have derived from very small, histologically undetected melanocytic lesions, especially that the pigs included in the study were black.

Controlled breeding and the early observation of a high number of Sinclair pigs allowed the cited authors to reach some conclusions regarding the SMD incidence and evolution. The high incidence of melanomas in Sinclair swine is also influenced by a hereditary component, which is also found in the case of human malignant familial melanoma, where the hereditary factor would act as either a polygenic factor, or with the involvement of an autosomal gene.

One of the most interesting observations was that with the advancement of age, regressions occur, which can result in the disappearance of cutaneous melanomas. It is known that human melanoma is one of the few tumors that can regress and, although the regression mechanism is not known, certain serum or plasma factors are associated with melanoma regression *in vivo* and *in vitro*. In both human and swine tumors, regression has been associated with a high number of tumor-infiltrative lymphocytes. These lymphocytes have been identified as CD8+ cytotoxic T cells and display granulated lymphocyte morphology. These are the lymphocytes that react to Melan-A/MART (melanoma antigen recognized by T cells), a cytoplasmic marker that is weaker or not evident in the population of melanoma cells remaining after partial remission of human melanomas. The observations on SMD in dwarf Sinclair pigs suggest the possibility of controlling and fighting the development of melanoma, a phenomenon acquired with age. The findings and suppositions in SMD increase the attraction of the study of this type of tumor, regarding host-tumor interactions during the development, growth and regression of melanoma.

SMD in Sinclair swine and the elucidation of the genetic mechanisms involved in the transmission and regression of neoplasms need to be further studied, making this unique model extremely attractive, through the possibilities offered by the study of melanoma.

RUMINANT MELANOMA (RUMINANT MELANOTIC DISEASE, RMD)

Melanoma in cattle, sheep and goats has been less studied, probably due to the low incidence of this neoplasm in ruminants. Cattle melanomas represent about 5% of the cases of tumors found in this species, and of all cases, 92% are benign. The melanoma cases found in young cattle made COTCHIN (1972) assume that this is a congenital disease. The author noted a higher incidence of RMD in Aberden Angus cattle. Congenital melanomas in calves are more rarely reported in the literature. These tumors have no preferential location, they can be found as naevi or pediculate tumors (trunk, limbs, jaws, abdomen).

MILLER et al. (1995) reported two cases of neuroectodermal melanoma, which, like in children, is characterized by a preferential maxillomandibular location and which contained a high number of melanic cells and "like" cells surrounded by connective tissue. LAMBOTTE et al. (1989) described the first case of congenital melanoma in a Blanc-Blue-Beige calf, mentioning that congenital melanoma in calves has no preferential location. It can be found as a naevus or pediculate tumor. In a newborn calf, the authors described a pediculate, black, glabrous tumor formation of 4.5 kg in the thorax, in the upper posterior region of the shoulder. Six months after resection, the tumor did not relapse. Histological findings included: hyperkeratosis, the spinous layer presented vacuolized cells, with large intracellular spaces, acanthosis and spongiosis. In the proximity of the epidermis, there was an intensely colored area, with cells overloaded with melanin. In the depth of the tumor, cells were polyhedral, with variable melanin amounts, arranged as cords or disorderly. The authors established the diagnosis of benign cell melanoma.

BOURGOIN (1972) noted a predisposition to malignant melanocytoma in the African Angora goat. KIRCHER et al. (1974) reported a case of malignant melanoma in the mandibular sinus of a sheep. In a synthesis study, NOBEL et al. (1979) found of 709 cases of neoplasms in domestic animals, 5 melanomas:

2 in horses, 2 in dogs and 1 in cattle. DRIEUX and PRAT (1968) described a melanoma case in a cow killed in a slaughter house. History showed the absence of particular clinical signs, just a slight hesitation to rise from the decubitus position. At the *post mortem* examination, sternebrae and the body of dorsal and lumbar vertebrae were black; small gray-reddish areas appeared in the cross section. The hypophysis had the size of a nut, with intense black pigmentation, and its consistency was higher than normal.

Traces of the glomerular area and some cords from the fasciculate area of the adrenal gland were histologically identified. The rest of the tissue was neoplastic. Two types of cells were identified: some fusiform cells with small nuclei containing scarce chromatin, arranged as a palisade, and large cells, with irregular arrangements, clearly delimited cytoplasm and large nuclei. Both cellular types contained intracytoplasmatic melanin granules. The neuronal and glandular portions were impossible to identify in the hypophysis. The development of fasciculated tissue with numerous necrotic foci predominated in the hypophysis. Fusiform bundle cells dissociated and dislocated the hypophyseal cords and acini. Fusiform cells were loaded with melanin. Among acidophilic glandular cells, there were voluminous cell elements, isolated or grouped into small clusters, with the cytoplasm loaded with melanin granules. Based on the histological description, the presence of a mixed melanoma may be assumed, in which fusiform cells were dominant, but epithelioid cells were also present.

In another cow killed in a slaughter house, BABA et al. (1983) reported a malignant melanoma located at the level of the cerebellum. The neoplasm was formed by tumor masses with the size of an orange, intense black color and dense elastic consistence. The subject also presented black tumor formations in the subcutaneous connective tissue and myocardial tissue. Histologically, the cerebellar melanoma was formed by fusiform cells and round cells in the cytoplasm of which there were numerous melanic pigment granules. Thorough microscopic examination could only be performed after the bleaching of the sections. Malignant melanomas in domestic ruminants are, as it has been mentioned, neoplasms that are more rarely reported in these species. The short life and killing for economic reasons could explain the lack of data in this field.

EXPERIMENTAL MODELS IN THE STUDY OF MELANIC TUMORS

The finding of an experimental animal model, even the ideal animal in a well defined experimental scheme, is essential for cancer research.

In order to clarify some essential problems that might explain the development of melanoma in man, it is necessary to find some experimental schemes in animals or cell cultures. The unanimous acknowledgement that the activity of phenotypically defined premalignant cutaneous lesions can progress directly into melanoma in predisposed individuals has led to the investigation of possible genetic influences on this progression. There are still difficulties regarding the establishment of the genetic or molecular base for the malignant transformation of a benign naevus-like cutaneous lesion, or of a melanocytoma into a melanoma, in humans.

The possibility to isolate and culture human melanocytes in successive generations, with the involvement of environmental agents that activate the genetic components of the transformation, has opened the perspective for the understanding and explanation of the complex phenomenon of melanocyte malignization and the appearance and development of melanomas.

The dwarf Sinclair pig, used as an experimental model for the study of melanic tumors, has led to some clarifications and has opened perspectives in the elucidation of some issues regarding the role of the genetic factor and ecological factors in the initiation and development of melanocytic melanomas.

The advantage of this experimental model is the resemblance in certain ways to human melanocytoma and melanoma, such as the involvement of a genetic factor, the transformation of melanocytomas into melanomas, and spontaneous regression that sometimes is completely unexplainable. The Sinclair pig is considered to be the animal that can be used as a unique model for the study of the aspects of development, growth and regression of melanic tumors.

The hereditary factor in the appearance of melanocytomas and melanomas in Sinclair pigs has been verified by controlled breeding, such as the breeding of males and females that both have melanic tumors. This has resulted in the obvious increased incidence of cutaneous and visceral melanomas. A hereditary component influences the incidence of melanomas in Duroc-Jersey pigs.

Selective breeding of pigs with exophytic melanomas has enhanced the incidence of congenital melanomas in newborns, accurate estimates being difficult since some neoplasms develop during the course of life, and others can regress completely during intrauterine growth. MANNING et al. (1974) concluded that in the regression of melanomas, although it has not been proved to be immunologically mediated, the lysis of neoplasms is accompanied by inflammatory infiltrate, which suggests the intervention of immunological factors.

HOOK JR. et al. (1979) noted in Sinclair pigs a peculiarity similar to that of human melanomas, i.e. melanic tumors decrease with age, which suggests a control from the part of the host and the possibility of favorably influencing neoplasm regression. The fact that heritability is not related to sex in congenital cutaneous melanoma in Sinclair pigs makes it much similar to human melanoma.

The microscopic structure and developmental stages have allowed the classification of lesions according to the criteria established for primary melanoma. Melanotic naevi and melanomas appear in Sinclair pigs since birth up to the adult age, with the mention that hyperplastic melanocytic naevi have been identified even in 11 week-old fetuses. Melanocyte hyperplasia occurs in the dermis and can be accompanied by perifollicular aggregates of macrophages loaded with pigment, images that are similar to that of early human melanoma. Melanophages develop early in areas adjacent to hyperplastic foci.

Lesions in the regional lymph nodes, brain, meninx, lung, liver, urinary bladder, heart, spleen, adrenal glands, ovaries, muscles, bones, stomach and intestine have been described in both fetuses and adults, in animals with solitary or multiple exophytic primary lesions.

The death of pigs with melanomas occurs during the first 6 weeks of life, and after this age, primary lesions and metastases regress. The regression of the primary tumor is associated with an increase in the activity mediated by the host leukocytes. The *in vitro* reactivity of leukocytes seems to be correlated with the tumor volume. The reactivity of leukocytes has been significantly higher in pigs with a lower tumor growth rate and a smaller tumor volume.

BEATTIE et al. (1988) monitored the effect of cellularly mediated immunity on the development of the tumor by the stimulation with three mitogens of lymphocytes from pigs with tumors and control pigs. The mitogens induced a significantly higher blastogenic response of lymphocytes in control pigs compared to pigs with melanomas. In conclusion, there is also an immunodepressive effect in pigs with tumors, at least in the phase of early tumor development.

The reduction in the tumor volume occurs with the decrease in the number of melanophagic cells. This suggests that immunological mechanisms mediated by manipulable host cells are associated with tumor regression. It remains to be established whether the response is initiated by the host or the tumor cells.

Regarding the hereditary character of cutaneous melanoma in Sinclair pigs, there are two plausible inheritance patterns, one autosomal dominant as a result of an autosomal mutation that has occurred in the original Sinclair pig population, and the other as a multifunctional mode by which the Sinclair pig has accumulated, by chance, a high number of polygenes for cutaneous melanoma. As a work hypothesis, the cited authors emphasize the importance of the establishment of cell lines, from cells or naevi, during the development of preneoplastic cells and of cells programmed to undergo a transformation *in vitro,* which could allow a study of the possible clonal evolution types that has not yet been performed. Cultured cells from metastatic lesions have an epithelioid to fusiform-epithelioid aspect that can subsequently suddenly change into fusiform cells. This sudden change of the phenotype can result from a genetically programmed alteration as an expression of genes responsible for phenotypic differentiation. This aspect has also been noted in human melanomas at early and intermediate stages.

The involvement of oncogenes in melanic neoplasms is achieved by the transfer of the oncogenic *c-Ha-ras* activity towards the non-metastatic K-1735 melanoma, causing the accelerated growth of tumor cells *in vivo* and enabling them to form colonies in the lung, after intravenous injection, but without conferring the capacity to metastasize from a primary tumor developed subcutaneously.

In conclusion, the transfer of the *c-Ha-ras* gene activity to non-metastatic melanic cells induces a significant acceleration of the growth potential, but does not confer the capacity for invasion and dissemination from the subcutaneously developed primary tumor.

There are some uncertainties regarding the extrapolation of results from veterinary oncology to human oncology. The use of cutaneous pig melanoma studies in human oncology requires some reserves, such as the relatively low incidence of familial melanoma, since the majority of human melanoma cases are sporadic and not congenital. It is not clear whether the activation of oncogenes is associated with the initiation or promotion of cutaneous melanoma in Sinclair pigs.

The authors suggest that the cutaneous melanoma model of the Sinclair pig is not limited to familial or inherited neoplasms, and it can serve as a model of the host-tumor interaction for regulating experiments, also acting in sporadic or non-familial cancer. Cutaneous melanoma in Sinclair pigs is an excellent model for the study of molecular changes related to the evolution of human benign cutaneous lesions towards melanomas.

In conclusion, recent progress in the maintenance of human melanocytes *in vitro* and the fact that swine melanoma offers an excellent animal model for the isolation of specific development factors, involved in the initiation, development, metastasizing and regression of human melanoma, will significantly improve the conditions for the study of melanomas.

The experimental study of melanic tumors was and remains a scientific passion justified, on the one hand, by the interest in the elucidation of some aspects of tumorigenesis and, on the other hand, by the enigmas of the black tumor. Thus, in 1978, WARREN et al. transplanted tumor tissue in the hamster pouch, monitoring the "rooting" of the graft, the growth of the tumor and the initiation and development of blood vascularization. The authors found that the tumor extension took place only after a vascular relation was established between the implanted tumor and the host. It is estimated that this experimental model can be used, being so far insufficiently evaluated.

The transplantation of human tumors in immunodepressed nude mice is an important experiment for the biological study and treatment of human cancer. CORNIL et al. (1989) experimented the possibility of the subdermal implantation of human melanoma, in order to study melanogenesis and metastasis in nude mice. The authors examined the possibility of the development of melanoma after the intradermal injection of melanic cells, being known that melanocytes normally have dermoepidermal junctions. Results show that subdermal inoculation, although it is not an ideal orthotopic site for the implantation of melanoma, could be more favorable and suitable than deep subcutaneous inoculation. The experiment shows that the extension of results to other types of melanomas, as well as to other skin tumors, such as basal cell or squamous cell carcinoma, will allow the finding of new ways for the study of malignant tumors and the reactions of different tissues or organs in order to elucidate the progress of the extension and metastasizing of the neoplasm

An interesting experiment was performed by BOMIRSKI et al. (1988), who managed, by successive transplantations over almost 25 years, to maintain a spontaneous skin melanoma, in the golden hamster. The authors identified 5 variants that, although having the same origin, differ in terms of differentiation, malignancy, intermediate metabolism, number of chromosomes and properties of the cellular surface. These melanomas, once stabilized, have a high phenotypic stability through the tens of years of passage. One of the tumor lines lost its capacity to synthesize melanin. After the first transfer to cell cultures, this amelanotic melanoma regained differentiation, but lost to a large extent its proliferative capacity. This process is reversible by the reimplantation of tumor cells in the hamster. The observations of the mentioned authors lead to the conclusion that tumor cells can lose a part of their malignancy spontaneously or as a result of mild experimental manipulations. By generalizing the above, it may be concluded that the observations and experiments can be used in the strategy and control of tumor disease in humans and animals.

PRINCIPLES OF DIAGNOSIS IN MELANIC TUMORS

Histological diagnosis does not pose particular problems. Rapid growth, metastases, especially in the lung and lymph nodes, will be considered for malignant forms. Melanomas are frequently bordered by inflammatory reaction.

Difficulties are encountered in the case of amelanotic melanomas, as well as between malignant and benign forms.

The following histological criteria are considered:

- Presence of melanic cells under a dispersed, ungrouped form in the depth of the epidermis;
- Melanocyte pleomorphism;
- Sporadic presence of tumor cells in the superficial epidermis;
- Presence of an inflammatory infiltrate, in the absence of trauma or infection and presence of melanocytes in the lumen of lymphatic or blood vessels;
- Presence of abnormal mitoses and cytonuclear atypias.

Additional diagnostic techniques, more traditional histochemical techniques, are taking second place to immunohistochemistry (IHC), incorporating the use of monoclonal and polyclonal antibodies. Other methods include electron microscopy (EM) and, more recently, in situ hybridization (ISH).

Histochemistry: more specific to melanin-producing cells is the enzyme tyrosinase, or dihydroxyphenylalanine oxidase (DOPA), which is responsible for converting tyrosine to melanin. This test is more applicable to non-pigmented melanomas, where the above chemical reaction results in the formation of brown/black pigment in the cytoplasm of neoplastic cells, confirming the presence of DOPA oxidase/tyrosinase in those cells. The major disadvantage of this test is that it requires fresh tissue.

Electron microscopy: melanosomes indicate that the cell in which they occur is actively producing melanin pigment, but they must be differentiated from compound melanosomes, which are really secondary lysosomes extruded by melanocytes and taken up by keratinocytes and macrophages. True melanosomes usually have cross-striations, whereas secondary melanosomes do not.

Immunohistochemistry: canine melanomas were 100% vimentin positive, 92% Melan-A positive, 98% S-100 positive, and 89% NSE positive; Melan-A is a useful marker for melanocytic derivation in tumors of uncertain lineage. In diagnosing melanomas in cats, Melan-A and S-100 are useful markers for melanocytic neoplasms.

The researches of SMITH et al., 2002 on the pathogenesis of melanoma have created a source of additional information for the development of future diagnostic methods. Much work is currently underway in human medicine to try and identify specific tumor markers, i.e. proteins that are associated with malignancy. Applicable detection techniques encompass IHC, ISH, polymerase chain reaction (PCR), and reverse transcriptase-PCR, as well as fluorescent IHC. Serological markers, including cytokines, cell adhesion molecules, and melanoma-inhibitory activity, have also been investigated as potential sentinels of melanoma.

Differential diagnosis is made with other malignant cutaneous tumors, such as epidermoid carcinomas and fibrosarcomas, but also with skin melanosis, brown or black tissue pigmentation.

THERAPEUTIC CONSIDERATIONS

In malignant melanoma, like in other forms of neoplasms, the following treatment methods are used: the surgical method, radiotherapy and chemotherapy. Immunotherapy has been recently experimented with encouraging results, especially in the case of malignant melanoma.

Surgery is the oldest method in the armamentarium of the fight against cancer. This method has been improved in time, benefiting from more and more efficient techniques and equipment.

Radiotherapy appeared at the beginning of the century and has developed especially after the introduction of high-penetration rays. In tumors, ionizing radiation stops cell division and induces the death of cancer cells.

Chemotherapy, including hormone therapy, prevents cell division and stops proliferation. Local cisplatin treatment of mouth melanoma in dogs is an alternative to surgery and radiotherapy. Results are encouraging, producing a complete positive response, with the prolongation of survival and improvement in the quality of life. In contrast, tumors with other locations and large sizes were difficult to control and required complex treatment. The use of cimetidine, 2.5 mg/kg*per os,* every 8 hours, for 2 months up to one year, in three horses with multiple melanomas, resulted in a significant decrease in the number and size of melanomas (50–90%). The progress of the disease was stopped in 2 horses and controlled in the third horse. In principle, the early removal of all melanic neoplasms will be initiated, when they are easy to approach. Early removal is indicated especially for dog and horse melanomas, when these have a cutaneous location. Mouth tumors are difficult to approach by usual surgical methods; this is why the cryosurgical method is indicated. Surgical or cryosurgical treatment will always be supplemented with radiotherapy.

Immunotherapy, a recently introduced method in the fight against cancer, which is also less harmful for the sick organism, needs to be extended, but in the first place experimented and improved in animals. Bispecific anticancer immunotherapy by oncolysis is an immunological method for the treatment of malignant tumors by regulated and prolonged hyperimmunization, due to autologous (autovaccines) or homologous preparations (stock vaccines). This bispecificity involves the specific antigenic strengthening of "weak" tumor antigens (inactive alone) by a "strong" viral antigen (inactive alone) that is capable of enhancing the "foreign" character of tumor antigens

All organisms with tumors oppose or tend to oppose an immune resistance to the development of malignant tumors. Cancer immunotherapy tends to ensure a maximal antigenic mismatching between the tumor and the normal tissues of the patient and stimulate as much as possible its immune defense system.

According to LEFTERIOTIS et al. (1980), the various modalities of anticancer immunotherapy can be synthesized as follows:

1. Non-specific stimulation of the immune reaction.
2. Active immunization, with or without an adjuvant, including:
   a. Immunization with tumor cells, cell fractions or purified autologous tumor antigens;
   b. Immunization with tumor cells, cell fractions or purified homologous tumor antigens.
3. Passive immunization by:
   a. Serum or immunoglobulins of patients in remission or patients who have rejected their own tumors;

   b. In the case of leukemia, heterologous antilymphocyte serum (directed against patients' malignant lymphoid cells).
4. Adoptive immunization, carried out:
   a. With lymphocytes of normal individuals;
   b. With lymphocytes of a normal individual immunized against the patient's tumor;
   c. With lymphocytes of patients in remission or who have rejected their own tumors;
   d. In the case of leukemia, with bone marrow transplant from a healthy individual to a patient in whom leukemic cells have been destroyed.
5. Immunogenetically coded messages:
   a. With homologous or heterologous lymphocytes that have responded to tumor cell immunization;
   b. With nucleic acids (RNA) extracted from messanger cells.
6. Suppression of favoring antibodies:
   a. Relative immunodepression by drugs (suppression of antibody formation without the suppression of the cellular immunity reaction)
   b. Antibody depletion (drainage of the thoracic canal and reinjection of lymphocytes).
7. Combination of radio-, chemo- and immunotherapy:
   a. Tumor mass reduction before immunotherapy;
   b. Selective immunodepression of humoral immunity.
8. Interferon and interferon-inducible double-stranded RNA.

The phenomenon of viral oncolysis or the lysis of tumor cells by a virus, by means of stimulating tumor cell immunogenicity, as well as immunity obtained following oncolytic viral treatment, made the object of researches performed by MOORE, KOPROWSKI, BOONE and LINDENNMANN 30 years ago. Experiments performed in mice showed that subjects who survived the oncolytic inflammation virus developed postviral oncolytic immunity and resisted to a new dose of the same tumor. Malignant melanoma is a favorable form for investigations in the direction of neoplasm immunity. The lysis of malignant melanocytes by the vaccine virus releases masked antigens that, reinjected in the patient, accelerate or trigger antibody production.

Oncolytic vaccine has been experimentally used in humans, in the case of advanced malignant melanoma forms. Melanoma in animals may exist under a latent form, sometimes for years, or it can evolve rapidly, by generalization, metastasizing and death. In dogs, pigs and horses, spontaneous regressions have been described, preceded by the depigmentation of the area, which suggests the intervention of a cellularly mediated factor.

The use of oncolytic viruses in animal melanoma remains an extremely attractive issue that is open to the future. Some attempts made by LEFTERIOTIS et al. (1980) in dogs and horses seem to be promising, especially that these species present favorable reactivity to the vaccine virus. In addition to the immediate interest in the treatment and salvation of these animals, we consider as a major desideratum the possibility of experimenting new treatments on animals in the hope of transferring results to humans. The perspective is open for the extension of oncolytic viruses to other neoplasms, such as breast carcinoma in dogs, serving in this way as an experimental and research basis for human oncology.

The literature, as well as our own experience, remarks the high incidence of benign forms (80%) with cutaneous location in cattle. In this case, melanocytoma is treated surgically, by exeresis with local anesthesia. Cicatrization occurs within two weeks, without recurrences or metastases.

PERIANAL GLAND TUMORS

Perianal gland tumors are also known as *hepatoid gland tumors* or *circumanaloma.* These tumors appear almost exclusively in male dogs, exceptionally in females, while in other species they are practically unknown. Incidence is high at an advanced age, over 8 years, and the Cocker Spaniel breed seems to be more sensible, the incidence ratio between males/females being 8.8/1. Some authors mention the following breeds as being more susceptible to develop perianal gland tumors: English Bulldog, Beagle and German Shepherd. In the etiology of these tumors, hyperandrogenism of testicular origin is involved, which is proved by the possibility of neoplasm prevention by castration, and the excision operation should be accompanied by castration.

Location is in the skin around the anus, but other locations may be occasionally found: the skin around the prepuce, the tail, lumbar, sacral skin. Tumor growths may be solitary or multiple.

The following forms are morphologically distinguished: hyperplasia, adenoma and carcinoma (adenocarcinoma). It should be envisaged that these three forms are evolution states of the same process, which makes difficult to establish the exact consequences and prognosis.

Perianal gland hyperplasia is characterized by the evidencing of nodular masses that are more obvious than normal and form a fold in perianal skin, with a tendency to expansion. The lesion obviously has a hyperplastic and not a tumoral character, the differentiation from the normal gland being difficult to make.

Perianal gland adenoma (hepatoid gland adenoma, circumanal gland adenoma) macroscopically appears as prominent, single or multiple nodular formations, sometimes reaching 10 cm in diameter. These nodules frequently become infected and ulcerated. The adenoma in section has a lobular structure, the lobes being separated by a fine connective stroma.

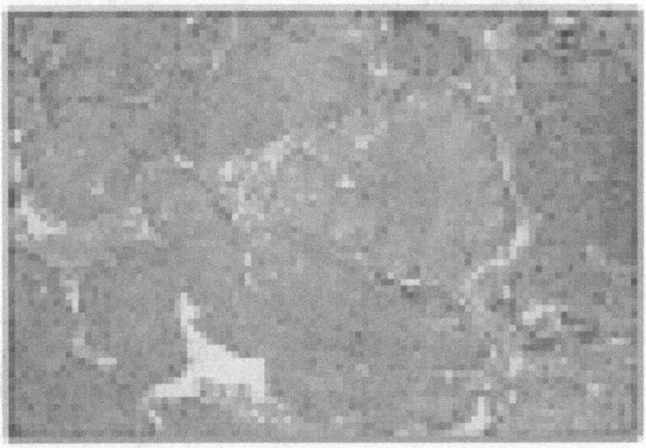

Fig. Hyperplasia of hepatoid gland.

Histologically, structure is little differentiated from the normal aspect, having a more obvious lobular differentiation, with polygonal or round cells arranged as cords or nests. Cells are large, with abundant and finely granulated acidophilic cytoplasm. The lobe periphery is marked by the presence of small, basophilic basal replacement cells and numerous blood capillaries. Mitoses are sporadic. In lobular structures, round and spiral formations occur, which are squamous metaplasias. Ulceration, hemorrhage, necrosis, and secondary infection with inflammatory cell infiltrates are common. A proportion of cases will show foci of ductal differentiation (squamous metaplasia) and/or foci of cytoplasmic lipidization (sebaceous differentiation). Some authors define perianal gland adenoma as a microscopically benign tumor and clinically, as local invasive adenoma.

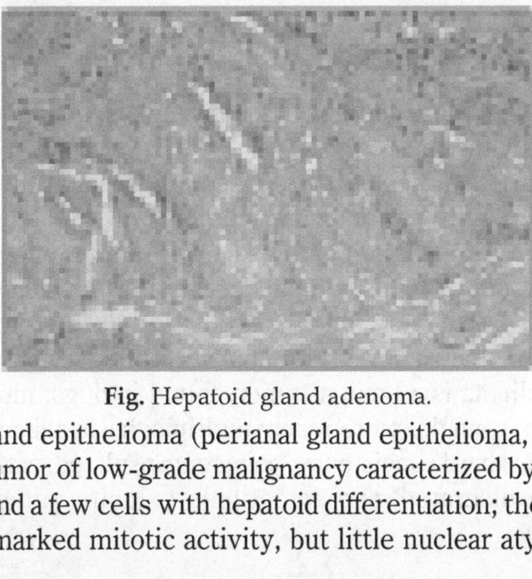

Fig. Hepatoid gland adenoma.

Hepatoid gland epithelioma (perianal gland epithelioma, circumanal gland epithelioma), a tumor of low-grade malignancy caracterized by a preponderance of basaloid cels, and a few cells with hepatoid differentiation; the basaloid reserve cells may show marked mitotic activity, but little nuclear atypia.

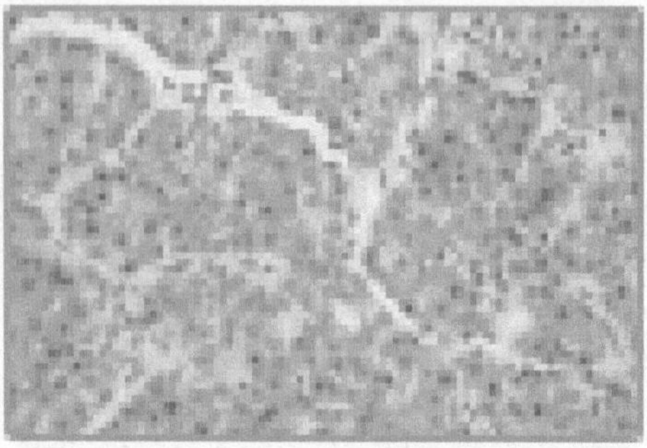

Fig. Hepatoid gland epithelioma.

Perianal gland adenocarcinoma (hepatoid gland carcinoma, circumanal gland carcinoma) is characterized by a high growth rate with a local invasion aspect; it can ulcerate and metastasize in locoregional lymph nodes, and more rarely, metastases develop in viscera.

Histologically, it is characterized by cell pleomorphism, sometimes with the almost exclusive presence of large polyhedral cells and few or no small replacement cells. There are also adenocarcinomas in which replacement cells are dominant. Malignancy is increased by poor cell differentiation, nuclear and cytoplasmic heterogeneity, the presence of mitoses and infiltrative growth in the stromal connective tissue or even in the small lymphatic formations.

Perianal gland tumors generally have a benign evolution, after excision they only recur in exceptional cases, when neoplastic remnants are left or when they generate tumors in other glands from that region.

Malignant forms grow rapidly, developing into large formations that ulcerate, metastasize in the locoregional lymph nodes, as well as in the pelvic canal or even at distance, and extremely rarely in the lung, liver, kidney and bone.

HAIR FOLLICLE TUMORS

Hair follicle tumors are generally rare in animals, the only species in which they are frequent being the dog.

The classification of these tumors takes into account the structure of hair follicles, with a generally low malignant tumor growth potential.

The animal literature has borrowed from human oncology, where hair follicle tumors are multiple and complex, a classification adapted to veterinary oncology. Thus, the following are distinguished: trichoepithelioma; tricholemmoma; pilomatrixoma and calcinosis circumscripta.

Trichoepithelioma is a benign tumor, found in dogs, more rarely in cats. The tumor, which is single, more rarely multiple, has a subcutaneous location, a 1–10 cm diameter, hard consistency, being encapsulated, with a multilobulated aspect; it can be ulcerated, and sometimes exhibits mineralization foci or multiple cysts.

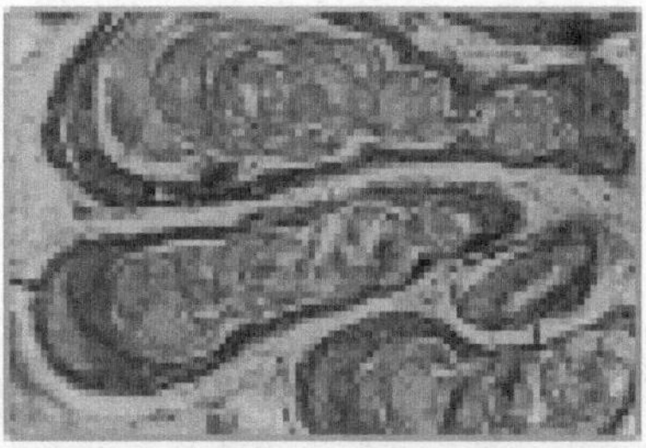

Fig. Trichoepithelioma, abrupt keratinization.

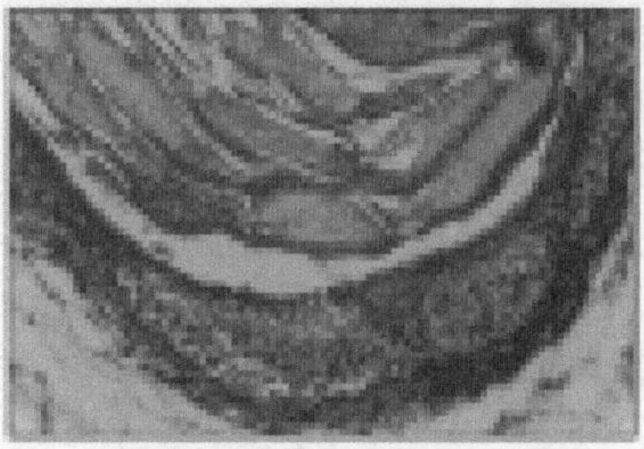

Fig. Trichoepitelioma.

Incidence is estimated to be 2–3% in dogs and 1.5–4% in cats, according to PRIESTER (1973, cited by MADEWELL and THEILEN, 1987), and according to NIELSEN and AFTOSMIS (1964) 4.5% of all cutaneous tumors in dogs, occurring at the age of 5 years, regardless of breed.

Histologically, the predominance of basal cells is found, which delimit the formation of microcysts with horny content. The transition from basal cells to squamous cells is neat, without an intermediate area. Sometimes, due to the more abundant proliferation of basal cells, the tumor resembles a basalioma, but in trichoepithelioma there will always appear horny cysts and even hair remnants. Infections through ulceration, as well as calcification, create perifocal inflammatory reaction.

Malignant trichoepithelioma (matrical carcinoma), a malignant follicular tumor with matrical and inner root sheath trichogenic differentiation that often metastasizes to regional lymph nodes and lungs. This tumor consists of a multilobulated dermal and subcutaneous mass that often has central cystic degeration and necrosis. The peripheral cells are basophilic with hyperchromatic nuclei and little eosinophilic cytoplasm is observed; characterized by the presence of occasional cells with intracytoplasmic trichohyalin granules and foci of abrupt matrical keratinization; lymphatic invasion can be identified.

Tricholemmoma is a rare hair follicle tumor; in dogs, it is structurally similar to trichoepithelioma, the main difference being the character of the basal cell cytoplasm with a tendency to resemble follicles. In tricholemmoma, the cytoplasm of the most external cells contains glycogen deposits, which differentiates them from the other cells of the hair root sheath.

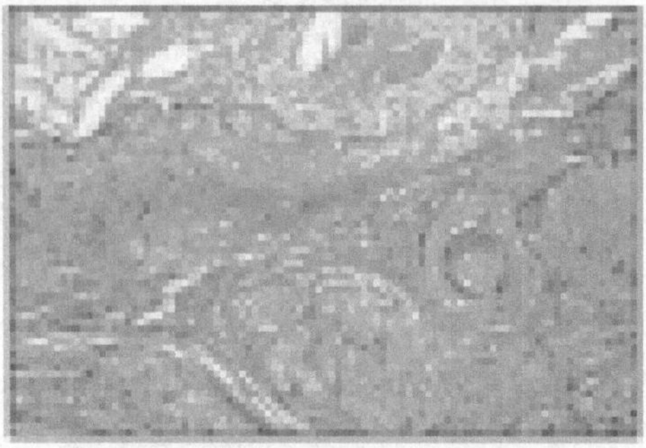

Fig. Malignant trichoepithelioma. Muyltilobulated hair follicle neoplasm with central cystic degeneration and necrosis.

Inferior tricholemmoma is composed of islands of neoplastic cells separated by fine fibrillar collagenous stroma with a prominent eosinophilic basal lamina zone. The peripheral cells palisade and have abundant pale eosinophilic cytoplasm, while the central cells have an abundant deeply eosinophilic cytoplasm. The tumor cells show little pleomorphism or mitotic activity.

Isthmic tricholemmoma is composed of epithelial aggregates, sometimes having an association with the epidermis. There are cords of epithelial cells, which show central tricholemmal keratinization, and large islands may appear cystic. Tumor cells have a moderate amount of lightly eosinophilic, glassy cytoplasm with small euchromatic nuclei. These tumors may be heavily melanized.

Trichoblastoma, a benign neoplasm derived from or reduplicating the primitive hair germ of embryonic follicular development; it was previously classified as a basal cell tumor.

Ribbon trichoblastoma, a tumor that occurs as a dermal and subcutaneous tumor, rarely showing any association with the overlying epidermis. The tumor consists of long cords of cells that may branch and join together, with nuclei arranged in a palisade perpendicular to the long axis of the column. Nuclei are euchromatic to hyperchromatic, and the number of mitoses is variable, but may be high. The surrounding collagenous matrix may be quite extensive and appear hyalinized.

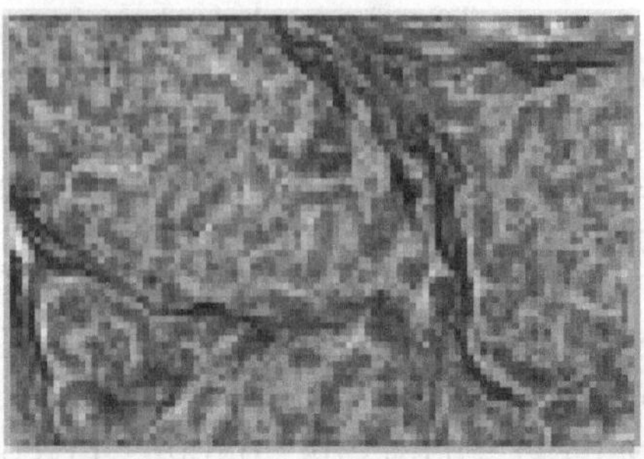

Fig. Trichoblastoma, ribbon type. Undulating ribbons of basaloid cells.

A variation of the above pattern consists of cords of cells radiating from a central island of densely packed cells, which have more abundant eosinophilic cytoplasm – medusoid pattern. Foci of incomplete trichogenesis are occasionally observed in this neoplasm.

Trabecular trichoblastoma is a multilobulated intradermal tumor. The epithelial cells at the periphery of the lobules palisade and appear more elongated with euchromatic nuclei, little cytoplasm, and indistinct cell borders. There is little interlobular collagenous stroma.

Granular cell trichoblastoma is composed of cells with abundant, finely granular eosinophilic cytoplasm and eccentrically located small hyperchromatic nuclei.

Spindle cell trichoblastoma, most frequently found in the cat, is often seen as a well-circumscribed dermal mass in which an attachment to the normal epidermis may be found. There are trabeculae and lobules of small basaloid epithelial cells with sparse fibrous stroma; cells have ovoid nuclei, inconspicuous nucleoli, scant cytoplasm, and few mitoses. The tumor cells may appear elongated, resulting in a whorled appearance of the neoplasm. Melanin may be present in melanophages and within tumor cells.

Pilomatrixoma (pilomatricoma, necrotizing and calcifying epithelioma of Malherbe) is an extremely rare tumor, being only found in dogs, with an incidence of approximately 3%, in old subjects from the Terrier and Poodle breeds, with no differences between sexes.

The tumor is found in the dermis and subcutaneously, as a solitary formation, in the limbs, shoulders and head. Macroscopically, it is frequently solitary, dense, well circumscribed, non-adherent to tissues, 1–10 cm in diameter, and it frequently ulcerates.

Multiple microcysts are histologically found, and eosinophilic epithelial cells can be identified, similar to immature hair cortex cells, with keratin content. Cysts are lined by multiple layers of intensely basophilic, dark cells. The proliferation of basal cells is found, which suddenly become eosinophilic, without a nucleus, but maintaining their cell membrane, aspect due to which they have been termed "mummified cells" or "ghost cells".

The mineralization of these tumors as well as infection determine a typical infectious pyogranuloma aspect with perifocal reaction.

Pilomatrixoma has all the characteristics of a benign tumor: it grows slowly, it is not invasive, it does not produce recurrences and metastases, but may have moderate mitotic activity.

Malignant pilomatricoma (pilomatrix carcinoma), a malignant follicular tumor with exclusively matrical differentiation. The tumor occupies the dermis and subcutis. These tumors invade the subcutaneous tissue, and there is often a desmoplastic response. Ossification may be present in the primary tumor and metastases. Differentiation from malignant trichoepithelioma is based on the lack of trichohyalin granules within the cell cytoplasm and/or the presence of foci of calcification and ossification. This diagnosis is best made when lymphatic invasion can be identified.

Calcinosis circumscripta is macroscopically similar to a hair follicle tumor; it is found in the subcutaneous tissue in dogs, but locations in the tongue and the paravertebral connective tissue have also been reported. In section, nodular formations have a necrotic aspect, with mineral granulations. Histologically, basophilic mineral granulations are found, delimited by macrophagic cellular reaction, and giant cells, lymphocytes and plasmocytes, all delimited by fibrous connective tissue.

Etiologically, these calcifications are supposed to be the consequence of microtrauma or the consequence of apocrine gland hyperplasia, with dystrophic mineralization and abnormal apocrine secretion.

CUTANEOUS SMOOTH MUSCLE TUMORS

LIU and MIKAELIAN describe cutaneous smooth muscle tumors of vascular and arrector pili origin in dogs and cats as typically solitary lesions that infrequently invade into the subcutis. Immunohistochemistry for á-SMA, and to a lesser extent desmin, may aid in their diagnosis, particularly in poorly differentiated tumors. In most cases, surgical excision is curative, and no metastases have been reported.

These authors describe cutaneous smooth muscle tumors: hamartomas, piloleiomyomas, piloleiomyosarcomas, angioleiomyomas, and angioleiomyosarcomas.

Cutaneous smooth muscle hamartomas and neoplasms are infrequently described in dogs and cats; they have been reported most often in the dog and ferret.

Hamartomas of arrector pili muscle were present on the dorsal lumbar region, in dogs. Histologically, they were solitary dermal lesions, unencapsulated and well demarcated. They were composed of abnormally large or thick but well-differentiated smooth muscle bundles, which encompassed dermal appendages. There was no evidence of cytologic atypia or mitotic figures. The overlying epidermis was moderately acanthotic, and a few hyperplastic sebaceous glands were entrapped in all hamartomas.

Piloleiomyomas, in dogs, the dorsal trunk was the most common site, followed by the limbs and head; in cats, the thigh and dorsal muzzle were affected. Tumors were solitary, nodular, well-demarcated, unencapsulated, expansile dermal masses, which showed a connection to arrector pili muscles in most dogs, but not in cats. They were composed of narrow to broad, interlacing bundles of well-differentiated smooth muscle cells, which were narrow to plump and spindloid, with elongated cigar-shaped nuclei containing sparse, finely granular chromatin. The cytoplasm was homogeneous to slightly fibrillar and deeply acidophilic. Mitoses were not observed. Piloleiomyosarcomas, in dogs, the dorsal trunk was the most common site, involving the limbs and muzzle; in the cat, they occurred on the thigh. Tumors varied from well circumscribed to poorly circumscribed and invasive. Neoplastic cells showed a connection to arrector pili muscles, either entrapped or displaced dermal appendages. The canine tumors were composed of well-differentiated smooth muscle but, unlike benign lesions, they were invasive and displayed a mitotic rate of > 2 per 10 high power fields; other tumors included poorly differentiated areas. These regions were hypercellular and contained streaming and interlacing bundles of narrow spindle cells with pale eosinophilic to amphophilic cytoplasm. Neoplastic cells occasionally showed nuclear palisading and packeting in the least-differentiated regions. These cells displayed mild anisokaryosis. Some tumors had areas of central coagulation necrosis, lymphoplasmocytic inflammation, mild mastocytic infiltration, hemorrhage, ulceration, pyogranulomatous inflammation centered on ruptured hair follicles, or mild myxomatous stromal degeneration.

Angioleiomyoma was diagnosed on a front digit of a domestic shorthaired cat; it was nodular, well delineated, unencapsulated, expansile, and moderately cellular. It was made of closely packed interlacing bundles of neoplastic smooth muscle fibers originating from the wall of blood vessels. Neoplastic cells were plump to fusiform, with indistinct cell borders and a moderate to large amount of fibrillar acidophilic cytoplasm. The nucleus was elongated to cigar shaped with blunt ends. Anisokaryosis and anisocytosis were mild; mitoses were not found. Angioleiomyosarcomas were diagnosed in dogs and cats, located on the face, limbs and body; they had an expansile or invasive growth pattern. Neoplastic cells in angioleiomyosarcomas were also smaller and had an increased nuclear/cytoplasmic ratio. Neoplastic cells blended into or seemed to originate from the media of blood vessels in all tumors.

PLASMACYTOMA

Solitary plasmacytoma has a cutaneous-mucous location, which is not identical to that of intraosseous malignant multiple plasmacytoma. Solitary cutaneous plasmacytoma is a benign tumor, common to dogs, extremely rare in cats, being reported in no other species. In dogs, it appears at adult and old age, under the form of spherical nodules, being located in legs, the ear duct and lips. Lesions have also been identified in digestive mucosae such as the gums and the rectum.

Histologically, the presence of plasmocytes with a classical cellular structure is found; pleomorphic cells may appear, grouped in the form of a solid or lobulated tumor, with fine connective filaments. Cells are basophilic, with eccentric nucleus, sometimes binucleated or multinucleated cells with mitoses. The benign evolution makes tumor recurrences exceptional.

4

Types of Cancer and Treatment

INJECTION-SITE SARCOMAS IN CATS

WHAT IS AN INJECTION-SITE SARCOMA?

An injection-site sarcoma is a tumor of the connective tissues in the cat. The most common cell type affected is the fibroblast, giving the tumor the name fibrosarcoma. Other tumor types have been described (osteosarcoma, chondrosarcoma, malignant fibrous histiocytoma). The tumor types behave similarly and are therefore treated the same way. The tumors are often located between the shoulder blades, in the hip region, and in the back legs.

A cause and effect relationship has been established for certain vaccines, including rabies and feline leukemia. However, other chronic injections such as lufeneron (Program) also have been thought to cause these tumors. It is likely there is a genetic predisposition to the development of these tumors in certain cats; however, the exact genetic problem has not yet been identified.

How is it diagnosed?

Needle aspirates are useful in increasing your veterinarian's suspicion of cancer. However, sometimes it is difficult to differentiate an injection-site sarcoma from a vaccine reaction or granuloma on aspirates. Sometimes these tumors do not exfoliate or "give up" their cells easily, making a diagnosis by needle aspirate difficult. A surgical biopsy is often necessary to make a definitive diagnosis of injection-site sarcoma.

Once an injection-site sarcoma has been diagnosed, other diagnostic tests are indicated prior to considering treatment. Chest x-rays are necessary to determine whether there is any evidence of metastasis (spread) to the lungs. Additionally, routine bloodwork should be performed to assess the overall health of your cat. In most cases, a CT scan may be recommended to determine the extent of your cat's tumor and to appropriately plan surgery and/or radiation treatments.

SURGERY

Surgery is the mainstay of treatment for injection-site sarcoma. Surgical removal needs to be wide and deep, as these tumors grow by sending "finger-like" projections of tumor cells among the tissues. The edges of the biopsy are evaluated to determine whether the margins are "clean" (no evidence of tumor cells at surgery margins) or "dirty" (tumor cells extend to surgery edge). Cats with "clean" surgery margins often do better, but tumors have been noted to regrow, even with "clean" margins. In most instances, surgery alone is not curative and other treatments, such as radiation and chemotherapy are recommended as an adjunct to surgery.

RADIATION THERAPY

Radiation therapy may be performed either before or after surgical removal of the tumor in order to treat any residual tumor cells that may remain after surgery. Radiation therapy is performed on a daily basis, Monday through Friday, for 4 weeks. A brief anesthetic episode is necessary so that your cat remains still for treatment. Radiation therapy is well tolerated by animals. Radiation therapy combined with surgery has been shown to significantly improve survival over surgery alone.

CHEMOTHERAPY

Chemotherapy may be recommended for recurrent tumors or tumors that may behave aggressively based on biopsy characteristics. Two chemotherapy drugs, doxorubicin and carboplatin, have shown activity against injection-site sarcomas. Both drugs are given intravenously every 3 weeks for 5-6 treatments. Chemotherapy is extremely well-tolerated in cats. Doxorubicin in combination with surgery may be better than surgery alone.

WHAT IS THE PROGNOSIS?

Injection-site sarcomas in cats are very difficult to treat. Cats with tumors on the legs, where an amputation can be performed, appear to do better than cats with tumor on the trunk of the body. Treatment combining surgery, chemotherapy and radiation therapy significantly increases the tumor free time to 18-24 months. Cats with tumors that have grown back after repeated surgeries before additional treatments are tried are more difficult to control. Thus, the best time to treat with chemotherapy and radiation therapy is the very first time the tumor occurs. Metastasis (spread) to other parts of the body is reported between 10-25%. This potential increases when the tumor regrows after repeated surgery.

CAN INJECTION-SITE SARCOMAS BE PREVENTED?

We do not know which patients are predisposed to these tumors or all of the events that go into development of these tumors. The risk of not vaccinating for certain diseases may be much higher than the risk of tumor development. The best advice is to vaccinate discriminately and make educated decisions about which diseases your cat should be vaccinated for.

LYMPHOMA IN DOGS

WHAT IS LYMPHOMA?

Lymphoma is a cancer of white blood cells, which circulate through the blood and especially through a special circulatory system called the lymphatic system (made up of a complete network of lymphatic vessels and lymph nodes). This system is as extensive throughout the body as the venous and arterial bloodstreams. Normally the function of the lymphatic system is to filter out debris from dead cells and bacteria and produce antibodies against foreign substances encountered throughout the animal's life. In dogs, lymphoma is a rapidly progressive disease that, if left untreated, results in death within a relatively short time, usually a few weeks from the time of diagnosis.

WHAT CAUSES LYMPHOMA?

What causes a healthy cell to be transformed into a tumor cell is still largely unknown. Much research has been conducted in recent years concerning lymphoma. In cats and cows it is associated with infection by the feline leukemia and bovine leukemia viruses. In dogs a virus that causes lymphoma has not been identified.

WHAT ARE THE CLINICAL SIGNS?

Lymphoma is recognized in several clinical forms (generalized, intrathoracic, skin, gastrointestinal, and leukemia). The most common sign is painless enlargement of the lymph nodes. Because of the disease affects the whole body, any organ system can be affected and non-specific signs such as depression, weakness, weight loss, loss of appetite, diarrhea or vomiting, and difficulty breathing can also be seen. Often dogs with lymphoma are slightly anemic. The liver and spleen can also be involved and enlarge in size.

CAN LYMPHOMA BE CURED?

We are unable to cure canine lymphoma. However, the vast majority of dogs can be rendered free of any evidence of disease (placed into complete remission) for many months. Their quality of life during this period is generally excellent.

WHAT DOES TREATMENT CONSIST OF?

Because lymphoma affects every organ system in the dog, the entire body must be treated. This leaves chemotherapy as our most effective means of killing the rapidly growing and dividing cancer cells. Some drugs are given intravenously and some orally. Many times a combination of drugs are given.

ISN'T CHEMOTHERAPY MORE HARMFUL THAN THE DISEASE?

No. Any chemotherapy drugs we recommend provide benefits by their actions against the cancer cells that are expected to far outweigh any potential side effects associated with their use. We will try to outline the side effects you might see following chemotherapy and give you recommendations as to what can be done to monitor and prevent them. We would never recommend a treatment plan that is expected to be worse than the disease itself.

WHAT IS LIFE EXPECTANCY WITH TREATMENT?

It is impossible to predict how long your pet will live. We know from the hundreds of dogs that have been treated, that 75-80% will achieve a complete remission with a return to an excellent quality of life. We also know that this disease always comes back. The range of time it stays away the first time varies from 4 months to 2 years. The average is 6-8 months. At that time, about half of the dogs can be placed into remission again. Generally the second remission is shorter than the first. Sometimes a third, fourth, and even fifth remission can be achieved, but each is generally shorter than the last as the cancer cells become resistant to the chemotherapy just as bacteria become resistant to antibiotics. On average, with treatment, patients will live for approximately 1 year.

However, every dog is an individual, and some will live much longer than this. Sometimes, additional diagnostic tests that tell us more about the location or type of lymphoma will also give us more information to help predict prognosis. Ultimately, the only way to know how well any specific patient will respond is to treat him or her.

HOW WILL I KNOW WHEN IT IS TIME TO SAY GOOD-BYE?

We will help you with that difficult decision in any way we can. We will also do everything we can to make sure that it is not for many months to come. Nobody knows your pet better than you do and they will let you know when their quality of life is no longer acceptable. It is rarely anything specific. Trust yourself. Don't hesitate to ask all of your questions. We will work together as a team and the most important members are you and your pet.

NASAL TUMORS IN DOGS

Nasal tumors make up approximately one percent of all cancers seen in dogs. It is thought that long-nosed breeds or dogs living in urban environments are at higher risk for the development of nasal tumors. Nasal tumors are seen less commonly in the cat.

The majority (two-thirds) of nasal tumors are carcinomas. Sarcomas make up the remaining tumor types. Lymphoma may also be seen in the nasal cavity; it occurs more commonly in cats than in dogs. Nasal tumors occur more commonly in older animals.

WHAT ARE THE SIGNS OF A NASAL TUMOR?

Symptoms may include an intermittent and progressively worsening nasal discharge or bleeding from one or both nostrils. Facial deformity can be seen in some cases. On rare occasions, animals may present with only neurological signs due to direct invasion into the brain cavity. Transient improvement in clinical signs is often noted with symptomatic treatments such as antibiotics, antihistamines and steroids, which often delays diagnosis.

HOW ARE NASAL TUMORS DIAGNOSED?

Other diseases of the nasal cavity such as fungal infections, foreign bodies, and allergies have symptoms similar to nasal tumors. Systemic bleeding disorders caused by low platelet numbers or clotting factor disorders may also manifest as nasal bleeding. Thus, a definitive diagnosis of intranasal cancer requires a tissue biopsy. Routine bloodwork and clotting times will rule out bleeding disorders and assess your pet's overall health prior to biopsy of the nasal cavity.

Computerized tomography (CT) is the ideal way to diagnose and plan treatment of intranasal cancer. CT can indicate the presence of a tumor and the extent of the tumor, especially as it relates to invasion into the brain cavity and around the orbit of the eye. A tissue biopsy can be planned based on the location of the tumor identified from the CT. A nasal biopsy is easily procured by passing a forceps up the nostril and into the tumor. Mild to moderate bleeding is expected and usually subsides within minutes. You may notice increased discharge from the nose that may be blood tinged for several days following the biopsy. Some dogs may have increased nasal congestion or noise for several days following the biopsy.

Although a CT scan is ideal, occasionally, x-rays of the nasal passage and/or rhinoscopy may be utilized to evaluate the extent of disease and to obtain a biopsy. Once a diagnosis of a nasal tumor is made, x-rays of the lungs are recommended to evaluate for metastases. Needle aspirates of the local lymph nodes may also be recommended if the lymph nodes are abnormal in size or consistency. Your doctor may also recommend an abdominal ultrasound to complete staging.

HOW ARE NASAL TUMORS TREATED?

Dogs and cats with nasal tumors usually present with a relatively advanced stage of the cancer in a critical location near the brain and eyes. Invasion into bone often occurs early and curative surgery is not possible. Chemotherapy alone yields only a 30% response rate and responses are usually short-lived.

Radiation therapy is the treatment of choice in dogs and cats with intranasal cancers. Stereotactic radiosurgery is a procedure that consists of a high dose of radiation delivered in multiple arcs that target the center of the tumor, minimizing damage to normal tissue around the tumor. It is often a one-time treatment of radiation done here at the University of Florida Small Animal Hospital. If the tumor is close to or invading important structures such as the eyes or the brain, however, it may be necessary to use Stereotactic Radiation Therapy (SRT). The difference between the two is that instead of delivering one large dose, SRT would consist of several (up to three) smaller doses to minimize side effects to the important structures such as the eyes and brain. Radiation therapy will result in improvement or resolution of clinical signs in the majority of patients.

Side effects are usually minimal but can include change in fur color, peeling of the skin and ulceration. Ulcers and burn-like side effects are temporary and can usually be well managed with antibiotics and anti-inflammatory medications. Depending on how close the tumor is to the eyes, the tear production can be compromised, resulting in keratoconjunctivitis sicca (dry eye).

WHAT IS THE PROGNOSIS FOR NASAL TUMORS?

Nasal cancer in dogs and cats is not curable. However, patients undergoing radiation therapy can achieve clinical remission with good quality of life.

For cancers other than lymphoma, radiation therapy will result in remission times ranging from 9 – 15 months, with an average of 12 months. Dogs and cats with significant extension into the brain have a poorer prognosis, with average survival times of only 4-6 months, even with radiation therapy.

Lymphoma of the nasal cavity responds very well to radiation therapy. Response is often complete and may be permanent. Survival times in patients with nasal lymphoma following radiation therapy are often 2 years or longer. Some patients with nasal lymphoma, however, may develop systemic lymphoma months to years after radiation treatment.

OSTEOSARCOMA IN DOGS

WHAT IS OSTEOSARCOMA?

Osteosarcoma is a malignant cancer of the bone. In dogs, it usually occurs in the large and giant breeds, such as Great Danes, St. Bernards, Dobermans and Rottweilers, but any breed can be affected. Osteosarcoma typically affects middle aged to older dogs, but young adult dogs are also susceptible. Osteosarcoma most commonly affects one of the bones of the limbs such as the shoulder, carpus or wrist, and stifle or knee. Occasionally, osteosarcoma can originate in the bones of the skull, ribs or spine. The cause of this tumor is unknown.

HOW IS IT DIAGNOSED?

Your pet may experience persistent lameness of the affected leg and/or swelling in the region of the affected bone. X-rays of the bone will alert your veterinarian to the likelihood of osteosarcoma; however other diseases such as bacterial or fungal infections of the bone, bone injuries, and other types of tumors can mimic the appearance of osteosarcoma on the x-ray. Thus, a fine needle aspirate of the bone or a bone biopsy must be performed to definitively diagnose osteosarcoma. These are relatively quick and simple procedures used to obtain a sample of the tumor for laboratory analysis. Some dogs will experience increased discomfort of the leg for 2-3 days following the biopsy. Pain medications are often prescribed to help lessen the discomfort. In rare instances, fracture of the diseased bone may occur following the biopsy procedure.

Once osteosarcoma is diagnosed, x-rays of the lungs are recommended, as this tumor has a high potential for metastases (spread) to the lungs. Occasionally, an abdominal ultrasound to detect spread to abdominal organs and/or a technetium bone scan, to detect tumor spread to other bones, may also be recommended before therapy of osteosarcoma is considered. Routine bloodwork is always recommended to evaluate your pet's overall health and evaluate a specific enzyme (alkaline phosphatase) that may help in determining your pet's prognosis with treatment.

HOW IS IT TREATED?

Treatment of osteosarcoma usually involves both local treatment of the primary tumor affecting the bone, and systemic chemotherapy to help delay or prevent the onset of metastases.

1. *Surgical amputation:* Amputation of the affected limb is often considered for treatment for osteosarcoma. The purpose of this procedure would be to remove the tumor, which would in turn alleviate the pain and source of potential fracture. Even large and giant breed dogs can function well after an amputation and most owners are pleased with their dog's mobility and quality of life following surgery. For some dogs, where amputation is not an option, limbsparing may be considered
2. *Stereotactic radiosurgery:* This procedure consists of a high dose of radiation administered to the exact location of the bone lesion. It is a one-time treatment of radiation done here at the University of Florida Small Animal Hospital. This procedure would require a CT scan with pin placement under anesthesia. The purpose of this procedure is to

alleviate bone pain and irradiate the tumor cells without removal of the leg. The major risk of this procedure is bone fracture at the sight of the bone lesion. If fracture occurs, then amputation of the affected leg could be pursued or possible surgical fracture repair

3. *Surgical limbspare procedure:* Limbsparing involves removal of the cancerous bone and replacement with a bone graft, which is fixed to the remaining normal bone with a bone plate. The purpose of this procedure is to alleviate bone pain and to remove the tumor. Limbsparing is usually recommended for dogs with severe pre-existing orthopedic or neurological problems that would not be functional with an amputation. Limbsparing is most successful in dogs with tumors located in the carpus or wrist, where the tumor has not invaded more than half of the bone and has not invaded into the surrounding tissues. The risks involved include infection of the bone or skin/tissue in the surrounding area, which may require repeated doses of antibiotics and/or surgery with hospitalization and possible tumor recurrence.
4. *Fractionated palliative radiation:* Palliative radiation may be used to decrease pain and improve function of the affected leg. Radiation therapy is usually administered in 2 consecutive daily treatments. Radiation therapy offers pain relief for 60-70% of dogs. Pain relief and improved function may be seen immediately after the first treatment or it may take several treatments to see the effects. In those dogs experiencing a decreased pain response with radiation therapy, the effects may last anywhere between 1-6 months, with 3 months being average. Fracture is also a risk with this treatment. This type of treatment is usually reserved for very painful dogs or dogs with metastatic disease.
5. *Chemotherapy:* This can be done in conjunction with any of the above treatment options. Chemotherapy would consist of treatments administered intravenously every 3 weeks. With Osteosarcoma, the most common drug used is carboplatin which is given intravenously every 3 weeks for a total of 4 to 6 treatments. Side effects are usually mild and can include gastrointestinal side effects (anorexia, nausea, vomiting, diarrhea) and white blood cell suppression. More severe side effects are uncommon and occur in less than 5% of patients being treated with chemotherapy, but may require hospitalization.
6. *Pamidronate (a bisphosphonate drug):* Bisphosphonates are a class of drugs given intravenously over 2 hours that help to slow or prevent bone lysis thereby strengthening bone. These drugs are commonly used in humans in diseases such as osteoporosis and bone metastasis. These drugs have been shown to provide pain relief from

osteosarcoma in up to 50% of dogs. These drugs are usually used in combination with carboplatin and/or radiation therapy, but may occasionally be used alone.

WHAT IS THE PROGNOSIS?

Osteosarcoma is not a curable cancer; however, it is treatable. Local disease control in combination with systemic chemotherapy can give your pet many months to years of good quality of life.

Because of the high potential for metastases, amputation or local treatment alone is purely palliative. Amputation or radiation alone of the affected limb will eliminate pain and offer good quality of life for your pet; however the development of metastases occurs on average in 3-4 months. With the addition of chemotherapy, we can extend your pet's life for an average of 10-13 months. Approximately 20-25% of dogs will live 2 years.

ORAL SQUAMOUS CELL CARCINOMA IN CATS

Squamous cell carcinoma (SCC) is the most common oral malignancy in the cat, arising from either the jaw bones or the tongue. Squamous cell carcinoma is an aggressive cancer in the cat and is often not diagnosed until the tumor is advanced.

WHAT ARE THE SYMPTOMS?

Owners may notice a mass in the cat's mouth. Tumors that occur in the back of the mouth or under/on the tongue are rarely seen until signs of drooling, weight loss, halitosis (bad breath), difficulty eating, and bloody discharge from the mouth are noted. Loose teeth can also be a symptom of oral cancer in the cat.

HOW IS IT DIAGNOSED?

A tissue biopsy is necessary to definitively diagnosis SCC. X-rays of the jaw are helpful in determining the presence of bone destruction, but often underestimate the extent of involvement. Routine bloodwork and chest X-rays are recommended to assess your cat's overall health prior to anesthesia and biopsy.

CAN IT BE TREATED?

Oral SCC is a very aggressive cancer in the cat. Severe and extensive bone involvement is common. Most cats present with advanced disease, making surgical removal impossible.

Radiation and chemotherapy have been used to treat oral SCC in the cat with little success. Although these tumors may shrink initially with treatment, the tumors often regrow rapidly after treatment is completed. Most cats have enough difficulty eating at the time of diagnosis that feeding tubes may be necessary if radiation therapy is to be pursued.

WHAT IS THE PROGNOSIS?

The prognosis of oral SCC in the cat is extremely poor. The 1 year survival rate is less than 10%, even with combinations of radiation therapy and chemotherapy. Pain medications, such as piroxicam and buprenorphine, may be helpful in reducing discomfort associated with the tumor. However, most cats are euthanatized due to their inability to eat and drink and poor quality of life within 1-3 months of diagnosis.

HEMANGIOSARCOMA IN DOGS

WHAT IS HEMANGIOSARCOMA?

Hemangiosarcoma (HSA) is a malignant tumor of the cells that line blood vessels. It is much more common in dogs than other species. HSA occurs most frequently in large breed dogs, especially German shepherd dogs, Labrador retrievers, and Golden retrievers. In dogs, the primary site is usually the spleen. Other sites include liver, heart, kidneys, bladder, muscle, and subcutaneous tissues. HSA is a very aggressive cancer, with high potential for wide spread and early metastases to other tissues, especially liver, lungs, and peritoneum. The cause of HSA is unknown.

WHAT ARE THE SYMPTOMS?

Clinical signs of HSA are often related to rupture and hemorrhage of the tumor, including weakness, abdominal distension, pale mucous membranes, and collapse. Some patients may suffer sudden death due to rupture of a mass in a critical location or severe and acute blood loss into a body cavity. Dogs with tumors involving the right atrium of the heart may present with arrhythmias, muffled heart sounds, and signs of heart failure. Some patients may have intermittent episodes of weakness with recovery within hours to days.

HSA may occur as cutaneous, subcutaneous or deep masses invading or originating from the muscle. These locations may cause lameness, a hard swelling within the muscle, or edema of the affected region.

HOW IS IT DIAGNOSED?

Many dogs with the splenic form of HSA will present to the veterinarian for rupture of the tumor and bleeding within the abdomen. An abdominal tap will usually reveal free blood that does not clot. Abdominal ultrasound is useful in evaluating the presence of liver metastases prior to consideration of surgery. Aspirates and biopsies of splenic masses often reveal only blood and are diagnostic of cancer in only a small percentage of cases. Surgical removal of the spleen (splenectomy) and associated masses with histological evaluation is necessary for diagnosis. Echocardiography may be recommended to evaluate for heart masses. Approximately 10-20% of dogs with splenic HSA will have right atrial involvement. Diagnosis of cutaneous and subcutaneous HSA requires a tissue biopsy.

Chest X-rays are important in the evaluation of possible lung metastases prior to consideration of surgery. Additionally, many patients with HSA suffer from clotting abnormalities, anemia, and low platelets; therefore assessment of blood work, including clotting times is imperative before surgery.

HOW IS IT TREATED?

Surgery

Surgery remains the primary method of treatment for HSA. The surgery should be as aggressive as possible to remove all locally affected tissue. For splenic HSA, splenectomy is required. At the time of splenectomy, all suspicious lesions within the liver or elsewhere within the abdominal cavity should be biopsied. In dogs with atrial HSA, surgical exploration of the chest cavity may be considered, but most cardiac HSA are not amenable to complete surgical removal. For HSA of bone or subcutaneous HSA involving the leg muscles, amputation of the limb may be necessary. Many patients with HSA are prone to arrhythmias and bleeding during the surgery. Blood transfusions and anti-arrhythmic drugs are often necessary during and immediately after surgery.

CHEMOTHERAPY

Because of the high potential for metastatic disease, chemotherapy has been considered an appropriate adjuvant to surgery. Combination chemotherapy using a doxorubicin-based protocol is most commonly recommended. While chemotherapy may extend disease-free intervals, long-term control rates remain very low.

Very few studies have been conducted to evaluate immunologic or biologic therapy in HSA.

RADIATION THERAPY

Palliative radiation therapy may help decrease symptoms of external masses, but will not impact disease at other sites.

WHAT IS THE PROGNOSIS?

Median survival times with surgery alone range between 30 to 120 days. The addition of chemotherapy may double survival times. Although survival times are short, the quality of your pet's life during this time should be excellent. New treatment options are constantly being explored.

Cutaneous HSA with invasion into subcutaneous areas or muscle have an average survival time between 6-11 months. Invasive cutaneous tumors warrant adjuvant chemotherapy.

LUNG TUMORS IN DOGS

WHAT ARE LUNG TUMORS?

Lung tumors are relatively rare in dogs, accounting for only 1 % of all cancers diagnosed. The average age at diagnosis is 10 years with no sex or breed predilection. Metastatic cancer to the lungs is much more common than primary lung cancer.

The majority of primary cancers of the lung are carcinomas with adenocarcinoma being the most common. Although rare, other primary lung tumors include benign adenomas, squamous cell carcinoma, sarcomas, and lymphomatoid granulomatosis. Lung tumors have moderate to high potential for metastases. Metastases occur to other lung lobes, pleura (lining of the chest cavity), bone, and brain.

HOW ARE LUNG TUMORS DIAGNOSED?

Most dogs with lung tumors present with coughing, exercise intolerance or other respiratory signs. Occasionally, patients will present for more vague, non specific signs such as loss of appetite, weight loss, or lethargy. Chest x-rays usually demonstrate a solitary lung mass. Ultrasound guided aspirate or biopsy can be performed for definitive diagnosis, but is not necessary for consideration of surgical removal.

Abdominal ultrasound may be recommended to evaluate for a possible primary or secondary mass. A CT scan is often recommended to better evaluate for secondary lung metastases or feasibility of surgical removal. Routine blood work is helpful in evaluation of your pet's overall health prior to surgery. At the time of surgery, the regional lymph nodes are biopsied to determine whether the cancer has spread.

WHAT IS THE TREATMENT?

Surgery

Surgery is the mainstay of treatment for dogs with lung carcinoma, providing no metastatic lesions are observed in other lung lobes. Removal of the affected lung lobe is usually by means of a lateral or sternal thoracotomy. Dogs generally tolerate this surgery well and are discharged from the hospital within 1-4 days following surgery. Additional therapy (chemotherapy) may be recommended depending on the grade of the tumor, size of the tumor, presence of tumor in lymphatics and/or blood vessels.

CHEMOTHERAPY

Chemotherapy is often recommended for high grade tumors, large tumors, or tumors that show evidence of lymphatic or vascular invasion. Little information has been published on the efficacy of chemotherapy for dogs with primary lung carcinomas. Drugs with documented activity against lung carcinomas include platinum compounds (cisplatin, carboplatin), gemcitabine, and vinorelbine. Chemotherapy alone, without surgical removal of the mass, does not often yield significant decrease in the size of the mass or significant improvement in survival time.

WHAT IS THE PROGNOSIS?

For dogs with small, low grade (well differentiated) tumors without lymph node involvement, the average survival time is 16 months or longer with surgery alone. Dogs with high grade (poorly differentiated) tumors with lymph node involvement have an average survival of 4 months, even with surgery. Because many patients will ultimately develop metastatic disease following surgery, chemotherapy is commonly recommended.

SOFT TISSUE SARCOMA IN DOGS

WHAT IS A SOFT TISSUE SARCOMA?

Soft tissue sarcomas are a group of malignant cancers that arise from the skin and subcutaneous connective tissues, such as fat (liposarcoma), muscle (rhabdomyosarcoma, leiomyosarcoma), cartilage (chondrosarcoma), fibrous connective tissue (fibrosarcoma), nerves (schwannoma, malignant peripheral nerve sheath tumor, neurofibrosarcoma) and the "pericytes" of small blood vessels in the subcutis (hemangiopericytoma). These tumors are often considered collectively because of their similarity in clinical behavior.

Soft tissue sarcomas may arise from any anatomic site. They tend to appear discrete and well encapsulated, but are actually very invasive into surrounding tissues. As such, local regrowth of the tumor is common after conservative surgical removal. Soft tissue sarcomas are graded as low, intermediate, or high grade. Most soft tissue sarcomas are low to intermediate grade, and have a relatively low chance of spreading to other places (metastatic rate of less than 25%). High grade sarcomas have a higher potential for metastasis (25-40%)

HOW ARE SOFT TISSUE SARCOMAS DIAGNOSED?

As a general rule, these tumors do not readily shed their cells; however, a fine needle aspirate is an easy, non-invasive test that can often confirm the presence of a sarcoma. A biopsy may be necessary if fine needle aspirates are non-diagnostic. Further, biopsies can also be used to classify the specific type or grade of soft tissue sarcoma.

Once a diagnosis of a soft tissue sarcoma is made, staging is recommended to rule out spread of disease and evaluate your pet's overall health. Staging for sarcomas typically involves routine bloodwork, chest x-rays, and evaluation of regional lymph nodes. An abdominal ultrasound may also be recommended.

WHAT IS THE TREATMENT?

Surgery

Surgery is the mainstay of treatment for soft tissue sarcomas. Surgical excision must be wide and deep in order to remove all of the tumor tissue. When tumors are excised surgically with "clean" surgical margins, no further treatment may be necessary. If the tumor was not removed with adequate margins, a second surgery may be recommended to excise more tissue and help ensure adequate removal of all tumor cells.

RADIATION THERAPY

In some instances, aggressive surgery is not possible without severe disfigurement or loss of function. In cases where aggressive surgery is not possible, or despite an aggressive resection where tumor cells remain at the margins, radiation therapy is used to prevent or delay regrowth of the tumor. Radiation therapy is extremely well-tolerated in dogs. Side effects are transient and limited to the site where radiation therapy is performed.

Radiation therapy may also be used for large tumors that cannot be surgically removed. Radiation therapy for measurable tumors is not considered to be as effective as radiation therapy for microscopic disease after surgery. These tumors do not rapidly regress after radiation and "control" may be defined as a slowly-regressing (6 months or longer) tumor, or a tumor that stays stable in size. In some instances, the tumor may regress enough to make surgical removal possible.

CHEMOTHERAPY

Chemotherapy is often recommended for high grade sarcomas to prevent or delay the onset of distant metastases. Doxorubicin has been shown to be the most active chemotherapy for soft tissue sarcomas. It may be given alone, or combined with other chemotherapy drugs. Chemotherapy is very well-tolerated in dogs.

WHAT IS THE PROGNOSIS?

Soft tissue sarcomas that are low to intermediate grade and can be removed completely with aggressive surgery have an excellent long term prognosis. Control rates for low grade soft tissue sarcomas that are excised with less than adequate margins and followed by radiation therapy are also very good.

Approximately 85% of these patients are tumor free three years after treatment. By comparison, the majority of dogs with incompletely excised soft tissue sarcomas that receive surgery without follow-up radiation therapy will have their tumors regrow by 1 year after surgery. Many of these tumors often regrow within the first few months after surgery.

For high-grade sarcomas, the long-term prognosis is more guarded. Chemotherapy is indicated to help delay the onset of metastasis; however, the median survival time for these patients is approximately one year.

The best time to treat a soft tissue sarcoma is the very first time it occurs. Tumors that regrow after an initial surgery are often more aggressive in their behavior. This makes the potential for metastases greater and our ability to control the tumor locally, even with adjuvant radiation therapy, much more difficult.

5

Tumors of Joints in Animals

BONE AND JOINT TUMORS

The complex structure of the bone, as well as its origin, by either mesenchymal condensations, intramembranous ossification, or the gradual replacement of cartilaginous tissue by bone tissue, endochondral ossification, are also reflected in the oncology of this organ. Bone is composed of different tissues, including bone, cartilaginous and fibrous tissue, blood vessels, nerves and bone marrow. All these elements have cancer potential, alone or in various associations. The difficulties of an accurate classification have made veterinary and human pathologists correlate clinical, radiological, morphological and histopathological data in the diagnosis of bone tumors, in order to define each individual bone neoplasm.

Primary skeletal tumors are frequent in dogs and rarer in cats. Primary osteosarcomas in dogs are more frequent than benign tumors, while in cats the two forms seem to be equally distributed, and in horses and cattle, benign forms outnumber malignant forms. In other species, bone tumors are sporadic. Bone tumors are rare in cats, but they appear in old females. Of 24 primary bone tumors, LIU et al. (1974) diagnosed 4 juxtacortical osteosarcomas, 3 chondrosarcomas, 1 osteoma and 1 osteoid osteoma. Locations were in the long bones of the extremities and in the skull. Bone dysplasias and bone cysts have also been diagnosed, and other lesions located in bones have been identified, fibrosarcomas, squamous cell carcinomas, lymphosarcomas, rhabdomyosarcomas and one case of melanoma, hemangiosarcoma, reticular cell sarcoma, meningioma and *myositis ossificans,* all being malignant tumors. Primary bone tumors in cats were diagnosed by LINNABARY et al. (1978), classified into chondrosarcomas, osteosarcomas and fibrosarcomas. Bone tumors, especially osteosarcomas, are found with a high incidence in large dog breeds.

The classification of bone tumors, as well as positive diagnosis, is difficult. The origin of these tumors may be different, from the bone skeleton, from the possibility of appearance in connective tissues, whose plasticity is well known.

Thus, tumors may originate from the connective tissue proper, from blood vessels, cartilage and various cell types, such as bone cells, osteoclasts, osteoblasts or their precursors. The possibilities of embryogenic transformation are added, including myxomatous, cartilaginous and bone transformation.

A classification of bone tissue and cartilaginous tissue tumors depending on histological structure recognizes, in spite of al inherent difficulties, the neoplasms specific for these tissues, as well as the possibility of cancerization of other structures, such as blood vessels, connective tissue or periarticular tissues.

The anatomohistological classification includes:

- Osteoma;
- Osteoid osteoma;
- Osteosarcoma;
- Juxtacortical osteosarcoma;
- Fibrosarcoma;
- Liposarcoma;
- Osteoliposarcoma;
- Chondroma;
- Osteochondroma;
- Chondrosarcoma;
- Multilobular bone tumor;
- Bone marrow tumors: myeloma;
- Vascular tumors; hemangioma; hemangiosarcoma;
- Mixed bone tumors;
- Metastatic bone tumors;
- Tumor-like bone lesions: solitary bone cyst; aneurysmal bone cysts;
- Articular and adnexal structure tumors: synovial sarcoma; fibroxanthoma; malignant giant cell tumor of the connective tissue.

Histological Classification of Bone and Joint Tumors of Domestic Animals (Slayter et al. 1994)

I. Benign Tumors
   A. Osteoma
   B. Ossifying fibroma
   C. Myxoma of the jaw
   D. Osteochondroma
   E. Feline osteochondromatosis
   F. Chondroma
   G. Hemangioma
II. Malignant Tumors
   A. Central
      1. Osteosarcoma
         a. Poorly differentiated

- b. Osteoblastic
  - i. Nonproductive
  - ii. Productive
- c. Chondroblastic
- d. Fibroblastic
- e. Telangiectatic
- f. Giant cell type

2. Chondrosarcoma
3. Fibrosarcoma
4. Hemangiosarcoma
5. Giant cell tumor of bone
6. Multilobular tumor of bone

B. Peripheral
1. Periosteal chondrosarcoma
2. Periosteal fibrosarcoma
3. Maxillary fibrosarcoma (dogs)
4. Periosteal osteosarcoma
5. Parosteal osteosarcoma

C. Joint tumors
1. Synovial sarcoma

D. Miscellaneous tumors
1. Liposarcoma
2. Malignant mesenchymoma
3. Others

E. Tumors of bone marrow
1. Myeloma
2. Malignant lymphoma

III. TUMOR-LIKE LESIONS
- A. Fibrous dysplasia
- B. Solitary bone cyst
- C. Juxtacortical bone cyst
- D. Epidermoid cyst of the phalanx
- E. Myositis ossificans
- F. Villonodular synovitis

Diagnosis in bone tumors is based on the correlation of data obtained following clinical, radiological and histopathological examination. It is important that clinical data should include the species, breed, age, position in the skeleton and number of affected bones, the duration and evolution of the tumor, as well as the general state of the subject.

Radiological examination cannot establish in all cases a positive diagnosis, since changes are not pathognomonic. Some categories of bone diseases can be radiographically established: congenital, evolutional, metabolic and nutritional, traumatic, infectious, neoplastic diseases. Histological examination completes and finally defines positive diagnosis.

Benign bone tumors, as well as some tumor-like lesions should be differentiated from fibrous lesions.

Extraskeletal bone tumors are generally rare, osteomas, osteosarcomas and chondrosarcomas being described in dogs and cats. In 1999, THOMSEN and MYERS reported in a 14-year-old dog an osteosarcoma located in the submandibular salivary gland, with metastases in the lungs, kidneys and mandibular lymph nodes. Histologically, the tumor contained large pleomorphic, osteoid-producing cells and numerous multinucleated giant cells. Neoplastic cells were immunoreactive to vimentin and actin, but not to keratin, desmin and protein S-100. POOL (1978) suggested that ossifying fibroma and fibrous dysplasia could be intermediate stages of osteoma.

In the etiology of bone cancer, the following factors are recognized: ionizing radiation, viruses, chemical carcinogens, foreign bodies, skeletal abnormalities and predisposing genetic factors.

An epidemiological study performed by DORN in 1969 revealed that approximately 98% of bone tumors in dogs are malignant and 85% of primary skeletal neoplasms are histologically classified as osteogenic sarcomas, while the rest of 15% are: chondrosarcomas 10%; hemangiosarcomas, fibrosarcomas, lymphomas, reticular cell sarcomas and myelomas 3%; osteomas and chondromas 2%.

OSTEOMA

Osteoma is a benign tumor that develops on the bone surface, being poorly delimited, and arises by subperiosteal ossification, sometimes being difficult to differentiate from exostoses even microscopically.

Osteomas may be found in all species, but more frequently in horses and cattle. In dogs, bone tumor formations, which are multilobular osteomas and chondromas, are found in the head bones.

Osteomas grow slowly, and sometimes recur after incomplete excision. They are frequently located in the mandibular bones, nasal sinuses, facial and cranial bones. In cats, they may be located in the limbs, sternum, ribs and skull. STEINBER and GEORGE (1989) describe in two goats the appearance of a mandibular osteoma as a result of subluxation with contralateral mandibular deviation, and the development of an osteoid-producing sarcoma following a comminuted humeral fracture, respectively. The authors also remark the appearance of bone tumors at the site of fractures in other species, especially in dogs.

Macroscopically, osteoma is clearly delimited, being covered with connective vascular tissue. Sometimes the tumor may reach impressive sizes, having a high density; it has the aspect of dense bone tissue, with fibrous connective tissue, being well anchored in the bone tissue in which it develops. Histological structure is characterized by trabecular growth, with bone structures perpendicular to the surface of the tumor formed by osteoblasts, and by osteoclastic modeling, lamellar bone is formed. The osteoma contains fat connective tissue and hematopoietic islands, the general aspect of the tumor being that of compact bone. Architecture and the predominance of one structural component or another determine wide variations from one osteoma to another. Thus, osteoma-like morphological aspects, ossifying fibromas or fibrous dysplasias can be microscopically evidenced. Some medullary bone developments do not present periosteum. In active osteomas, the periosteum is well differentiated and composed of osteogenic and fibrous layers. In some osteomas, the periosteum is poorly differentiated, being almost exclusively fibrous.

In general, osteomas have a slow development, without metastasizing, and malignant transformation has not been reported.

OSTEOID OSTEOMA

Osteoid osteoma is a benign tumor, rarely found in animals, which has been reported in the 10th thoracic vertebra in cats, under the form of a small lesion, of osteoid tissue and connective fibrocellular tissue. Radiologically, the central area is transparent, and the periphery radiopaque. Histologically, it is characterized by intense cellularization and good vascularization of osteoid tissue and immature bone. The tumor is delimited by bone reaction.

The lesion is similar to human osteoblastoma.

OSTEOSARCOMA

Osteosarcomas are dynamic structures whou radiological aspect can change in a short time their radiological aspect. Depending on the radiological aspect, there are: osteoblastic, osteoid and mixed osteosarcomas. Osteosarcomas may originate from the bone tissue proper or they can be osteomas of periosteal origin.

Histologically, pleomorphic osteosarcomas can be identified, formed by ovoid or fusiform cells, with a high mitotic index. *Compound osteosarcomas* are those in which tumor cells produce osteoid and bone proliferations, as well as malignant cartilage. *Undifferentiated osteosarcomas* are those in which tumor cells do not form spicules or trabeculae with fibrillar matrix.

Osteosarcomas represent the most frequent histological type of primary bone tumors, over 80% in dogs, and over 50% in cats. Osteosarcoma is the most common bone tumor, representing 5–6% of all tumors of the canine species.Osteosarcoma has the highest incidence in large or very large breeds, in which the risk of developing this tumor can be 5 to 30-fold higher compared to all the other tumors.

Osteosarcomas in dogs develop in the metaphyseal region of long bones, followed in decreasing order by the radius and humerus bones (especially in Boxer dogs), while other locations are sporadic. Mixed tumors of chondrosarcoma and fibrosarcoma type are frequently diagnosed in dogs. In other species, horse, ox and sheep, osteosarcomas are frequently located in the head.

Bone tumors and extraskeletal cartilaginous tumors are extremely rare in dogs and cats.

The mean onset age of osteosarcoma in dogs is 7.6 years, with age ranges between 1 and 15 years. Rib tumors occur more frequently in adult-young dogs, while cranial location appears at a more advanced age. In cats, old individuals are frequently affected, with a mean age of 10.6 years.

The more affected dog breeds are the Boxer, Danish Dog, Saint-Bernard, German Shepherd and Irish Setter breeds. No breed predisposition has been reported in cats.

Sex does not seem to influence the incidence of osteosarcomas, although there are observations accrediting supporting a higher frequency in female cats.

Macroscopically, an increased volume of the affected bone is found, and in section, congestion, edema appears, with the presence of osteofibrous tissue. Concomitant muscular atrophy develops, and regional lymph nodes are enlarged and hard. Fracture at the tumor site is not an uncommon complication, sometimes very soon after the development of the tumor, but in most cases, in advanced forms. Rhinorrhagia frequently appears when osteosarcoma is located in the nasal bones.

Osteosarcomas in cats are located in the posterior limbs (tibia and femur); they develop slowly, and metastases occur late, after the diagnosing of the disease, with pulmonary and renal locations. Juxtacortical osteosarcoma in cats is usually located in the long bones of the extremities and cranial bones, frontal bone and mandible. Histologically, depending on the dominance of a certain tissue, osteosarcomas can be classified as: *osteoblastic, chondroblastic* and *fibroblastic.* In practice, there are also combinations of these forms, and some authors maintain that these variants depend on the site and number of the sections examined. Sometimes, the exact definition of the sarcoma type is difficult.

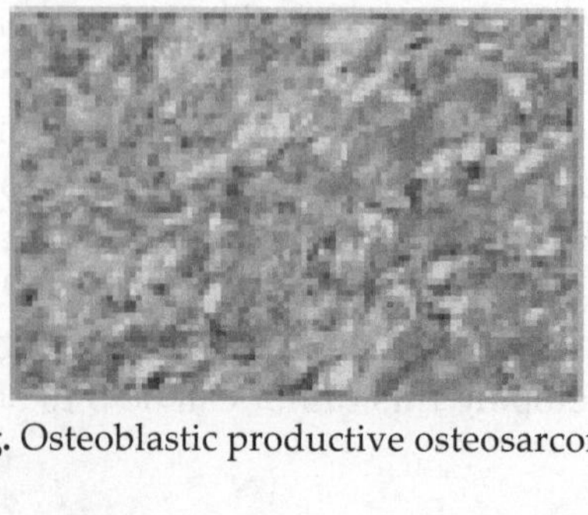

Fig. Osteoblastic productive osteosarcoma.

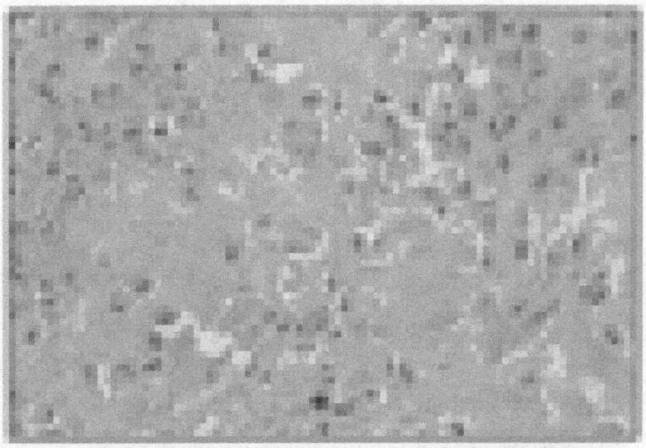

Fig. Osteoblastic productive osteosarcoma.

Fig. Fibroblastic osteosarcoma.

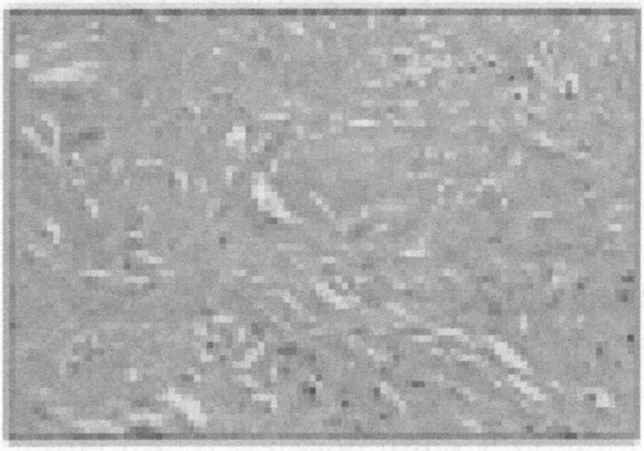

Fig. Osteoblastic osteosarcoma, productive, poorly differentiated.

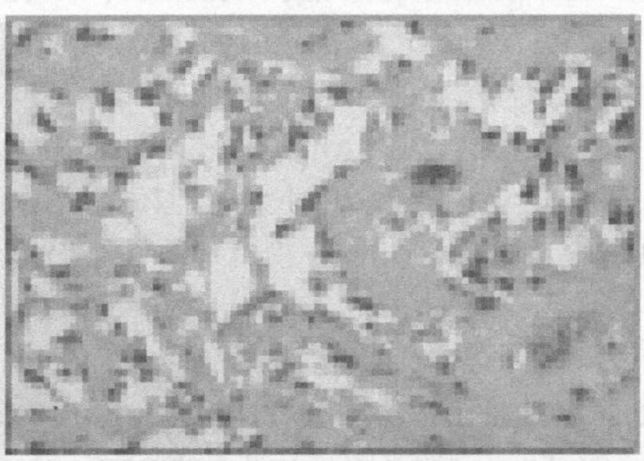

Fig. Osteoblastic osteosarcoma, productive, well differentiated.

Fig. Chondroblastic osteosarcoma.

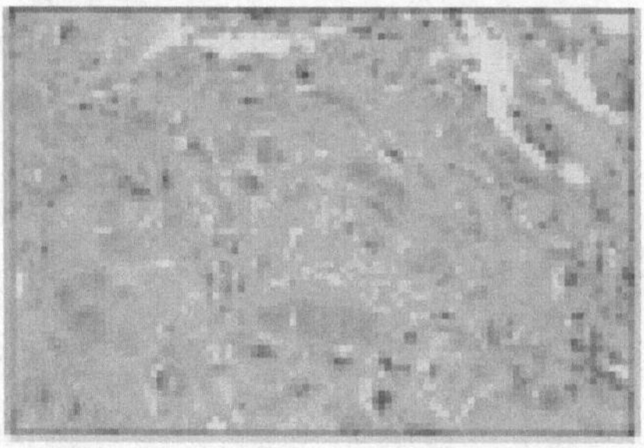

Fig. Chondroblastic osteosarcoma.

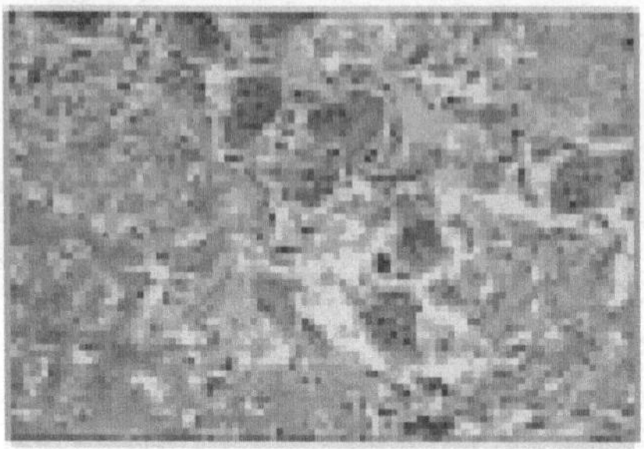

Fig. Osteosarcoma, osteoblastic, productive, osteoclasts.

According to POOL (1978), *osteosarcomas in dogs* could be grouped into two categories. A first category is characterized by the destruction of bone and its replacement by a mineralized tumor matrix. An intense osteolytic process takes place, and the tumor tissue is soft, intensely vascularized. It is a highly cellularised *pleomorphic osteosarcoma,* intensely cellularized, and a *simple osteosarcoma* with reduced tumor bone production.

Due to the intense vascularization, these tumors have been labeled as *telangiectatic osteosarcomas,* being included in the group of osteolytic tumors. NORRDIN et al. (1989) diagnosed in two sibling dogs telangiectatic osteosarcomas; in one subject, the isolated cells stimulated with parathyroid hormone showed a high osteoblastic differentiation, while in the second, cells did not respond to hormonal stimulation. Histologically, simple osteosarcomas are of fibroblastic type, with osteolysis and the replacement of bone tissue by neoplastic fibrous stroma, with reduced tumor bone. The second group of osteosarcomas includes tumors in which neoplastic bone is formed, sometimes in association with neoplastic cartilage proliferation. Productive *simple osteosarcomas* of this group may be classified as mixed osteoblastic osteosarcomas, with high density, so that they are difficult to dissect. *Compound osteosarcomas* are characterized by bone but also cartilage proliferation, and are frequently found compared to simple osteosarcomas. They have a slow development, which allows the delimitation of tumors by periosteal or fibrous tissue, but in time they can cross tissues, reaching the articular cavity.

LOSCO et al. (1984) propose the term *multilobular osteosarcoma* for tumors developing in the head bones. Histologically, it has benign characteristics, but due to its invasive character and the early and massive metastasizing in the lung, it is recognized as being potentially malignant. CALLA et al. (1989) describe for the first time a multilobular osteosarcoma with primary mandibular location and pulmonary metastasis. Other authors also report multilobular osteosarcomas located in the head bones, mentioning a mean age of 7.5 years, a mean weight of 28.9 kg and no sex predisposition. The authors found, in 16 subjects with multilobular sarcomas, a survival rate of 21 months, of 12 subjects, following adequate treatment, 7 had recurrences and 7 developed metastases.

Extraskeletal osteosarcomas and *chondrosarcomas* have been diagnosed with locations in different tissues or organs: eye, adrenal gland, spleen, kidney, liver, ileum, stomach, testicle, vagina, subcutaneous connective tissue of the limbs, thyroid, esophagus, lung, aorta, etc.. THOMSEN and MYERS (1999) reported an extraskeletal osteosarcoma located in the submaxillary salivary gland, in a 14-year-old dog, with the presence of metastases in the lungs, kidneys and bilateral submaxillary lymph nodes. Histologically, osteosarcoma was formed by multinucleated giant cells and large osteoid-producing pleomorphic cells. The mean age of dogs with extraskeletal osteosarcoma was 11 years, and of dogs with chondrosarcoma, 14 years. Osteoblastic osteosarcoma was the most frequently diagnosed histological type, followed by fibroblastic, fibrous histiocytic, chondroblastic and mixed osteo-chondroblastic osteosarcomas. Distal metastases were present in 7 of 11 dogs with extraskeletal osteosarcoma and were absent in dogs with chondrosarcoma. Osteosarcoma metastases in dogs develop in 75% of cases in the lung, as primary tumors. Simple osteosarcoma metastases usually have a solid to soft and fleshy consistency, with no calcification tendency. Primary telangiectatic osteosarcomas produce soft, cystic secondary hemorrhagic tumors. Voluminous metastases generally tend to become hard and bony.

In cats, osteosarcoma appears as a tumor with multiple foci, with osteocartilaginous hyperplasia of the periosteum of the apical and appendicular skeleton, with malignant potential, especially at an advanced age. In some felines, osteosarcomas with multinucleated giant cells are reported. The association, in some cases, of malignant osteochondroma with the presence of thymic lymphosarcoma makes DOIGE (1987) suppose a possible viral etiology.

The advanced cases of primary sarcomas can be recognized as malignant tumors on radiological and clinical examination, but the diagnosis of certainty is established by histological examination. Incipient osteosarcomas, chondrosarcomas and bone fibrosarcomas have no radiological pathognomonic aspects.

The histology of osteosarcomas can be characterized by POOL's statement (1978), according to which there are no two identical osteosarcomas and a microscopic description characteristic of osteosarcoma is impossible. This has made difficult to establish criteria for a histological classification of bone tumors. The author remarks the possibility that a subdominant tissue aspect might be much more malignant than the main aspect, which makes prognosis difficult, but underscores at the same time the importance of the histopathologist's experience.

Malignant osteosarcoma cells produce fibrillar stroma, tumor osteoid and tumor bone, sometimes neoplastic cartilage. Each of these components may be dominant in a certain tumor, so that an osteosarcoma can mimic a fibrosarcoma or a chondrosarcoma. Osteoid can have an irregular arrangement in the malignant fibrillar stroma, among pleomorphic or fusiform tumor cells, it can be organized as irregular spicules. In canine osteosarcomas, but mostly in feline osteosarcomas, extensive giant cell areas may be found, and these cells may be confused with osteoclasts. However, in osteosarcomas, osteoid appears homogeneous, poorly delimited, and mitoses are present in osteoblastic cells. These neoplastic cells are enlarged and each cell is connected to other cells by numerous cytoplasmic processes.

Electron microscopically, osteosarcomas include anaplastic blast cells, osteoblasts, chondroblasts and fibroblasts. The proportion of these cells varies from one neoplasm to another. Some osteosarcomas contain groups of multinucleated cells, giant cell tumor.

GIANT CELL BONE TUMOR

It is a rare neoplasm in dogs, cats and cattle, but frequent in humans, with an expansive osteolytic aspect in long bones.

In dogs and cats, it develops in metacarpal bones, ribs and axial skeleton, it has an expansive character, with the destruction of the bone cortex, but it does not produce metastases.

Histologically, two types of cells are described that are found in these neoplasms, numerous mononuclear stromal cells with histiocytic or fibroblastic aspect, with frequent mitoses, and multinucleated giant cells, similar to osteoclasts. Hemorrhage and cavernous vascularization are frequently found, which leads to the confusion with aneurysmal cystic tumors or telangiectatic osteosarcoma. Due to hemorrhage, necrosis and secondary reactions, histological diagnosis requires the examination of various tumor portions.

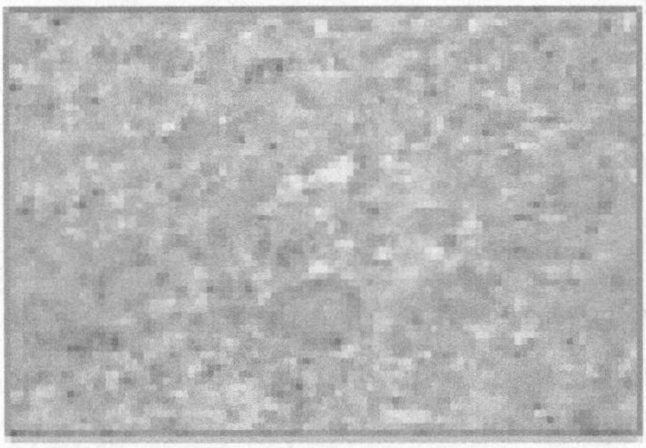

Fig. Osteosarcoma, giant cell type.

The origin of neoplastic giant cells is supposed to be in malignantly transformed osteoclasts; these structural elements have also determined the designation of this neoplasm as osteoclastoma. Extraskeletal giant cell bone tumors have been diagnosed in man and extremely rarely in animals.

JUXTACORTICAL OSTEOSARCOMA

Juxtacortical or parosteal osteosarcoma originates in the periosteum, being a periosteal sarcoma with trabecular bone structure, frequently mature and lamellar, with mixed fibrous and cartilage tissue. The aspect of differentiated tissues does not exclude the malignant character, represented by pleomorphism and cellular mitoses. However, compared to typical osteosarcoma, the malignancy of juxtacortical osteosarcoma is lower.

In 1971, JACOBSON reported 31 bone tumors, of which 14 cases were in dogs aged between 6 months and 13 years, belonging in particular to large and giant breeds, with a number twice as high in males as in females. In cats, 11 cases of bone tumors were diagnosed, with a mean age of 7 years, age ranges between 1 and 13 years. The author mentions three tumors in horses, one tumor in a pig and another one in a cat. Statistics do not mention the number of juxtacortical osteosarcomas.

Periosteal proliferations were identified by POOL (1978) in cats, as multiple or solitary superficial bone excrescences, with slow development, from several months to several years. The tumor margins contain cartilage structures, while in the tumor mass, medullary spaces with fibrous elements and hematopoietic tissue are found. In humans, these neoplasms produce pulmonary metastases and result in the patient's death.

Radiologically, the tumor has clear-cut margins, without the involvement of the cortex and metaphyseal spongy bone.

Histologically, in humans, stromal, fibrous and sometimes cartilage structures are identified, which rather resemble reactive tissue than neoplastic proliferation. Cell pleomorphism is low or even absent and mitoses are absent.

In animals, in parosteal osteosarcomas, cell pleomorphism is present, cellularized, with mitotic activity. Sometimes, this tumor has the appearance of an osteochondroma.

Juxtacortical osteosarcoma develops slowly, in time the tumor becomes aggressive, it can invade the cortex and even the medullary cavity. Sometimes, pulmonary metastases occur.

FIBROSARCOMA

Fibroblastic bone tumors may develop from the periosteal tissue or the stromal tissue of the medullary canal. The medullary location is less frequent, compared to periosteal incidence.

Medullary fibrosarcoma has a high incidence in large size dogs, reaching 5% of bone neoplasms. The tumor is also rare in other species. It is located in long bones, having an invasive and destructive character, producing osteolytic lesions. Unlike fibrosarcomatous osteosarcoma, medullary fibroma develops rare or no metastases. In dogs, multicentric primary neoplasms have also been found, sometimes located in the mandible and the spine.

Macroscopically, in early or less aggressive forms, the bone cortex is not affected, the white-gray tumor fills the medullary canal. In advanced and aggressive forms, bone is destroyed, and a soft, compact tumor mass appears. By osteolysis, the neoplasm can invade the articular cavity.

Radiological examination cannot establish diagnosis, since medullary fibrosarcoma cannot be differentiated from other bone tumors, such as simple osteosarcomas, pleomorphic osteosarcomas, undifferentiated sarcomas, chondrosarcomas, etc.

Histologically, medullary fibrosarcomas have a moderately or well-differentiated aspect. Dense collagen foci, similar to osteoid substance, may be sometimes identified. Differential diagnosis will be made with fibroblastic osteosarcomas. Sometimes, medullary fibrosarcomas are highly cellularized, anaplastic, composed of large fusiform cells, with large hyperchromatic nuclei and numerous mitoses.

Metastases, even in anaplastic forms, are absent.

Periosteal fibrosarcoma is located in the majority of domestic animal species in the head bones, especially in the maxilla and mandible, sometimes in long bones. Macroscopically, the tumor appears as a non-homogeneous soft mass, attached to the bone. Although present, metastases are rare, with pulmonary location. In horses, fibrosarcomas of the nasal sinuses and dental arches are found. In dogs, more rarely in cats, maxillary, mandibular and other head bone locations require differential diagnosis with fibromas that develop in gums, in the oral and labial mucosae, as well as in the palatal mucosa, which are relatively frequent. Periosteal fibrosarcoma determines osteolysis, penetration in the bone cortex and results in fractures.

Histologically, periosteal fibrosarcoma is characterized by a well differentiated, collagenous stroma, hyperchromatic cells, cellular pleomorphism and rare mitoses.

Undifferentiated sarcoma is a tumor with pleomorphic structure, lacking a specific histological structure.

LIPOSARCOMA

Bone liposarcoma originates in the fat cells from the medullary canal that are malignantly transformed. The tumor is rare in animals, being reported in the long bones of young dogs. Medullary liposarcoma is a gray fat tissue mass that replaces bone marrow and induces cortical osteolysis.

Histologically, the presence of lipoblasts is found, large cells with eccentric nucleus and lipid vacuoles in the cytoplasm. Fibrosarcomatous islands or undifferentiated sarcoma with fusiform cells may be detected.

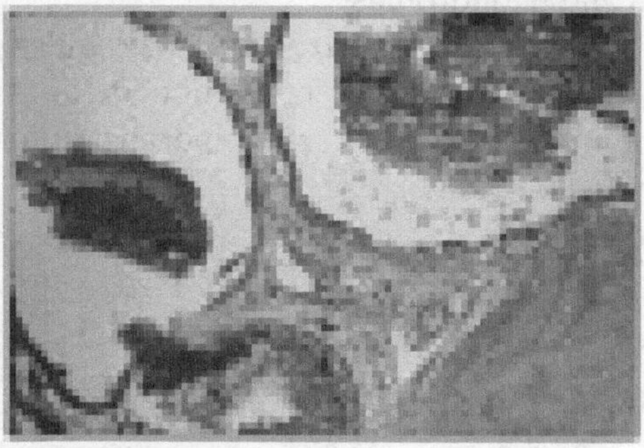

Fig. Cavernous hemangioma, lombar vertebrae.

Metastases have been reported in the liver and lymph nodes.

OSTEOLIPOSARCOMA (MALIGNANT MESENCHYMOMA)

Osteoliposarcoma or malignant mesenchymoma is a tumor formed by both osteosarcomatous and liposarcomatous tissue. Generally, it is rarely found in dogs and it has not yet been reported in other species.

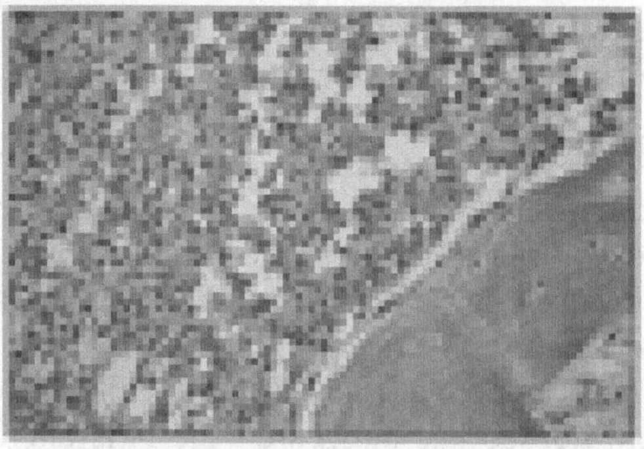

Fig. Hemangiosarcoma, rib.

CHONDROMA

Chondroma is a benign tumor, found in dogs, cats, cattle and sheep. It is characterized by nodular formations, delimited by fibrous tissue, located in long bones and ribs. In section, it has a lobular structure, in which the cartilage tissue is traversed by connective trabeculae, with ossification or mineralization areas. In some tumors, myxomatous masses, with hemorrhage and necrosis, may appear. JUBB, KENNEDY and PALMER (1992) distinguish, depending on location, the enchondroma, which is a tumor present as cartilage tissue in the medullary canal, and the ecchondroma, which is a benign cartilaginous tumor that appears in other areas of the skeleton. Primary bone chondromas grow by expansion, primary benign cartilaginous tumors are rare and sometimes difficult to diagnose in animals, and their malignant transformation is exceptional. Chondrosarcomas in animals appear *de novo*. The author distinguishes primary bone enchondromas and chondromas.

Histologically, they are characterized by an irregular lobular structure of hyaline cartilage tissue, with the presence of characteristic mature chondrocytes. Enchondral ossification and mineralization foci, and sometimes myxomatous tissue nodules may be seen.

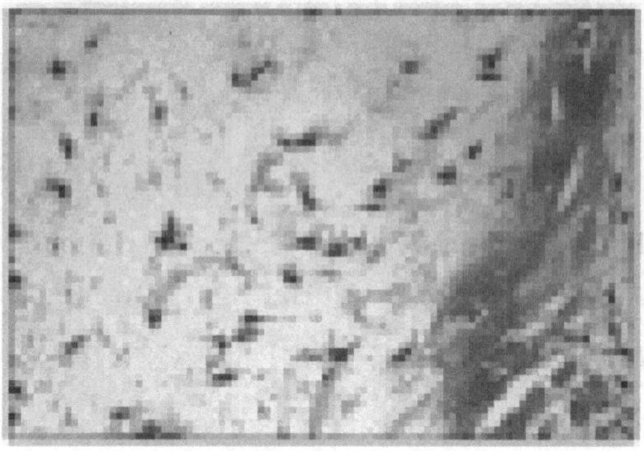

Fig. Chondroma.

Chondroma develops slowly, and malignant transformation has not been demonstrated. In the case of development in the medullary canal, the tumor produces bone deformity and even fracture.

OSTEOCHONDROMA

Osteochondroma is a benign tumor that appears under the form of osteocartilaginous exostoses, as either single or multicentric lesions. The tumor has been reported in cats, dogs and horses, and also in humans. It is frequently located in long bone metaphyses, ribs, vertebrae and the scapula. The multicentric formation can be malignantly transformed. The familial character present in dogs and horses has suggested the existence of a predisposing hereditary factor.

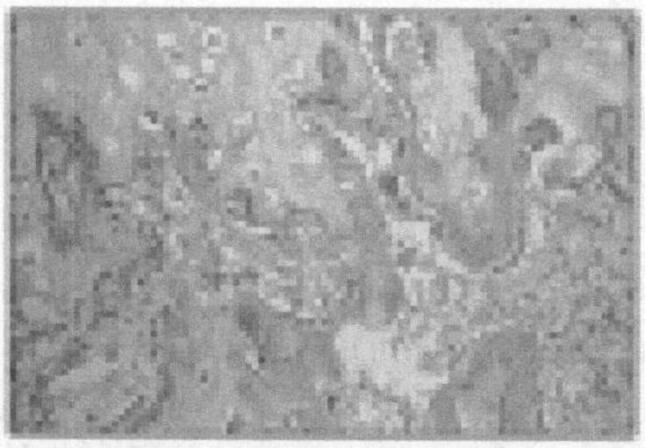

Fig. Osteochondroma.

In cats, the presence of cartilaginous exostoses with positive serological reaction for type C virus and the presence of type C viral particles in cartilage cells suppose the implication of feline leukemia virus (FELV) and feline sarcoma virus (FSV) in the appearance and the multicentric development of osteomas in this species.

CHONDROSARCOMA

Chondrosarcoma is a malignant tumor, in which malignant cells produce neoplastic chondroid and fibrous matrix. This neoplasm can develop in the medullary bone area or subperiosteally. Periosteal chondrosarcomas are rarer in animals; two patterns exist that are distinctively different: the first contains cytologic features of low to intermediate-grade malignancy with lobulation and resembles central chondrosarcoma, the second pattern lacks lobulation, has less anaplasia, and frequently contains nonneoplastic bone formed through endochondral ossification.

Chondrosarcoma in dogs has a significantly higher incidence compared to osteosarcoma.

Chondrosarcomas may originate directly from normal cartilage structures, except for articular cartilage, which become malignant, like in the case of the neoplasms of the nasal conchae, or they may result from the malignization of benign tumors, such as chondromas.

The incidence of chondrosarcomas has been frequently reported in adult animals. The tumor has been studied in particular in dogs, where it represents less than 10% of all bone sarcomas, affecting in the first place large breeds, with no sex preference, the Boxer and German Shepherd breeds being more frequently affected. In sheep, tumors have been frequently found at adult and old age.

Chondrosarcoma is frequently located in flat bones, especially in cattle and horses. In sheep, tumors have been reported in sternocostal structures, scapular cartilage and the coxal tuberosity, sometimes also in long bones. In cats, the most frequent locations have been in the scapula, vertebrae and limbs, but also in the head bones.

Macroscopically, the chondrosarcoma developed on the bone surface, periosteally, appears under the form of large masses, with lobulated surface, firm or hard consistency, relatively well delimited. In section, the tumor tissue has a hyaline cartilage aspect, with multiple small nodules of soft to dense tissue, white-gray color, with bluish iridescence. Sometimes, consistency is viscous, mucoid, especially in nasal chondrosarcomas. In the tumor mass, calcified or fibrous cartilage-like nodules may appear.

Central chondrosarcomas, developed in the medullary canal of long bones, have a symmetric development and gradually penetrate the bone cortex, sometimes reaching below the periosteum. In this case, the periosteum produces an extremely dense, ivory, reactive bone belt around the neoplasm.

Histologically, a differential diagnosis between chondrosarcoma and chondroma is difficult to establish, and requires the correlation of clinical and morphological data. Microscopically, the following characteristics of increased malignancy are estimated in the case of chondrosarcoma: cells with pyknotic nuclei, numerous cells with two pyknotic nuclei and the presence of large cartilage cells with one or several nuclei, as well as heterochromatin bundles. The presence of mitoses, even in a small number, facilitates the chondrosarcoma diagnosis.

The level of chondrosarcoma malignancy can be assessed by 3 grades, depending on the histological structure of the tumor.

Grade I chondrosarcomas are formed by uniform tumor cells, most of which have double nuclei, and no mitotic figures. In general, mitoses are extremely rare or absent.

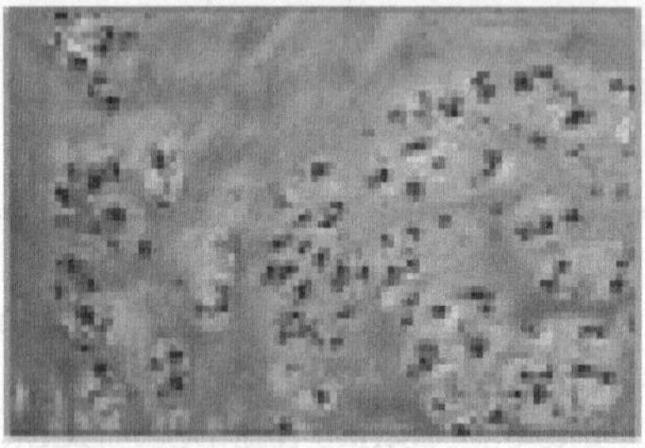

Fig. Chondrosarcoma, grade I.

Grade II chondrosarcomas are frequent, being composed of pleomorphic chondrocytes, with the presence of mitoses and metastases.

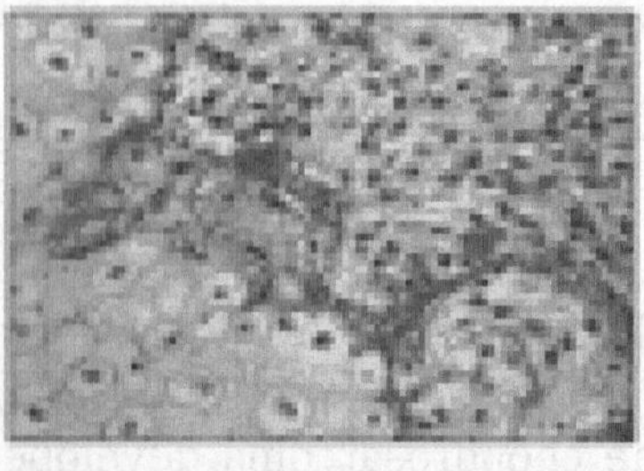

Fig. Chondrosarcoma, grade II.

In grade III chondrosarcomas, undifferentiated cells are dominant, with marked pleomorphism and numerous mitoses, and metastases are present in the lung. The malignant character is marked by the presence of fusiform chondroblasts.

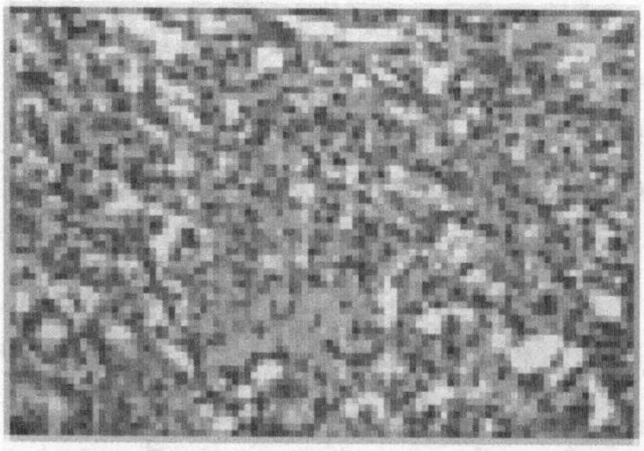

Fig. Chondrosarcoma, grade III.

The characteristic of these neoplasms is the presence of chondro-nucleoproteic matrix, with marked cellular anaplasia, especially when located in the nasal conchae, where intense cellularization is found. LINNABARY et al. (1978) report a chondrosarcoma in a cat, located in the nasal septum. Histologically, the authors find the presence of undifferentiated cells, bone, cartilage islands and numerous capillaries. In other locations, differentiation is more obvious and a lytic action of bone structures is noted.

MULTILOBULAR BONE TUMOR

The multilobular bone tumor is also known as chondroma rodens or cartilaginous counterpart of fibromatosis. This tumor is characterized by local aggressiveness, being formed by nodular masses of membranous bone tissue, with locations in the cranial bones and vertebrae. The affected species are cats, large size dogs, around the age of 1 year, and extremely rarely, horses.

The general aspect is that of an infiltrative fibromatous lesion, with calcification foci and metaplastic cartilage; it appears as a single cranial tumor, as a result of the pathological reaction of the periosteum.

The development of the tumor in the cranial bones, in the proximity of the brain and lacrimal bone, frequently in the temporo-occipital and zygomatic process area, determines clinical symptoms specific for these tissues. The tumor tissue appears under the form of dense sandy nodules in a fibrocartilaginous mass. In most cases, the size of the tumor is between 3 and 7 cm in diameter, and it can induce signs due to compressions on the origin of the cranial and brain nerves, exophthalmos and maxillary paralyses.

The explanation offered by POOL (1978) regarding the formation of this tumor in the cranial bones consists of the fact that ossification is intramembranous, where the periosteum serves as a framework on which osteoblasts deposit additional layers of bone tissue in order to give the cranial bone cortex a compact structure. In most tumors, thin fibrous tissue lamellae appear, which are comparable to the bone lamina of the developing skull, in puppies. Fibrous lamellae are covered with neoplastic osteoblasts, and in the mesenchyma, tissue spaces are completed by hyaline cartilage, hyalinized cartilage, and osteoid and/or chondroid matrix.

Some tumors are predominantly cartilaginous, other bony, but most neoplasms appear as a mixture of cartilaginous and bony lobules. The centers of cartilaginous lobules may undergo enchondral ossification, with intense vascularization and osteoclastic resorption of the ossified matrix.

The multilobular bone tumor, located in the cranial bones, is considered to be benign, although pulmonary metastases have also been reported, in the case of malignization.

Although it is a rare tumor, it is the second most frequent benign tumor, after osteochondroma, and the most common benign cranial tumor in dogs. It occurs in dogs aged between 15 months and 12 years, with a mean age of 7 years.

Histologically, it is characterized by multiple circular, oval or irregular nodules, with cartilage, bone or osteocartilaginous structure. These nodules are bordered by septa formed by fusiform cells and capillaries. In the center of lobules, enchondral ossification nodules, hyaline cartilage, osteoid substance and immature bone are formed. At the periphery of the lobules, fetal osteoblasts are identified.

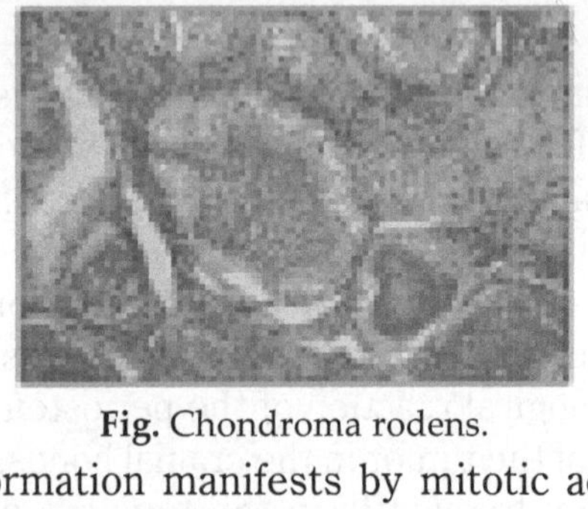

Fig. Chondroma rodens.

Malignant transformation manifests by mitotic activity, loss of lobular structure, necrosis and hemorrhage, and the excessive growth of one of the mesenchymal components. Malignization determines pulmonary metastases.

This tumor has a slow but progressive growth, and by its cranial location and compression on the nerve tissue, it may result in death.

BONE MARROW TUMORS

Myeloma. Malignant bone marrow tumors develop either as multiple or diffuse neoplasms. Macroscopically, they are characterized by round cells similar to poorly differentiated plasmocytes, including atypical multinucleated cells. The medullary lesion causes focal or diffuse osteolyses. Both in animals and humans, these tumors are associated with the presence of abnormal protein in the blood.

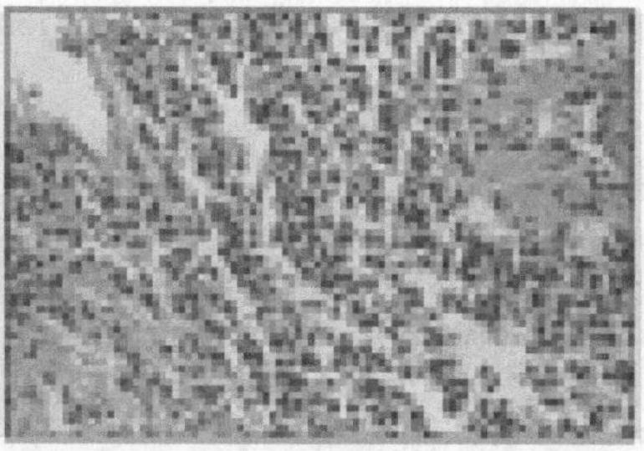

Fig. Mieloma.

Myeloma is found in dogs aged between 1 and 10 years, more rarely in pigs and horses, with the presence of nodular metastases or diffuse growths in different parenchymatous organs.

VASCULAR BONE TUMORS

Hemangioma is a benign tumor that may have the character of vascular or cavernous neocapillaries. In dogs, hemangioma has been reported in vertebrae, as a multiple tumor.

Hemangiosarcoma or malignant hemangioendothelioma is characterized by the irregular anastomosis of vascular channels, bordered by one or several layers of atypical endothelial cells. These cells are immature, being accompanied by poorly differentiated tissue masses or anaplastic tissue. Multiple bone or bone and soft tissue hemangiosarcomas have been diagnosed in dogs, horses, and less in other species. In the long bones of dogs, these neoplasms are rare, and there are difficulties in establishing whether the bone lesion is primary or secondary. It is supposed that these neoplasms develop from the small vessels of bone spaces.

Macroscopically, the tumor has the aspect of a hemorrhagic mass, similar to telangiectatic osteosarcoma. The character is that of intense malignancy, with early metastasis in the lung.

Histologically, both necrotic foci and hemorrhagic spaces are found in the neoplastic mass, along with capillaries with irregular lumen, bordered by immature endothelial cells, in a fibrous stroma. The neoplastic development has a lytic action on bone structures, followed by fractures. In the case of a less aggressive evolution, periosteal reaction that covers the osteolytic focus may occur.

MISCELLANEOUS BONE TUMORS

Miscellaneous bone tumors are frequently formed by neoplasms of the hematopoietic and lymphopoietic system. Sporadic juvenile leukosis in calves is associated with hemorrhagic and necrotic bone marrow lesions. Canine myeloma is associated with poorly differentiated histiocytic tumors.. Vascular bone marrow tumors may be associated with other neoplasms of the compact bone, so that osteoliposarcoma is associated with other mesenchymal neoplasms.

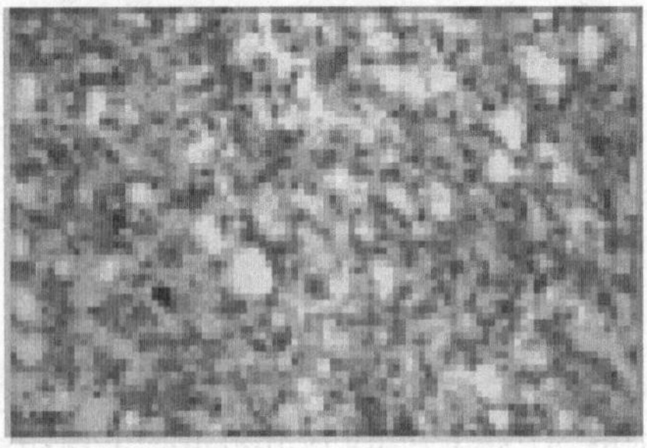

Fig. Liposarcoma, diaphyseal.

METASTATIC BONE TUMORS

Metastatic bone tumors, in dogs and cats as well as in other species, have a particular diagnostic importance, mimicking osteosarcomas or other malignant bone forms. Hematogenic metastases in farm animals are relatively rare. For companion animals, bone metastases of mamary, pulmonary and prostate tumors are well known. These metastases preferentially occur in ribs, vertebrae and long bones.

TUMOR-LIKE BONE LESIONS

SOLITARY BONE CYST

Solitary bone cyst may be defined as a clear liquid or sanguinolent collection, delimited by a connective capsule with multinucleated osteoclast cells. This connective structure may contain hyalinized or calcified fibrillar material. These cystic formations have been rarely found in the long bones of the dog. Numerous cases had a multilocular aspect.

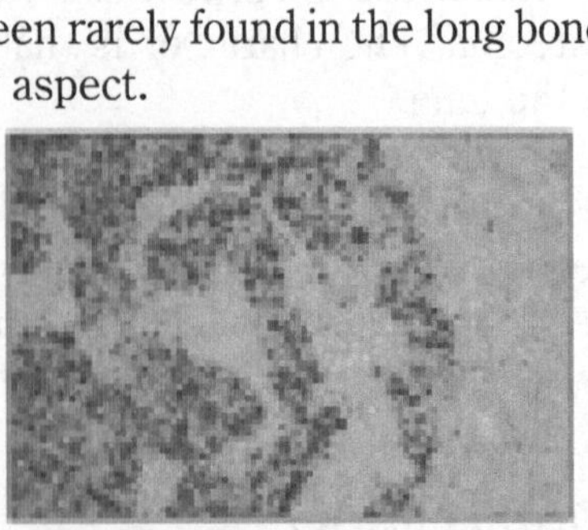

Fig. Osteoliposarcoma, radius.

Subchondral cysts have been noted in horses and swine, associated with osteochondrosis. They are located in epiphyseal cartilage or as hemorrhagic subarticular foci; at other times fibrous cysts contain synovial liquid. In horses, these lesions are located in the phalangeal, cuboid and other limb bones. In chronic degenerative arthrosis, cystic osteochondrosis may be diagnosed in both horses and cattle.

ANEURYSMAL BONE CYSTS

Aneurysmal bone cysts are expansive lesions with a multicellular aspect, derived from vascular bone marrow disorders. These formations are neoplastic lesions, but radiographically and clinically they may resemble tumors, sometimes being complications of bone neoplasms. Incidence is low, being diagnosed in dogs, cats and horses.

Macroscopically, aneurysmal cyst has an expansion tendency, deforming and corroding the bone, the osteolytic action determining the formation of fusiform or sacciform cavities, with the development of new subperiosteal bone. In section, content is red-blackish, with numerous fibrous flanges that contain osteoid bone substance, and numerous osteoclast giant cells, sometimes with endothelial delimitations. Occasionally, these hemorrhagic cysts with osteolytic lesions have a hemangioma, hemangiosarcoma or telangiectatic osteosarcoma appearance, which requires differential diagnosis.

Aneurysmal cysts appear as a result of traumatic lesions or they can be complications of circulatory bone marrow disorders due to medullary neoplasms.

OSSIFYING FIBROMA

Ossifying fibroma is a rare benign tumor, reported in horses and cattle, located in the mandible and the maxilla. The lesion is clearly delimited, being formed by fibrous tissue and hard or sandy bone spicules. Fusiform fibroblasts with osteoblastic changes are microscopically identified.

In severe lesions, the tumor includes fibrous dysplasia, osteoma, osteosarcoma, and in horses, fibrous osteodystrophy with a polyostotic aspect. The connective tissue cells from the ossifying fibroma are pleomorphic, with a high mitotic index and osteosarcoma aspect. In horses, there is a possibility of confusion with fibrous osteodystrophy, in which large osteoclasts are present freely in the connective tissue, which differentiates it from ossifying fibroma.

FIBROUS DYSPLASIA

The lesion is rare, with a fibrous bone aspect and frequently neoplastic development, being monostotic or polyostotic. It is found in horses, dogs and cats, especially in young or newborn individuals. Monostotic lesions are located in the maxillary sinus of horses, and in dogs, in the maxilla and infraorbital area. Polyostotic fibrous dysplasia with familial character has been reported in dogs. In cats, the lesion has a monostotic character, with the same locations as in humans, and histologically it seems to be a maturation defect in bone formation.

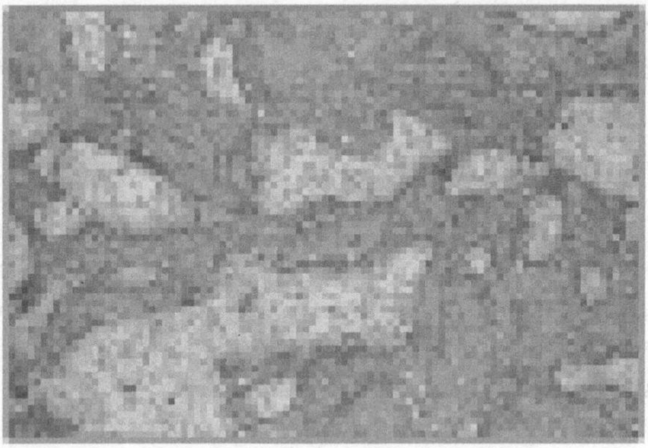

Fig. Fibrous displasia, ulna.

BROWN TUMOR IN HYPERPARATHYROIDISM

The brown tumor appears as a circumscribed lesion in humans, mimicking according to ASLEY (1978) osteoclastoma. It is accompanied by generalized osteoporosis or minimal decalcification, as a direct effect of the parathyroid hormone on bone, the transformation of the osteogenic potential of osteoblasts towards osteoclasia, respectively.

Microscopically, numerous osteoclast giant cells are found, arranged in groups, separated by well vascularized fibrous tissue and bone tissue islands. Hemorrhage is frequently noted. These lesions are rare in animals.

Other bone lesions, mentioned by MISDORP and VAN DER HEUL (1976), are: *myositis ossificans, epidermoid phalangeal cyst* and *infectious granulomatous proliferations.*

Myositis ossificans appears as an ossifying periostitis that extends to the connective tissue. It is characterized by fibrous and bone proliferations, sometimes with the presence of cartilage islands. This lesion has been reported in cats and pigs. Myositis ossificans similar to the human form is mentioned in the olecranon area in cats.

The epidermal cyst of phalanges contains a keratinized mass of squamous epithelium, being found in dogs.

Granulomatous infections such as tuberculosis in horses and coccidiomycosis in dogs could be radiologically confused with bone tumors, but histological examination establishes positive diagnosis.

JOINT AND RELATED STRUCTURE TUMORS

Articular cartilage tumors are not frequently found. Cartilage osteochondromas are cartilaginous or more frequently bone nodules, which develop in synovial structures following chronic cartilage irritations. The development of cartilage occurs by the metaplasia of connective tissue and bone components, by nodule metaplasia or by enchondral ossification with cartilage metaplasia. Synovial structure neoplasms are not frequent, but they have been diagnosed in dogs, cats, horses and cattle, especially in adult animals, in long bones.

SYNOVIAL SARCOMA

Synovial sarcoma is characterized by an acinar or linear structure, with endothelial cells, arranged in a mucoid mass and separated by reticular fibers or collagen formations, similar to fusiform cells with non-homogeneous arrangement in terms of density and hyalinization. In animals, unlike in humans, the extension and intensity of calcifications are reduced. Mucinous substance appears among fusiform cells.

Microscopically, synovial sarcoma is characterized by either a sarcomatous predominance, similarly to fibrosarcoma, or the dominance of the epithelioid component under the form of acini, being similar to a metastasized carcinoma. This biphasic character is not frequent in dogs; in contrast, epithelioid cells in this species have a linear arrangement and sometimes contain amorphous PAS positive material, similar to synovial fluid.

The fibrosarcomatous form is frequently predominant. In cats, synovial tumors contain multinucleated giant cells.

The histological examination of the connective tissue of the synovial sarcoma can identify hemangiopericytosis, malignant fibrous histiocytoma and giant cell tumors.

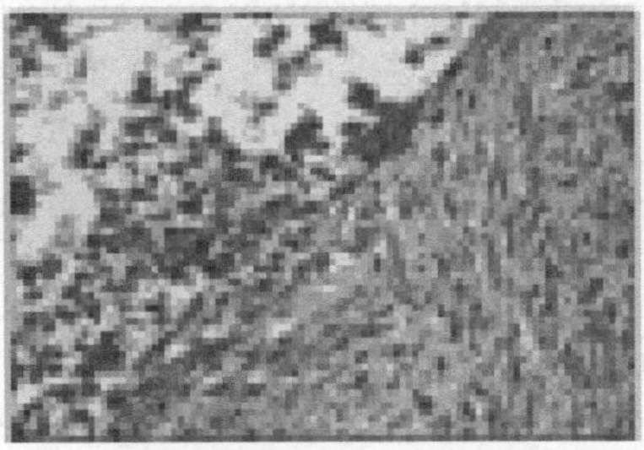

Fig. Synovial sarcoma.

The neoplasm has been diagnosed in dogs of different ages, and more rarely in cats and foxes. The tumor appears in the proximity of the cartilage of the extremities, having a local invasive character, and it frequently produces metastases in the lungs and adjacent lymph nodes. Villous papillary formations develop on the internal side of the synovial membrane.

FIBROXANTHOMA

Fibroxanthoma or fibrous histiocytoma is a benign tumor, with local or diffuse growth. The microscopic structure is characterized by the presence of fibroblasts, collagen, xanthomatous cells, multinucleated giant cells and hemosiderin deposits, with the frequent presence of calcifications and osteoid substance. In dogs, the tumor is located in the synovial membrane of the phalanges and the shoulder, similarly to human nodular tenosynovitis. In horses, it has been reported to have a villonodular pigmented synovitis appearance, similar to the human form. In dogs and cats, the tumor has been diagnosed in the proximity of or even within cartilage, sometimes having malignant characteristics, infiltrative growth and metastasizing.

MALIGNANT GIANT-CELL SOFT TISSUE TUMOR

This tumor of the connective tissue from the synovial fluid structure is characterized by pleomorphic fusiform cells, with collagen fibers and numerous giant cells disseminated throughout the tumor. Hyalinizations and cartilage material are found. The tumor has been reported in cats.

The standard treatment in bone neoplasms consists of the surgical excision of the tumors, or the extensive amputation of the bone and adjacent tissues. In general, the survival rate increases postoperatively, compared to unoperated subjects.

The association of surgical treatment with the administration of chemotherapeutic agents significantly increases the survival rate. Thus, following surgery, the mean survival time is about 3–4 months, death occurring by pulmonary metastases, while after treatment with liposome-encapsulated tripeptide muramyl phosphatidylethanolamine, mean survival was 222 days.

KRAEGEL et al. (1991) obtained much better results in 14 dogs treated with cisplatin, administered postoperatively, the mean survival time being 413 days. In a similar experiment in 22 dogs with appendicular osteosarcoma, following amputation or partial surgery of the affected limb, BERG et al. (1992) administered intravenous cisplatin (60 mg/m^2), starting with the first postoperative week, at 3 week interval. The authors reported a mean survival time of 46.6 weeks in 22 dogs, with variations between 1 year in 45.5% and 2 years in 20.9% of the treated subjects.

HEIDNER et al. (1992) experimented a complex formula, using in 12 dogs with appendicular osteosarcoma, 24–40 Gy radioactive cobalt 60 and two intraarterial cisplatin doses. Survival in subjects was between 6 months and over one year, with a mean of 4.9 months. The authors mention that therapy was well tolerated, being an alternative to amputation.

Therapy in cancer disease has made relatively rapid progress, but there are some principles that are respected, which do not exclude the latest discoveries and the possibility of including in current practice therapies that reflect the progress and conquests of science.

JOINT DISORDERS IN DOGS

Some joint diseases, such as arthritis, affect the joint membranes themselves. Other types of joint conditions affect the tendons, cartilage, bursae, and fluid within the joint. Joint disorders may be congenital (present at birth) or may be the result of injury to the joint, abnormal development, immune-related conditions, or infections.

KEY JOINTS IN A DOG'S LEG

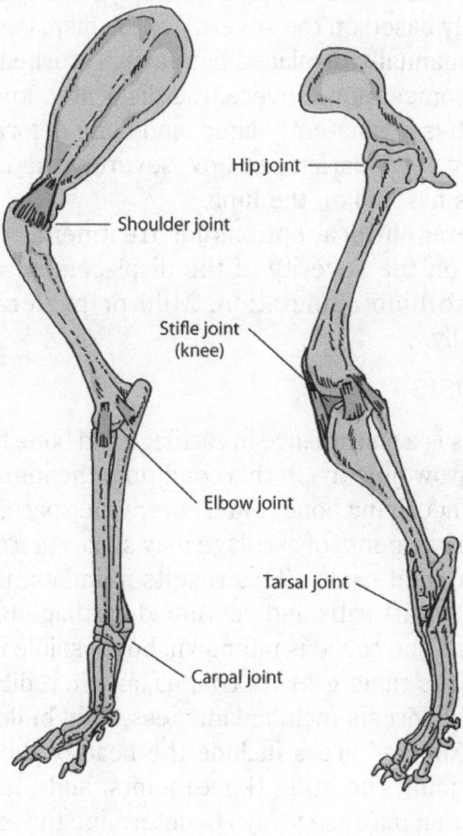

Fig. Legg-Calvé-Perthes Disease

This deterioration of the top of the femur (femoral head) seen in young miniature and small breeds of dogs is characterized by a lack of blood supply and destruction of blood vessels of the bone. The cause is unknown, although the condition may be hereditary in Manchester Terriers. The sudden loss of blood supply to the femur leads to collapse of the top of the bone. Signs include hindlimb lameness, wasting away of the thigh muscles, and pain during movement of the hip joint. Longterm cases have evidence of degenerative joint disease. X-rays can help identify characteristic changes of this condition and may be used to confirm the diagnosis. Treatment involves surgical removal of the affected femoral head and neck and physical therapy to stimulate limb usage. If these procedures are followed, most animals with this condition recover.

DISPLACEMENT OF THE KNEECAP

This hereditary disorder is caused by abnormal development of the kneecap (patella). Displacement of the kneecap is often associated with multiple deformities of the hindlimb, involving the hip joint, femur, and tibia. Affected animals are lame or walk with a skipping gait. Dogs of any age may be affected.

Signs vary widely based on the severity of the displacement. In mild cases, the kneecap can be manually displaced but easily returned to normal position. As displacement becomes more severe, the dislocated kneecap is more often out of place, the limb is consistently lame, and bone deformities may be seen. X-rays can help your veterinarian see how severely the kneecap is displaced and what effects this has had on the limb.

There are several surgical options for treatment. The type of surgery performed is based on the severity of the displacement and can range from minor procedures to limb amputation. Mild or moderately affected dogs generally recover fully.

OSTEOCHONDROSIS

Osteochondrosis is a disturbance in cartilage and bone formation of medium and large dogs that grow quickly. In this condition, the immature joint cartilage separates from the underlying bone. Fluid enters the space, and cysts may form under the cartilage. Fragments of cartilage may separate from the end of a bone and float loose in the joint cavity. This results in inflammation of the affected joint, and it can lead to arthritis and continued cartilage breakdown, severely affecting joint motion. The cause is unknown, but possible factors include high-growth diets, as well as rapid growth, trauma, and heredity.

Signs of osteochondrosis include lameness, fluid buildup within the joint, and joint stiffness. Affected areas include the head of the shoulder joint, the inside of the elbow joint, the stifle (knee) joints, and the ridges of the hock joints. Your veterinarian may use x-rays to determine the extent of the damage. Surgery using an endoscope can also be performed to identify cartilage or joint lesions. Treatment involves surgical removal of cartilage flaps or the free-floating fragments of cartilage known as joint mice. Dogs with degenerative joint disease may benefit from nonsteroidal anti-inflammatory drugs and joint fluid modifiers. Your veterinarian is best able to advise you on the appropriate use of any relevant medication. The outlook for recovery is excellent for the shoulders, good for the stifle joint, and fair for the elbow and hock (tarsal) joints. If the affected dog also has signs of degenerative joint disease, other joint conditions, or instability of the hock joint, the chances of recovery are reduced.

ELBOW DYSPLASIA

Elbow dysplasia is an abnormal development of the elbow joint in young, large, rapidly growing dogs. It involves abnormal bone growth, cartilage development, or joint stresses. It is considered to be one of the most common causes of osteoarthritis of the canine elbow.

Lameness can develop slowly between 4 and 8 months of age; however, some cases may not be diagnosed until the dogs are more than 1 year old. The joint may appear stiff or unable to move freely. Advanced cases develop osteoarthritis, fluid buildup within the joint, and a grating or crackling sound. Physical examination and the presence of the characteristic signs suggest the diagnosis, and x-rays can confirm it. Both elbows should be examined because the condition can develop in both at the same time.

Surgery should be performed before the degenerative changes of osteoarthritis occur. The outlook for recovery after surgery is good if degenerative joint disease has not developed in the joint. Aspirin or nonsteroidal anti-inflammatory drugs (as prescribed by your veterinarian) can reduce pain and inflammation. Joint fluid modifiers may also be helpful.

HIP DYSPLASIA

Hip dysplasia is an abnormal development of the hip joint in large dogs. It is characterized by a loose joint and subsequent degenerative joint disease (osteo-arthritis). Excessive growth, exercise, nutrition, and hereditary factors affect the occurrence of hip dysplasia.

The signs associated with hip dysplasia vary. Lameness may be mild, moderate, or severe, and is worse after exercise. The dog may walk with a "bunny-hopping" gait. A loose joint, reduced range of motion or stiffness of the joint, and a grating sound and pain during full extension and bending of the joint may be present. X-rays are useful in determining the degree of arthritis and planning treatments.

Both medical and surgical treatments are available. Dogs that have a mild case of hip dysplasia or that cannot undergo surgery due to health or owner constraints may benefit from other treatments. These include weight reduction, restriction of exercise on hard surfaces, physical therapy, anti-inflammatory drugs, and possibly joint fluid modifiers. Surgical treatments can include a range of procedures from reducing pain and arthritis to total hip replacement.

The outlook for recovery varies greatly and depends on the overall health, degree of dysplasia and joint damage, and environment of the animal. Surgery is generally beneficial if recommended and performed correctly. Dogs that do not undergo surgery may require lifestyle changes in order to be comfortable.

OSTEOARTHRITIS (DEGENERATIVE JOINT DISEASE)

The joint cartilage in freely moving joints may degenerate over time, leading to loss of joint movement and, in many cases, pain. This condition is characterized by thinning of cartilage, buildup of fluid within the joint, and the formation of bony outgrowths around the joint. Joint degeneration can be caused by trauma, infection, the body's own immune system, or malformation during development. This leads to inflammation of the joint membrane, continued cartilage destruction and inflammation, and abnormal joint function.

Signs of osteoarthritis include lameness, joint swelling, wasting away of muscle, and thickening and scarring of the joint membrane. Eventually enough damage can occur that a grating sound might be heard during joint movement. X-rays show increased fluid within the joint, soft-tissue swelling around the joint, the formation of bony outgrowths, hardening and thickening of bone beneath the cartilage, and sometimes a narrowed joint space.

Treatments can be either medical or surgical. Medical treatment may include the use of nonsteroidal anti-inflammatory drugs to reduce pain and inflammation. However, longterm use of these drugs in dogs can sometimes cause gastrointestinal problems such as lack of appetite, vomiting, and inflammation of the stomach and intestines. Corticosteroids also suppress inflammation, but they are usually given only for a short period in order to avoid adverse effects of continued use. Your veterinarian will prescribe appropriate medication based on your dog's signs, age, and overall health.

Surgical options include joint fusion, joint replacement, cutting of the joint, and amputation. The outlook for recovery depends on the location and severity of the joint disease.

Other treatments that might be considered include weight reduction, carefully monitored exercise on soft surfaces, and application of warm compresses to affected joints. Joint-fluid modifiers may help prevent further cartilage degradation.

SEPTIC ARTHRITIS

Infectious, or septic, arthritis is usually caused by bacteria that spread through the blood or enter the body as a result of trauma (with penetrating wounds) or surgery. Other causes of septic arthritis include rickettsia (Rocky Mountain spotted fever, ehrlichiosis) and spirochetes (Lyme disease).

Signs of septic arthritis include lameness, swelling, pain of affected joint(s), fever, listlessness, loss of appetite, and stiffness. X-rays may reveal increased fluid within the joint in early cases and degenerative joint disease in longterm conditions. Laboratory tests on fluid removed from the joint may be useful in confirming the diagnosis.

Treatment consists of antibiotics administered orally or intravenously, flushing of the joint cavity, and surgical removal of dead, damaged or infected tissue in severe cases.

IMMUNE-MEDIATED ARTHRITIS

Arthritis caused by the body's own immune system can cause inflammation of joints. It generally affects several joints. In some types of immune-mediated arthritis, joint cartilage and bone beneath the cartilage is destroyed. Rheumatoid arthritis andGreyhound polyarthritis are examples of arthritis that destroys joint cartilage and bone beneath the cartilage. Systemic lupus erythematosus is the most common form of arthritis that causes inflammation of the joint.

Signs include lameness, pain, and swelling in multiple joints, fever, a generalized illness, and persistent loss of appetite. These signs commonly come and go. In addition to signs, the diagnosis is aided by x-rays, biopsy of joint tissue, and examination of joint fluid (commonly called a joint tap).

Treatment involves anti-inflammatory medications and chemotherapeutic agents. The outlook for recovery is uncertain. Relapses are relatively common and the cause of the reactions is often unknown.

CANCEROUS ARTHRITIS

This type of arthritis is most commonly caused by a tumor known as a synovial cell sarcoma. It is the most common cancerous (malignant) tumor involving the joints. Signs include lameness and joint swelling. X-rays show soft-tissue swelling and a reaction around the bone. A biopsy reveals evidence of a soft-tissue tumor. Spread of the cancer to the lungs occurs in about 25% of animals; thus, amputation of the limb is usually recommended to prevent spreading.

JOINT TRAUMA

There are many types of joint trauma that can contribute to the development of joint disorders.

CRANIAL CRUCIATE LIGAMENT TEAR

Tearing of the cranial cruciate ligament of the stifle (knee) joint is usually caused by serious injury. However injuries are more likely to occur when the joint structure is already weakened by degeneration, the animal's own immune system, or defects in conformation (such as those seen in straight-legged dogs). Most injuries involve a tear in the middle of the ligament, although some result from bone separation at the origin of the ligament. A tear of this type can make the stifle joint unstable and can lead to cartilage injury, buildup of joint fluid, bony outgrowths, and hardening and thickening of the joint membrane.

Signs include those generally seen in joint disorders, such as lameness, pain, joint swelling, fluid buildup, and a grating sound when the joint is moved. In addition, the joint may appear to be abnormally loose. Partial cranial crucial ligament tears are characterized by a reduced ability to move the joint, especially bending it. Cartilage injury may be identified by a clicking sound during movement or when the joint is bent and extended. Xrays may show the injury and/or damage to the joint; testing of fluid removed from the joint may also be used to help diagnose the condition.

Both medical and surgical treatment options are available. Physical therapy, weight reduction, and nonsteroidal anti-inflammatory drugs ease discomfort from inflammation and degenerative joint disease. For active dogs, surgery to stabilize the knee joint is recommended. Physical therapy following surgery is critical for recovery. The outlook after surgery is good as long as degenerative joint disease has not progressed too far.

DISLOCATION AND FRACTURE OF THE ANKLE

Injury to the ankle (tarsus) is often seen in dogs that have been hit by a car. The ankle in dogs includes several bones that connect the lower leg to the foot.

Injuries may include fracture or dislocation of these bones or tearing of the ligaments that hold them together. Affected dogs will hold the injured leg up and refuse to put any weight on it. The foot may swing in unusual directions because of its loose attachment following the injury. The extent of the injury is confirmed by physical examination and x-rays. Treatment is surgery to repair the bones and ligaments. The outlook for recovery is good.

DISLOCATION OF THE ELBOW

Elbow dislocation is usually the result of trauma and is a common injury in dogs. Pain is variable, and the dog will usually hold the injured leg up and refuse to put any weight on it. Physical examination and x-rays are used to diagnose the condition. Treatment generally requires surgery, but the outlook for recovery is excellent.

DISLOCATION OF THE HIP

Hip dislocation is usually the result of injury or trauma that displaces the head of the femur from the socket of the hip joint. Signs of hip dislocation include lameness, pain during movement of the hip joint, and a shortened limb. X-rays are useful in confirming the dislocation and revealing the presence of fractures. Nonsurgical treatment involves forcefully moving the joint back into place (closed manipulation) and using slings to keep the hip in its normal position. Surgical treatment involves stabilization using sutures or pins. Surgical resection of the bones involved or total hip replacement may be performed if more conservative treatment has not succeeded. The outlook for recovery is usually excellent.

A NORMAL AND A DISLOCATED HIP JOINT OF A DOG

JOINT FRACTURES

The shoulder, elbow, carpal, hip, stifle, and tarsal joints are those most commonly involved in fractures due to injury. In young animals, the portion of the bone where growth occurs—called the growth plate and usually located at the ends of the bones—is weak compared with adjacent bones, ligaments, and joint membranes, making this area more prone to injury.

Signs of joint fractures include lameness, pain, and joint swelling. If the injury affects an active growth plate, limb deformities can result. X-rays are used to confirm and locate the fracture.

The goal of treatment is to allow the fracture to heal in proper alignment while maintaining joint and limb functions. This is usually done by holding the fracture in place internally with pins, wires, or screws in order to stabilize it. The outlook for recovery is good as long as damage to the joint is not severe.

PALMAR CARPAL LIGAMENT BREAKDOWN

Injuries sustained when falling or jumping can cause hyperextension, in which the limb extends beyond its normal range of motion. This produces excessive force on the wrist (carpus), which can cause tearing of the palmar carpal ligaments and fibrocartilage, leading to collapse of the joints. Signs include lameness, swelling of the carpal joint, and a characteristic stance in which the heel is touching the ground. For mild cases a splint or cast may be sufficient, but surgery is usually required. Surgery involves fusing the affected joints using a bone plate and screws, pins and wires, or an external system. The outlook for recovery is good.

BONE AND JOINT TUMORS IN DOGS

These notes are provided to help you understand the diagnosis or possible diagnosis of cancer in your pet. For general information on cancer in pets ask for our handout "What is Cancer". Your veterinarian may suggest certain tests to help confirm or eliminated diagnosis, and to help assess treatment options and likely outcomes. Because individual situations and responses vary, and because cancers often behave unpredictably, science can only give us a guide. However, information and understanding for tumors in animals is improving all the time.

We understand that this can be a very worrying time. If you do have any questions, please do not hesitate to ask us.

WHAT ARE THESE TUMORS?

Non-cancerous bone tumors are rare and mainly due to abnormal development. They include bone cysts and single or multiple lumps of bone in abnormal places (exostoses). Fibrous dysplasia is another developmental condition that causes bone swelling around areas of bone destruction. 'Craniomandibular osteoarthropathy' is a disease of the head bones that occurs in some terriers.

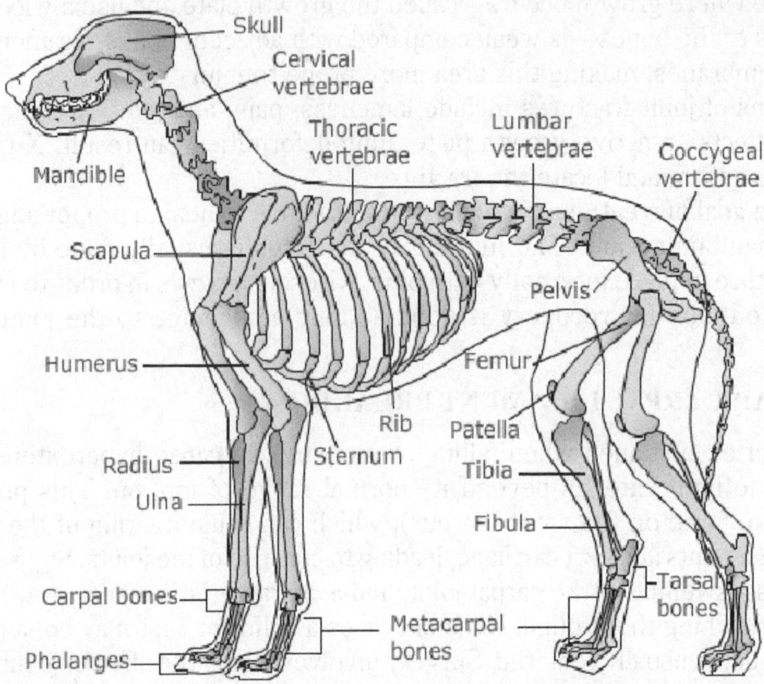

'Benign' (non-spreading, local) tumors are also rare and of several types. 'Ossifying fibroma' (osteofibroma, fibrous osteoma) is an expansile lesion in the jaw that destroys existing bone structure. Osteomas and osteochondromas are benign tumors in immature animals. Chondromas are also benign but some become malignant (chondrosarcomas) and infiltrate locally.

"In the dog, 90% of bone cancers are malignant (capable of spreading to other body sites)."

In the dog, 90% of bone cancers are malignant (capable of spreading to other body sites). Osteosarcoma is by far the most common, particularly in large dogs. In these animals, it is usually in a limb (appendicular skeleton). They often spread to other parts of the body (metastasize).

Rare types of malignant bone cancers include the locally aggressive 'giant cell tumor of bone' derived from bone marrow cells, and 'multilobular tumor of bone' (multilobular osteoma/chondroma of canine skull) which is slow growing. Joint tumors (synovial sarcomas) are rare. Spread of cancer (metastasis) from the mammary glands, liver, lung or prostate to the bones is relatively common.

WHAT DO WE KNOW ABOUT THE CAUSE?

The reason why a particular pet may develop this, or any cancer, is not straightforward. Cancer is often the culmination of a series of circumstances that come together for the unfortunate individual. Non-cancerous bone tumors such as cysts and exostoses are due to abnormal development. Craniomandibular osteoarthropathy has a genetic basis.

"A history of previous fracture at the site of osteosarcoma is not unusual."

We do not know the precise cause of malignant cancers of bones in dogs but abnormal bone cell growth and unusual hormone stimulation may be involved. A history of previous fracture at the site of osteosarcoma is not unusual. It is speculated that the excessive proliferation of cells that is required to heal a fracture gives greater opportunity for neoplastic mutation to occur. Body size and sex are also important in the development of bone cancers in dogs so genetic factors also have a role.

WHY HAS MY ANIMAL DEVELOPED THIS CANCER?

"Some animals have a greater tendency (genetic susceptibility) to cancer."

Developing cancer is often the result of chance misfortune. Some animals have a greater tendency (genetic susceptibility) to cancer. Some breeds have far more bone cancers than other breeds.

ARE THESE COMMON TUMORS?

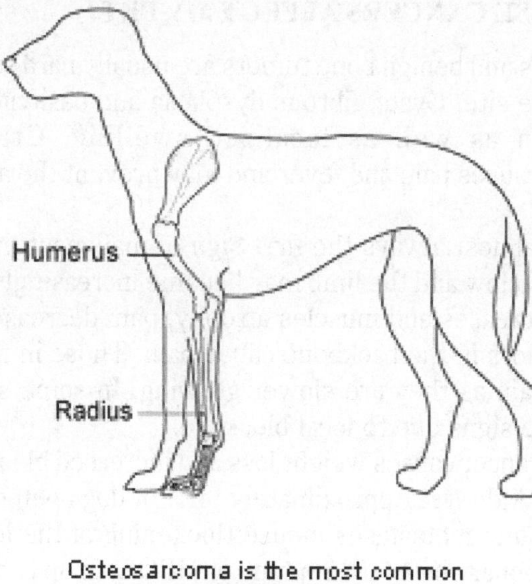

Osteosarcoma is the most common bone tumor in dogs

All the non-cancerous and benign bone tumors are rare.

Malignant bone cancer is not common but one survey found an incidence of 7.9 per 100,000 dogs. Osteosarcoma accounts for 80% of all canine bone tumors and chondrosarcoma for another 10%. The median age of affected dogs is seven years but tumors can occur in young dogs. The youngest recorded case is a 3 month old dog with a mandibular tumor. Rib tumors are common at an early age.

"Large, particularly giant breed, dogs are most frequently affected and 75% of tumors develop in the radius or humerus of the forelimb."

Large, particularly giant breed, dogs are most frequently affected and 75% of tumors develop in the radius or humerus of the forelimb. The most commonly afflicted breeds include Great Danes, German Shepherds, Rottweilers, Saint Bernards, Boxers and Irish Setters. Most cases are male dogs. In small dogs, most bone tumors are axial (bones other than in the limbs) and less than 50% of these are osteosarcoma. Large dogs with axial tumors are usually female.

Chondrosarcomas are the second most common primary tumor of bone and account for 5 to 10% of all bone tumors in dogs. The average affected dog is 8 years old (range, 1 to 15 years). The inside of the nose is the most common site but they also occur in the ribs and other bones.

Multilobular tumors of bone are uncommon but are usually in the skull. The average age of affected dogs is 7 years with a range of 1 to 12 years. Affected breeds are usually medium to large size and are rarely giant breeds. Giant cell tumors are very rare. Joint (synovial) tumors usually affect the stifle and elbow of large breeds such as the Rottweiler. They are rare and difficult to diagnose with certainty unless specialized techniques are available.

HOW WILL THESE CANCERS AFFECT MY PET?

Non-cancerous and benign bone tumors are usually hard swellings. Clinical signs vary with the site. Cysts, fibrous dysplasia and ossifying fibroma cause bone destruction as well as localized swelling. Craniomandibular osteoarthropathy causes pain and fever and may prevent the affected dog from eating.

Lameness is almost always the first sign of malignant limb bone cancer. Pain and swelling follow and the limb may become increasingly painful and hot. As the swelling increases and muscles atrophy, pain decreases but the bones may fracture. Tumors in the backbone cause pain. Those in the ribs and head often cause less pain as they are slower growing. In some sites such as the nose, there may be signs due to local blockage.

The growing cancer causes weight loss and increased blood calcium levels, which damage the kidneys. Approximately 90% of dogs will die of metastasis within one year. Lung metastases induce thickening of the lower limb bones (hypertrophic pulmonary osteoarthropathy or HPOA) in up to a quarter of dogs.

HOW ARE THESE CANCERS DIAGNOSED?

In large breed dogs, lameness associated with certain sites and typical clinical and x-ray (radiography) patterns are assumed to be cancer until proven otherwise. Sometimes, the accurate diagnosis also requires microscopic examination of a bone sample (biopsy), using a diagnostic technique called histopathology. This is the microscopic examination of specially prepared and stained tissue sections and is done at a specialized laboratory where a veterinary pathologist examines the sample.

When preparing the histopathology report, the veterinary pathologist usually adds a prognosis (what will probably happen). This may include information on the likelihood of local recurrence or metastasis (distant spread).

Sensitive MRI (magnetic resonance imaging) or computed tomography (CT) can demonstrate the extent of tumors.

WHAT TYPES OF TREATMENT ARE AVAILABLE?

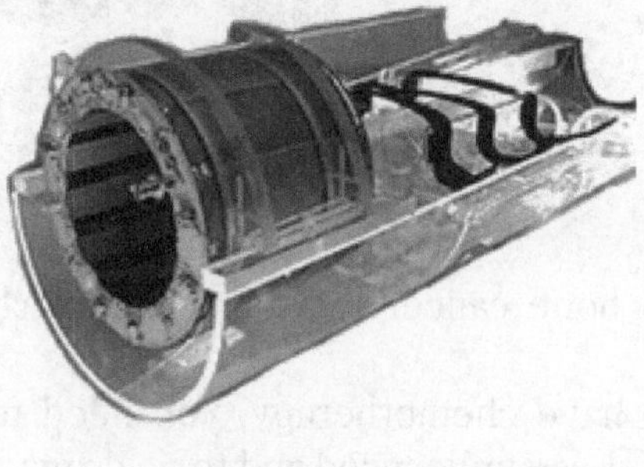

MRI Machine

The most common treatment is surgical. For benign tumors, removal of the lump is sufficient but for malignant tumors, surgery may be extensive and include amputation of the affected limb. The over-riding consideration in treating dogs with malignant cancer is that 90% will die with metastatic disease despite control of the primary tumor by amputation.

In some countries, chemotherapy has been used to prolong survival time. The drugs used for this are toxic to the animal, people and the environment so it has sometimes been combined with immunotherapy (using cells to stimulate the macrophages of the immune system). For these cancers, the optimal chemotherapy protocol is still uncertain. Radiotherapy can decrease the pain, and this treatment may be available at specialized referral centers.

CAN THESE CANCERS DISAPPEAR WITHOUT TREATMENT?

Cancer is a multi-step process so it may stop at some stages. The body's own immune system can kill cancer cells but it is rarely 100% effective. Rarely, loss of blood supply to a cancer will make parts of it die but the dead tissue will still need surgical removal.

HOW CAN I NURSE MY PET?

After surgery, your pet may need to wear an "Elizabethan collar" so he or she cannot interfere with the operation site. The incision site needs to be kept clean and dry. Report any loss of stitches or significant swelling or bleeding to your veterinarian. If you require additional advice on post-surgical care, please ask. Pain control in bone cancer is vital, particularly for aggressive bone tumors.

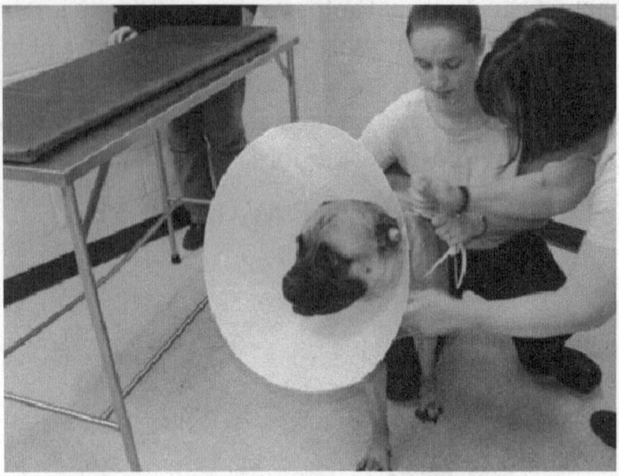

"Pain control in bone cancer is vital, particularly for aggressive bone tumors."

If your pet is to have chemotherapy, you need to understand the risks involved in the use of these unlicensed and toxic drugs. The safety precautions required to protect yourself, other people and the environment when handling and disposing of them will be explained if you consent to their use.

HOW WILL I KNOW HOW THE CANCER WILL BEHAVE?

"Your veterinarian will use the information obtained from the biopsy and subsequent histopathology report to predict how this cancer is likely to behave."

Your veterinarian will use the information obtained from the biopsy and subsequent histopathology report to predict how this cancer is likely to behave. The veterinary pathologist usually adds a prognosis that describes the probability of local recurrence or metastasis (distant spread).

A blood test (for alkaline phosphatase enzymes) can predict the behavior of some limb osteosarcomas in dogs. High blood levels of bone-specific alkaline phosphatase before surgery are associated with significantly shorter survival and disease-free intervals following surgery.

WHEN WILL I KNOW IF THE CANCER IS PERMANENTLY CURED?

'Cured' has to be a guarded term in dealing with any cancer.

Non-cancerous and benign cancerous bone swellings do not spread and many stop growing at maturity. If they interfere with the mechanics of the skeletal system, they can be cured by surgical removal.

Craniomandibular osteoarthropathy has an intermittent, progressive course lasting several weeks to a few months. If it prevents your pet from eating, you and your veterinarian will need to assess the whether the quality of his or her life is reasonable Osteochondromas may undergo malignant transformation to osteosarcoma after some time. Chondromas may recur following surgical removal and some progress to malignancy and infiltrate locally. Chondrosarcomas of the limbs may be cured by amputation, with average survival time of almost two years. Dogs with rib tumors survive an average of about three years but median survival of dogs with nasal tumors is only 210 to 510 days even with radiation therapy. Chondrosarcomas are too slow growing to respond to chemotherapy.

Osteosarcomas have a poor prognosis. High histological grade and high pre-treatment blood bone enzyme alkaline phosphatase are indicators of a particularly poor prognosis.

Less than 5% of dogs with long bone tumors have radiographically detectable metastasis to the lungs at the time of diagnosis. However, 75% will have metastatic lesions when they die and almost 90% will die within one year following amputation of the affected limb. Only 2% of affected dogs are likely to be alive two years after surgery. It is claimed that chemotherapy has improved the survival time of some dogs. Osteosarcomas of the axial skeleton metastasize less readily but survival remains approximately a year. Hypertrophic osteopathy (HPOA) can only be reversed if the lung disease can be cured.

Canine tumors affecting the head are locally aggressive rather than metastatic. Dogs undergoing removal of the lower jaw for a tumor have approximately a 70% one-year survival rate. Local recurrence leads to euthanasia. Multilobular tumor of bone behaves in a similar way. If complete removal is possible, the cancer may be cured but recurrence following surgery is the usual pattern of behavior. Giant cell tumors are locally aggressive but rarely metastasize. Amputation may be the treatment of choice but there is limited information on success rate because they are very rare. In man, benign-looking tumors have metastasized after radiotherapy.

BONE DISORDERS IN DOGS

Bone disorders can be developmental, infectious, nutritional, or due to bone tumors, trauma, or unknown causes.

DEVELOPMENTAL BONE DISORDERS

Developmental bone disorders appear in young animals when the bones do not grow correctly. They may be congenital (present at birth) or occur as the animal grows. Some of the more common causes include hereditary breed characteristics and dietary imbalances.

ANGULAR LIMB DEFORMITIES

Abnormal development of the bones of the foreleg (the radius and ulna) can occur after injury to growth plates. It may also be hereditary in some breeds (such as Bulldogs, Pugs, Boston Terriers, Basset Hounds, and Dachshunds). Irregular growth of the 2 bones leads to shortened limbs, bowing of the bones, partial displacement of the elbow joint, and bending or twisting of the carpus.

GROWTH PLATES ARE FOUND NEAR THE ENDS OF BONES IN YOUNG ANIMALS.

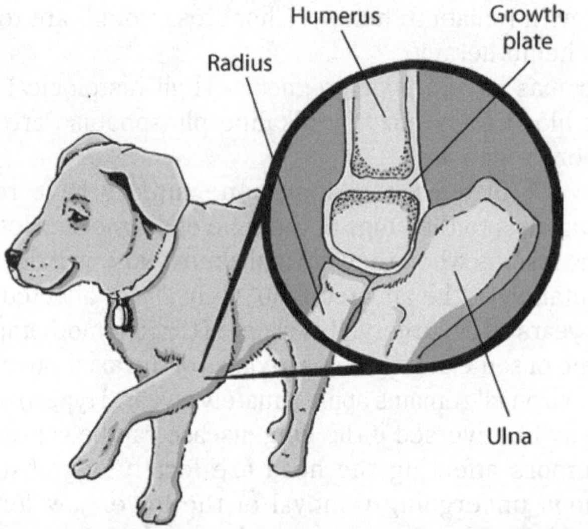

This condition results in lameness. Movement of the elbow or carpal joints is painful, and range of motion is reduced. X-rays may reveal the extent of bone deformity.

Treatment is based on correcting the position, shape, and length of the limb, and reestablishing normal joint movement. Surgical procedures include corrective surgery and stabilization with internal or external braces. The outlook for recovery is good as long as the limb deformities are not severe.

CRANIOMANDIBULAR OSTEOPATHY

Craniomandibular osteopathy is a bone disorder of growing dogs that affects the lower jaw (mandible) and the round bones behind the ears (tympanic bullae) of Terrier breeds. Certain portions of the normal bone are resorbed and replaced by immature bone. The cause is unknown, but it is probably genetic.

Signs may vary widely. They include mouth discomfort, weight loss, fever, and painful enlargement of the lower jaw. X-rays are used to confirm the diagnosis.

Treatment may involve the prescription of appropriate pain relievers or cortico-steroids to reduce inflammation and discomfort. A soft-food diet is usually recommended. The outlook for recovery is good, because the bone growth stops when the animal matures.

HYPERTROPHIC OSTEODYSTROPHY

This disorder affects the areas where growth occurs in the long bones of young, growing dogs, usually of large or giant breeds. The exact cause is unknown, although feeding puppies a diet that is very high in protein and/or calories may play a role.

Signs include pain and swelling in the radius and ulna, fever, loss of appetite, and depression. Affected dogs may be lame and reluctant to move. These signs may come and go. Deformities may develop in severely affected dogs.

Treatment is aimed at relieving pain; for example, nonsteroidal anti-inflammatory drugs may be prescribed. Supportive fluid care and dietary changes (as recommended by a veterinarian) may also be helpful.

OSTEOCHONDROMATOSIS (MULTIPLE EXOSTOSES)

Osteochondromatosis is an uncommon disorder of young dogs characterized by multiple bony growths (known as osteochondromas) that arise from the surface of the long bones, vertebrae, and ribs. Animals may have no signs, and diagnosis is confirmed by x-rays and physical examination of the growths. If lameness or pain develops, the masses can be surgically removed.

PANOSTEITIS

Panosteitis causes bone inflammation, primarily of the long bones, in young, rapidly growing dogs of large and giant breeds. The inflammation may involve single or multiple bones. It appears spontaneously and lasts only as long as the dog is growing, whether or not it is treated. The exact cause is unknown, although some factors thought to play a role include genetics (in German Shepherds), stress, infection, or the body's own metabolic and immune responses.

The condition generally affects dogs 6 to 16 months old. Animals are lame and feverish, have no appetite, and show signs of pain when the affected bones are handled. These signs may come and go. X-rays are used to confirm the diagnosis. Treatment is aimed at relieving pain and discomfort. Nonsteroidal anti-inflammatory drugs or corticosteroids as prescribed by a veterinarian can be used when signs are present. Although it is uncertain whether diets high in protein and calories or dietary supplementation play a role in the development of this condition, it is suggested that such diets be avoided in young, growing dogs.

RETAINED ULNAR CARTILAGE CORES

Retained ulnar cartilage cores is a disorder of the growth plate of the ulna in young large and giant dogs. Abnormal bone formation, in which bone does not harden appropriately, occurs. As a result, bone growth is restrained in the affected forelimb. The exact cause is uncertain, although diet may play a role.

Signs include lameness and angular limb deformities. X-rays are useful to confirm the diagnosis. Dietary supplementation should be stopped, and appropriate nutrition discussed with your veterinarian. Surgical division or removal of the bone may also be necessary to reduce limb deformation. The outlook for recovery is based on the severity of the condition.

OSTEOMYELITIS

Osteomyelitis is inflammation of the bone. The condition is most often associated with bacterial infection, although fungal diseases may also cause osteomyelitis. Factors contributing to infection include an inadequate blood supply to the bone, trauma, inflammation, bone damage, and the spread of an infectious agent through the bloodstream.

General signs of osteomyelitis include lameness and pain. Dogs may have pus-filled sores at the wound site, fever, persistent lack of appetite, and depression. Xrays, laboratory tests, and cultures to identify the source of infection can all help to confirm the diagnosis.

Longterm treatment with antibiotics, either injected or given by mouth, is the usual treatment. Additionally, flushing of the wound; removal of dead, damaged, or infected tissue; and removal of loose implants are recommended. Open or closed wound drainage and bone grafting can also be performed. In persistent cases, limb amputation may be necessary. The outlook for recovery varies based on the severity of the infection and on how long it has remained untreated.

HYPERTROPHIC OSTEOPATHY

Hypertrophic osteopathy is excessive thickening or growth of bone tissues of long bones in dogs occurring after tumors or infectious masses develop in the chest or abdominal cavity. The exact cause is unknown, but may be related to a reduced flow to blood to the bones.

Signs include lameness, long-bone pain, and signs of body cavity tumors. X-rays are used to reveal the primary tumors and bone reactions. Treatment includes chest or abdominal surgery to remove tumors and the surgical cutting of the nerve to block the associated bone changes.

NUTRITIONAL OSTEOPATHIES

Reduced bone mass, bone deformities, bony growths, fractures, and loose teeth (rubber jaw) are all conditions that can result from nutritional disturbances. These disturbances affect parathyroid hormone function and the metabolism of calcium and vitamins in the body. Specific causes may include an unbalanced diet resulting in an abnormally high level of parathyroid hormone (secondary nutritional hyperparathyroidism) or an abnormally high level of parathyroid hormone causing kidney damage (renal hyperparathyroidism), a deficiency of vitamin D, and excessive intake of vitamin A. Diagnosis is by blood tests, xrays, and identification of any underlying nutritional cause. Treatment is aimed at reversing the specific cause. Surgery is rarely needed.

BONE TUMORS

Skeletal tumors can be benign or malignant (cancerous). They can either begin in the bone or spread from other areas of the body. The most common primary bone tumor is osteosarcoma of the radius, humerus, femur, or tibia.

Signs include lameness, bone swelling, and fractures of the bone that are not caused by injury. X-rays of the affected limb can help confirm the diagnosis. Chest x-rays should be performed to look for any original tumors that may be spreading to the bones. A bone biopsy is required to confirm the diagnosis.

The outlook for recovery is guarded. Untreated animals rarely live more than several months. The recommended treatments are limb amputation and chemotherapy, which in many cases can double survival times. On average, dogs live for 5 months after amputation.

BONE FRACTURES

Bone fractures are often caused by car accidents, firearms, fights, or falls. Fractures can involve single or multiple breaks in the bone and may be open (also called compound) or closed. Open fractures have a wound or break in the skin that is associated with the fracture; closed fractures are those that do not produce an open wound. The shape and severity of the fracture depends on the force and type of the trauma.

FRACTURE TYPES CAN RANGE FROM INCOMPLETE TO COMPOUND

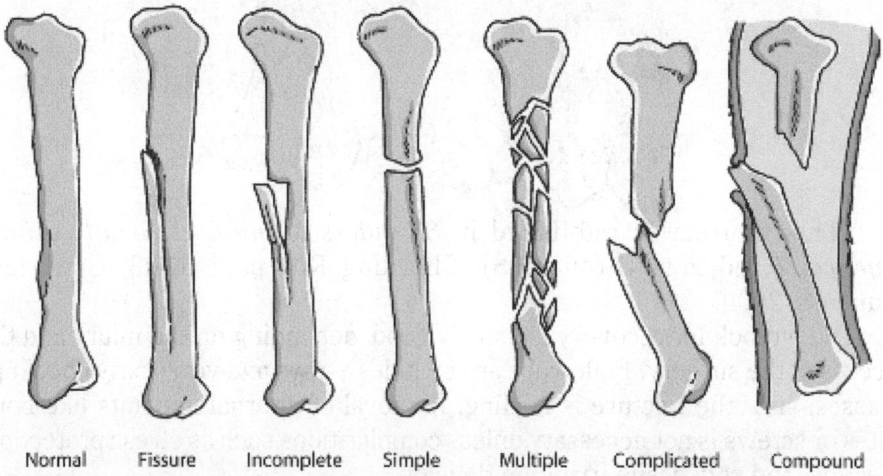

Signs of fracture are general and include lameness, pain, and swelling. X-rays are useful in determining the type and extent of the fracture. Treatment is based on the type of fracture, the dog's age and health, the owner's finances, and the surgeon's technical expertise.

Incomplete fractures in young, healthy dogs can be treated with external splints or casts. Other injuries are treated with bone plates, screws, orthopedic wires, or pins. Bone grafts are frequently used to help healing. Antibiotics are given to keep open fractures from becoming infected. Appropriate pain-relieving medication is used to reduce discomfort.

COMMON TREATMENTS FOR BONE FRACTURES INVOLVE THE USE OF BONE PLATES, SCREWS, ORTHOPEDIC WIRES, AND PINS.*

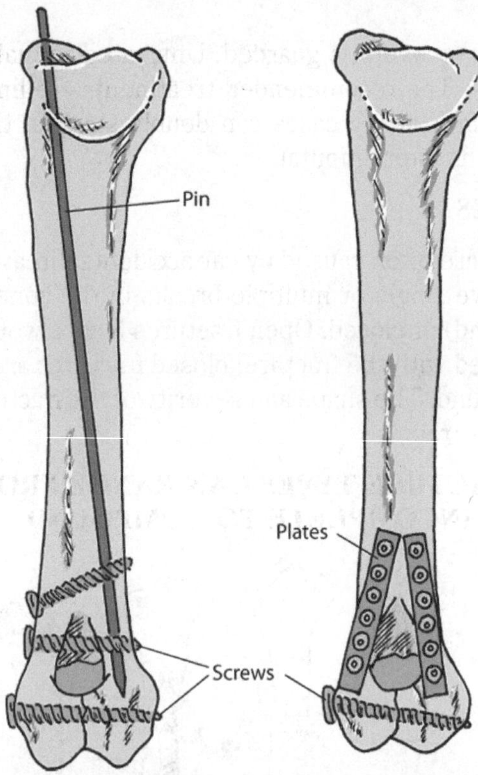

*This figure was published in *Saunders Manual of Small Animal Practice*, 2nd edition, Birchard SJ, Sherding RG, page 1108, Copyright Saunders, 2000.

The outlook for recovery is usually good, depending on the injury and the success of the surgery. Followup care includes x-rays and veterinary checkups to assess how the fracture is healing. Removal of internal implants like bone plates or screws is not necessary unless complications such as stress protection, infection, and soft-tissue irritation develop.

JOINT CANCER (SYNOVIAL SARCOMA) IN DOGS

Synovial sarcomas are soft tissue sarcomas – *malignant* cancers – that arise from the precursor cells outside the synovial membrane of the joints and *bursa* (the fluid-filled, sac-like cavity between joints that helps to facilitate movement). The synovial membrane itself is the layer of soft tissue that lines the surfaces within the joints.

Precursor cells have the ability to differentiate into one or two closely related forms: epithelial cells (skin cells) or fibroblastic (connective tissue) cells. Therefore, the tumor may have cancer resembling both cancers of the skin and of the connective tissue.

Synovial sarcomas are aggressive and highly locally invasive, spreading in greater than 40 percent of cases. They most often spread to the elbow, knee, and shoulder blade regions. They are more common in large-breed rather than small-breed dogs.

SYMPTOMS AND TYPES

- Limping
- Slowly progressive *lameness*
- Palpable mass
- Weight loss
- No appetite (anorexia)

Causes

- Unknown

DIAGNOSIS

You will need to give a thorough history of your dog's health and onset of symptoms. Your veterinarian will then perform a complete physical exam and order standard laboratory tests, including a biochemical profile, a complete blood count, a *urinalysis*and an electrolyte panel to rule out other non-cancerous causes for your dog's symptoms. Visual diagnostic techniques will show abnormalities around the affected joints. X-rays of the mass will show that the tumor is involved in both the bone and the joint. To obtain a definitive diagnosis, a *biopsy* of the soft and bony tissue of the tumor is necessary for histologic evaluation (microscopic analysis of the tissue sample). Using fine-needle aspirates (removal of fluid), the regional *lymph nodes* (i.e., the lymph nodes of the *groin*, armpits) should also be sampled and tested for evidence of*metastasis* (spread).

TREATMENT

The invasiveness of this type of *sarcoma* makes *amputation* (if possible) of the affected limb the treatment of choice. When appropriate, the patient should be treated with supportive chemotherapy. Pain medication will also be prescribed and administered as necessary.

LIVING AND MANAGEMENT

After surgery, you should expect your dog to feel sore. Your veterinarian will give you pain medication for your dog to help minimize discomfort, and you will need to set up a place in the house where your dog can rest comfortably and quietly, away from other pets, active children, and busy entryways. Trips outdoors for bladder and bowel relief should be kept short and easy for your dog to handle during the recovery period. Use pain medications with caution and follow all directions carefully; one of the most preventable accidents with pets is overdose of medication.

Your veterinarian will schedule follow-up appointments with you for your dog every two to three months for the first year after diagnosis of the synovial sarcoma. After the first year your dog may be seen by your veterinarian every six months for follow-up exams, and to affirm that the cancer has not recurred. X-rays should be taken at each visit to check for local recurrence and to confirm that the cancer has not spread to the lungs.

MUSCLE DISORDERS IN DOGS

Myopathies are diseases that primarily cause damage to muscles or muscle tissues. They may be present at birth or occur due to nutritional imbalances, injury, or ingestion of a poisonous substance.Myositides are diseases that produce a mainly inflammatory reaction in muscle. Common causes include infections, parasitic diseases, and immune—mediated conditions.

TYPE II MUSCLE FIBER DEFICIENCY

Type II muscle fiber deficiency is a muscle disease of unknown cause that is present at birth in some Labrador Retrievers. The condition appears to be inherited and has been known to occur in both yellow and black Labradors. Signs become noticeable at less than 5 months of age and include wasting away of muscle, stunted growth, and weakness. These signs progressively worsen until the animal reaches maturity (between 6 and 12 months), when they stabilize. Animals may have a normal life span. A definite diagnosis usually requires a urine test, muscle biopsy, and electromyography. There is no effective treatment.

FIBROTIC MYOPATHY

Fibrotic myopathy is an uncommon progressive disorder that leads to degeneration of the thigh muscles. The cause is not known, but German Shepherds appear to be predisposed to developing the condition. Affected muscles are characterized by a permanent, abnormal tightening. Normal tissues are replaced by thick connective tissue. Other signs include lameness without pain. Surgery is only sometimes helpful. The outlook for recovery is guarded because signs can recur.

MYOSITIS OSSIFICANS

Myositis ossificans is a muscle disorder in which noncancerous bony deposits appear in the muscles and connective tissue. The cause is unknown. The disorder frequently affects tissues near the hip joint in Doberman Pinschers. It may be related to a bleeding disorder (von Willebrand's disease) in these dogs. Surgical removal of the bony mass is usually helpful.

POLYMYOSITIS

Polymyositis is an inflammatory muscle disorder in adult dogs that affects the entire body. It may be associated with immune-mediated disorders such as lupus erythematosus or myasthenia gravis. The condition can have a sudden onset, or it may be recurring and progressive. Signs include depression, lack of energy, weakness, weight loss, lameness, muscle tenderness or pain, and wasting away of muscle. Corticosteroids are generally recommended for treatment; however, other drugs that suppress the immune system may also be used. The outlook for recovery is favorable, although signs sometimes reappear.

MASTICATORY MYOSITIS

Masticatory myositis is an inflammatory condition that affects the muscles used to chew. The exact cause is unknown, although the body's own immune response plays a role. In acute cases, muscles are swollen and the dog has difficulty opening the jaw.

In chronic cases, signs include persistent loss of appetite, weight loss, difficulty opening the jaw, and wasting away of muscle. Blood tests, electromyography, and a muscle biopsy may help to confirm the diagnosis. Although the condition sometimes improves on its own, treatment with corticosteroids given by mouth is recommended in most cases. Relapses are common, and longterm medication may be required.

MALIGNANT HYPERTHERMIA

Malignant hyperthermia is a disorder of skeletal muscle usually brought on by certain types of inhaled anesthesia and stress. It is characterized by an abnormal increase in metabolic rate. Although the condition is most common in pigs, it is also known to occur in some heavily muscled dogs (particularly Greyhounds).

Signs include rapid heartbeat, increased breathing rate, fever, muscle tightness and rigidity, and heart and lung failure. Signs develop between 5 and 30 minutes after exposure to the anesthetic agent. Treatment requires immediately stopping the anesthesia and administering oxygen. Fluid injections, corticosteroids, ice packs, and muscle relaxants are also used. The outlook is poor in severe cases.

EXERTIONAL MYOPATHY (RHABDOMYOLYSIS)

This muscle disorder of racing Greyhounds and working dogs is caused by overuse. It appears to be triggered by an inadequate supply of blood to the muscle after exercise or excitement. This can cause kidney disease and destruction of muscle cells. Signs include muscle pain and swelling that becomes noticeable 24 to 72 hours after racing or overuse. In severe cases, stiffness, deep or fast breathing, collapse, and kidney failure may occur. Urine tests are used to confirm the diagnosis. Treatment includes supportive care such as fluid injection, bicarbonate, body cooling, rest, and muscle relaxants. The outlook for recovery depends on the severity of the case.

MUSCULAR OR TENDON TRAUMA

Injury (trauma) of the muscles and the associated tendons can cause many forms of myopathy in dogs.

ABNORMAL CONTRACTURE OF THE SHOULDER MUSCLE

Abnormal tightening or contracture of the shoulder muscle (infraspinatus) is a muscle disease that affects one or both shoulders. It usually occurs after trauma in hunting or working dogs. Signs include noticeable lameness, pain, and swelling in the shoulder region. The lameness is temporary, but an abnormal gait develops 2 to 4 weeks after injury as the muscles thicken and tighten. Other signs include the pulling in of the elbow towards the body, pulling away of the foreleg from the body, and external rotation of the carpus and paw. The limb is moved in a circular motion with each stride of the leg. Treatment consists of the surgical removal of a portion of the muscle, including cutting of a tendon. Limb and joint functions usually improve immediately following surgery, and the outlook for full recovery is excellent.

INFLAMMATION OF THE BICEPS BRACHII TENDON AND ITS COVERING

Inflammation of the biceps brachii tendon and its covering (sheath) due to injury can occur in one or both forelimbs. It usually affects mature, large dogs. The injury may be direct, indirect, or due to overuse or migration of bone and cartilage fragments called joint mice.

Persistent lameness that may worsen over time is one sign of this condition. The lameness also worsens after exercise and improves with rest. The shoulder joint's range of motion is reduced, and the shoulder muscles may waste away. Applying pressure to the biceps tendon while bending and extending the shoulder joint causes severe pain.

X-rays and ultrasonography are often used to help confirm the diagnosis. Surgical inspection using an endoscope can help your veterinarian determine the extent of the injury.

In mild cases, the recommended treatments are rest and nonsteroidal anti—inflammatory drugs given by mouth. Severe cases can be treated with injections of an anti-inflammatory agent and rest. Chronic cases that are resistant to treatment of multiple corticosteroid injections or cases involving joint mice are treated through surgery. The outlook for recovery is good, although severe degeneration in longterm cases may cause lameness to persist after treatment.

QUADRICEPS CONTRACTURE (STIFF STIFLE DISEASE)

This serious thickening and tightening of the quadriceps muscles develops after improperly performed surgical operations on young dogs to repair fractures of the femur. The bone, the connective tissue surrounding the bone, and the quadriceps muscles bond together. This leads to extension and disuse of the limb, porous and easily broken bones (osteoporosis), osteoarthritis, and bone and joint deformations. The affected limb is overextended and often displaced. Surgery is usually required to remove the fibrous tissues that connect the bone and muscles and to restore motion in the stifle joint. Reconstruction of bone and soft tissues should be followed by application of bandages that restrict mobility of the joint. Physical therapy is usually required after surgery. The outlook for recovery is guarded. Careful surgical repair of bone fractures can help prevent this condition.

ACHILLES TENDON DISRUPTION (DROPPED HOCK)

This injury to the common Achilles tendon most often afflicts fully grown working and athletic dogs. It is usually the result of trauma. The tendon can be partially or completely ruptured or torn away from the bone. Signs include a severe lameness that keeps the animal from putting weight on the leg, overextension of the tarsus, and a stance in which the heel touches the ground. Swelling, pain, and torn or thickened tendon ends may also be seen. X-rays are useful for diagnosis and may reveal fragments torn from the bone. Surgical treatment can repair the torn ends and reattach the tendon to the bone. External splints are generally used after surgery to immobilize the joint for several weeks. The outlook for recovery varies, based on the duration of the injury, the success of the surgery, and the dog's expected performance when healthy.

MUSCLE TUMORS

Tumors that originate in the skeletal muscle can be benign or cancerous (malignant). Malignant tumors can spread and invade nearby muscle. They can also spread to other parts of the body.

Signs include localized swelling and lameness. The diagnosis is confirmed by taking a small tissue sample called a biopsy. The tumor generally must be surgically removed or the limb amputated.

BONE AND JOINT TUMORS IN CATS

These notes are provided to help you understand the diagnosis or possible diagnosis of cancer in your pet. For general information on cancer in pets ask for our handout "What is Cancer". Your veterinarian may suggest certain tests to help confirm or eliminated diagnosis, and to help assess treatment options and likely outcomes. Because individual situations and responses vary, and because cancers often behave unpredictably, science can only give us a guide. However, information and understanding about tumors and their treatment in animals is improving all the time.

We understand that this can be a very worrying time. If you have any questions, please do not hesitate to ask us.

WHAT ARE THESE TUMORS?

Non-cancerous bone tumors are rare in cats and are mainly due to abnormal development. They include bone cysts and single or multiple lumps of bone in abnormal places (exostoses). Fibrous dysplasia is another developmental condition that causes bone swelling around areas of bone destruction.

'Benign' (non-spreading, local) tumors are also rare and of several types. Ossifying fibroma (osteofibroma, fibrous osteoma) is tumor that develops in the jaw and destroys existing bone structure. Osteomas are benign tumors in immature animals.

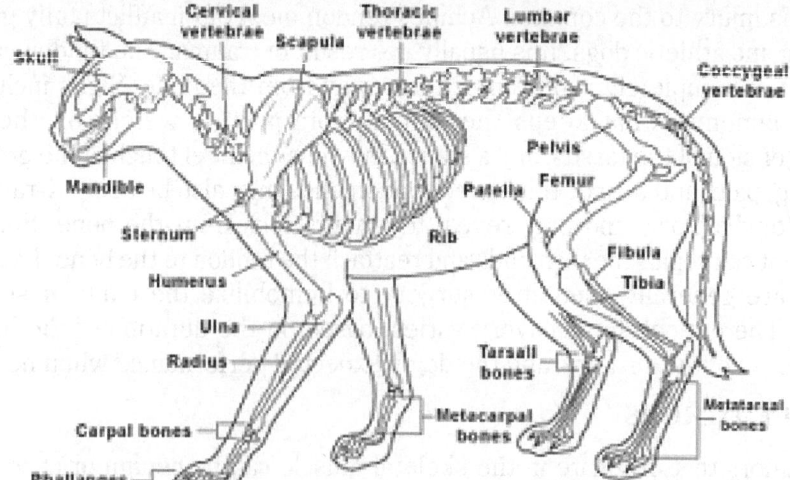

Cat Skeleton

'Feline osteochondromatosis' is a progressive tumor that occurs in skeletally mature animals and may develop on any bone. Viral particles (feline leukemia and sarcoma viruses: FeLV, FSV) are found in proliferating cells. Malignant change (transformation) of these cells and appearance of new lesions is common, imparting a poor prognosis in all cases.

Chondromas are benign but some progress to malignancy (chondrosarcoma) and infiltrate locally.

"In the cat, 90% of bone cancers are malignant (capable of spreading to other body sites)."

In the cat, 90% of bone cancers are malignant (capable of spreading to other body sites). Osteosarcoma is by far the most common. In cats, the tumors are most common in the non-limb (axial) skeleton. They often spread to other parts of the body (metastasize).

Rare malignant cancers include the locally aggressive 'giant cell tumor of bone'which is derived from bone marrow cells and the very rare 'multilobular tumor of bone'. Joint tumors (synovial sarcomas) are rare. Blood borne metastases of cancer from carcinomas of mammary, liver and lung origins are relatively common in bones.

WHAT DO WE KNOW ABOUT THE CAUSE?

The reason why a particular pet may develop this, or any cancer, is not straightforward. Cancer is often the culmination of a series of circumstances that come together for the unfortunate individual.

Non-cancerous bone tumors such as cysts and exostoses are due to abnormal development.

'Feline osteochondromatosis' contains viral particles (FeLV, FSV) in proliferating cells. Giant cell tumors in cats have also been associated with virus particles suggesting viral cause. There is also some evidence that some osteosarcomas in cats are virus associated. A history of previous fracture at the site of osteosarcoma is not unusual. It is speculated that the excessive proliferation of cells required to heal a fracture gives greater opportunity for neoplastic mutation to occur.

WHY HAS MY PET DEVELOPED THIS CANCER?

Some animals have a greater tendency (genetic susceptibility) to cancer.

ARE THESE COMMON TUMORS?

All the non-cancerous and benign tumors are rare.

Malignant bone cancer is not common but one survey found an incidence of 4.9 cases per 100,000 cats. Osteosarcoma accounts for 70% of feline bone tumors. The age range for cats with osteosarcoma is wide with rare cases less than a year, although the average age of cats with the tumors (10.5 years) is older than that of dogs. Sex and breed predisposition is uncertain. Chondrosarcomas are the second most common primary tumor of bone.

Multilobular tumors of bone are rare in cats. Sites include the skull and rib. Giant cell tumors are rare.

Synovial tumors usually affect the stifle and elbow. They are rare and difficult to diagnose with certainty unless specialized techniques are available.

HOW WILL THESE CANCERS AFFECT MY PET?

Non-cancerous and benign bone tumors are usually hard swellings. Clinical signs vary with the site. Cysts, fibrous dysplasia and ossifying fibroma cause bone destruction as well as localized swelling.

'Feline osteochondromatosis' is a syndrome of multiple growths and is progressive.

"Lameness is almost always the first sign of malignant limb bone cancer."

Lameness is almost always the first sign of malignant limb bone cancer. Pain and swelling follow and the limb may become increasingly painful and hot. As the swelling increases and the muscles atrophy, pain decreases. Bones affected by a malignant cancer may fracture. Tumors in the backbone cause pain. Those in the ribs and head often cause less pain as they are slower growing. In some sites such as the nose, there may be symptoms that are due to local blockage.

The growing cancer causes weight loss and increased blood calcium levels, which damage the kidneys.

HOW ARE THESE CANCERS DIAGNOSED?

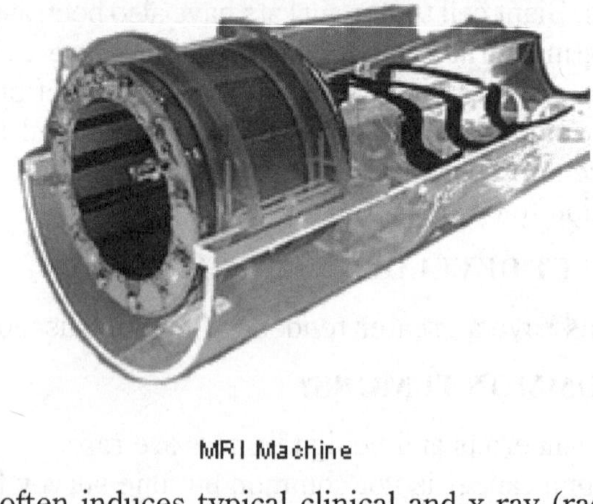

MRI Machine

Bone cancer often induces typical clinical and x-ray (radiography) signs but sometimes, the accurate diagnosis also requires microscopic examination of a bone sample (biopsy), using a diagnostic technique called histopathology. This is the microscopic examination of specially prepared and stained tissue sections and is done at a specialized laboratory where a veterinary pathologist examines the sample.

When preparing the histopathology report, the veterinary pathologist usually adds a prognosis (what will probably happen). This may include information on the likelihood of local recurrence or metastasis (distant spread).

Sensitive MRI (magnetic resonance imaging) or computed tomography (CT) can demonstrate the extent of tumors.

WHAT TYPES OF TREATMENT ARE AVAILABLE?

The most common treatment is surgical. For benign tumors, removal of the lump is sufficient but for malignant tumors, surgery may be extensive and include amputation of the affected limb. The over-riding consideration in treating cats with malignant cancer is that most will die from metastatic disease despite control of the primary tumor by amputation. There is no recommended drug therapy for cats with osteosarcoma but radiotherapy may reduce pain in some cases.

CAN THESE CANCERS DISAPPEAR WITHOUT TREATMENT?

Cancer is a multi-step process so it may stop at some stages. The body's own immune system can kill cancer cells but it is rarely 100% effective. Rarely, loss of blood supply to a cancer will make parts of it die but the dead tissue will still need surgical removal.

HOW CAN I NURSE MY PET?

After surgery, your pet may need to wear an "Elizabethan collar" so he or she cannot interfere with the operation site. The incision site will need to be kept clean and dry. Report any loss of stitches or significant swelling or bleeding to your veterinarian. If you require additional advice on post-surgical care, please ask. Pain control in bone cancer is vital, particularly for aggressive bone tumors.

HOW WILL I KNOW HOW THE CANCER WILL BEHAVE?

Your veterinarian will use the information from the biopsy and subsequent histopathology report to predict how this cancer is likely to behave. The veterinary pathologist usually adds a prognosis that describes the probability of local recurrence or metastasis (distant spread).

A blood test (for alkaline phosphatase enzymes) can predict the behavior of some limb osteosarcomas in dogs and possibly cats. High blood levels of bone-specific alkaline phosphatase before surgery are associated with significantly shorter survival and disease-free intervals following surgery.

WHEN WILL I KNOW IF THE CANCER IS PERMANENTLY CURED?

'Cured' has to be a guarded term in dealing with any cancer.

Non-cancerous and benign cancerous bone swellings do not spread and many stop growing as the cat reaches maturity. If they interfere with the cat's movement, they can be removed by surgical excision.

'Feline osteochondromatosis' is progressive and therefore has a poor prognosis. Chondromas may recur following surgical removal and some progress to malignancy and infiltrate locally. Chondrosarcomas of the limbs may be cured by amputation. Survival is longer for tumors in these sites rather than those occurring in the ribs.

In general, cats survive longer than dogs with osteosarcomas. One third of cats are alive five months after surgery for a limb tumor but survival for axial tumors is usually less than this, mainly because of the difficulty of surgical removal of these tumors. Feline tumors affecting long bones may metastasize but do so less readily than those affecting axial skeleton. The median survival time after amputation is four to five years.

Giant cell tumors are locally aggressive but rarely metastasize. Amputation may be the treatment of choice but there is limited information on success rate because they are very rare. In man, benign-looking tumors have metastasized after radiotherapy.

'Multilobular tumor of bone' is slow growing but locally aggressive. If complete removal is possible, the cancer may be cured but recurrence following surgery is the usual pattern of behavior.

Most joint tumors destroy local tissues and are difficult to remove surgically. Amputation appears to be the most effective form of therapy for joint tumors but only 25% of cases have been reported to survive more than a year following diagnosis.

ARE THERE ANY RISKS TO MY FAMILY OR OTHER PETS?

"Some of these tumors may be due to viral infection, but the tumors themselves are not transmitted from pet to pet or from pets to people."

Some of these tumors may be due to viral infection, but the tumors themselves are not transmitted from pet to pet or from pets to people.

The viruses implicated in some bone tumors may be transmitted between pets, occasionally from an infected queen to her kittens before birth but more commonly from close contact with infected cats, which shed the virus in saliva, urine and feces. If your cat is infected, he or she can pass the infection to other cats. The infection is not transmissible to people or other animals such as dogs.

6

Horse Disorders and Diseases

BLOOD DISORDERS OF HORSES

Blood cells form and develop mostly in the bone marrow, that is, the tissue located in the cavities of bones. Blood performs a variety of important functions as it circulates throughout the body. It delivers oxygen and vital nutrients (such as vitamins, minerals, fats, and sugars) to the body's tissues. It carries carbon dioxide to the lungs to be exhaled and waste products to the kidneys and liver to be eliminated from the body. It transports hormones, which are chemical messengers, to various parts of the body, allowing those parts to communicate with each other. Blood also includes cells that fight infection and platelets that control bleeding. There are 3 cellular elements of blood: red blood cells, white blood cells, and platelets. Basically, red blood cells supply the body with oxygen; white blood cells protect against infection; and, platelets start the formation of blood clots.

BLOOD IS A COMPLEX MIX OF PLASMA (THE LIQUID COMPONENT), RED BLOOD CELLS, WHITE BLOOD CELLS, AND PLATELETS

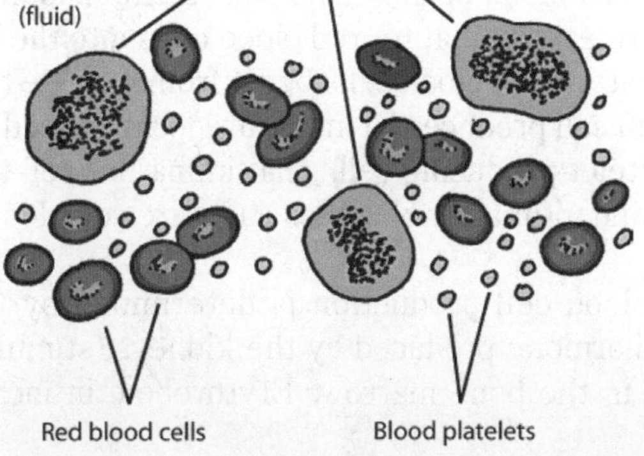

Blood disorders are quite diverse. They can occur as normal responses to abnormal situations; for example, a significant increase in the number of white blood cells in response to an infection or disease. They may also occur as primary abnormalities of the blood; for example, a deficiency of all cellular elements of the blood due to bone marrow failure. Furthermore, abnormalities may be quantitative (too many or too few cells) or qualitative (abnormalities in the way cells function). Like people, horses have different blood groups or types. Researchers have identified 8 blood types in horses: A, C, D, K, P, Q, U, and T. Horse owners and breeders will often have a record of their horse's blood type. This should be included in the medical records for your horse.

RED BLOOD CELLS OF HORSES

The main function of red blood cells (also called erythrocytes) is to carry oxygen to the tissues, where it is required for cellular metabolism. Oxygen molecules attach themselves to carrier molecules, called hemoglobin, which are the iron-containing proteins in red blood cells that give the cells their red color. Oxygen is carried from the lungs and delivered to all body tissues by the hemoglobin within red blood cells. Oxygen is used by cells to produce energy that the body needs. Carbon dioxide is left behind as a waste product during this process. The red blood cells then carry that carbon dioxide away from the tissues and back to the lungs, where it is exhaled. When the number of red blood cells is too low, this is called anemia. Having too few red blood cells means the blood carries less oxygen. The result is that fatigue and weakness develop. When the number of red blood cells is too high, this is called polycythemia. The result is that blood can become too thick, and impair the ability of the heart to deliver oxygen throughout the body. An animal's metabolism is geared to protect both the red blood cells and the hemoglobin from damage. Interference with the formation or release of hemoglobin, or with production or survival of red blood cells, causes disease.

The total number of red cells remains constant over time in healthy animals. Mature red blood cells have a limited life span; their production and destruction must be carefully balanced, or disease develops.

Production of red blood cells begins with stem cells in the bone marrow and ends with the release of mature red blood cells into the body's circulation. Within the bone marrow, all blood cells begin from a single type of cell called a stem cell. The stem cell produces an immature form of a red blood cell, a white blood cell, or a platelet-producing cell. That immature cell then divides again, develops further, and ultimately becomes a mature red blood cell, white blood cell, or platelet.

The rate of blood cell production is determined by the body's needs. Erythropoietin, a hormone produced by the kidneys, stimulates development of red blood cells in the bone marrow. Erythropoietin increases if the body lacks oxygen (a condition called hypoxia). In most species, the kidney is both the sensor organ that determines how much oxygen the body's tissues are receiving and the major site of erythropoietin production; so chronic kidney failure leads to anemia. Erythropoietin plays a major role in determining whether to increase the number of stem cells entering red blood cell production, to shorten maturation time of the red blood cells, or to cause early release of red blood cells. Other factors that affect red blood cell production are the supply of nutrients (such as iron and vitamins) and cell-cell interactions between compounds that aid in their production.

Anemia may be caused by blood loss, the destruction of red blood cells (hemolysis), or decreased production. In severe blood loss anemia, red blood cells are lost, but death usually results from the loss of total blood volume, rather than from the lack of oxygen caused by loss of red blood cells. Red blood cells may be destroyed by toxins such as red maple leaves, infections such as equine infectious anemia, and drugs such as phenothiazine tranquilizers. Hemolytic disease of newborn foals is caused by maternal antibodies against foals' red blood cells.

Decreased red blood cell production may be caused by primary bone marrow diseases (such as aplastic anemia or hematopoietic malignancy), kidney failure, or drugs such as phenylbutazone. Longterm debilitating diseases may also be associated with mild anemia. It is important to remember that anemia is a clinical finding, not a specific diagnosis. The prognosis and treatment depend on the underlying cause.

WHITE BLOOD CELLS OF HORSES

The function of white blood cells (also called leukocytes) is to defend the body against infection. There are 2 main types of white blood cells: phagocytes and lymphocytes.

Phagocytes

Phagocytes are cells in the bloodstream and tissues that surround and consume foreign particles, cell waste material, and bacteria. Their main function is to defend against invading microorganisms by surrounding and destroying them.

There are 2 types of phagocytes: granulocytes and monocytes. Granulocytes, primarily neutrophils, protect against bacteria and fungi. Others known as eosinophils and basophils are involved in allergic reactions. Monocytes become macrophages in the tissues and consume large foreign particles and cellular debris.

As with red blood cells, the production and number of phagocytes are tightly regulated by chemical messengers in the blood, including interleukins (chemicals found in white blood cells that stimulate them to fight infection).

Unlike the red blood cells, which remain circulating in the blood, the phagocytes use the blood's circulatory system as a pathway to the tissues. Because of this, the number of phagocytes in the blood can provide an indication of disorders in the body. For example, the number of neutrophils increases when inflammation is present anywhere in the body. The number of phagocytes produced as a result of these conditions varies from species to species. An abnormal response, such as an unusually low number of circulating white blood cells due to marrow failure, drugs, or toxins, can lower resistance to bacterial infections. Finally, those elements that produce phagocytes may become cancerous, resulting in a disease called myelogenous leukemia.

Lymphocytes

Lymphocytes are white blood cells that produce antibodies to attack infectious organisms. They are also responsible for rejecting foreign tissue or cancer cells. Lymphocyte production in mammals begins in the bone marrow. Lymphocytes then become 1 of 2 types: T cells or B cells. T cells are responsible for a variety of functions, especially fighting off viral infections and cancers. Most T cells remain in the bloodstream, but some are also present in the spleen and lymph nodes. B cells are responsible for producing antibodies that coat invading organisms or foreign substances. For example, bacteria coated by antibody are more easily recognized and removed by phagocytes. If lymphocytes are decreased or abnormal, the patient is immunodeficient and is at risk for a wide variety of infections.

Antibody molecules are called immunoglobulins. They fall into several classes, each of which has a different function. For example, one class is commonly found in the lungs and intestines; another is the first antibody produced in response to newly recognized foreign microorganisms; a third is the main antibody in the bloodstream; and a fourth is involved in allergic reactions.

Lymphocytes usually act appropriately to rid the body of foreign "invaders" that cause disease. However, sometimes lymphocytes do not react appropriately. One inappropriate response occurs when antibodies are produced against the body's own cells. Another inappropriate response of the immune system is allergy. When exposure to the allergen occurs, the reaction can be mild or life-threatening.

An increase in the number of lymphocytes in the bloodstream occurs in some animals as a response to the secretion of epinephrine (a hormone also known as adrenaline). A reduction in the number of circulating lymphocytes may be caused by corticosteroid hormones that are secreted in times of stress. Unusual lymphocytes may be seen in the blood in response to antigenic stimulation, such as vaccination. Malignant tumors starting in a lymph node (lymphoma) or lymphoid leukemia can occur.

HEART AND BLOOD VESSEL DISORDERS OF HORSES

The cardiovascular system includes the heart and the blood vessels—the veins and the arteries. The function of the heart is to pump blood. The right side of the heart pumps blood to the lungs, where oxygen is added to the blood and carbon dioxide is removed from it. The left side pumps blood to the rest of the body, where oxygen and nutrients are delivered to tissues, and waste products (such as carbon dioxide) are removed. In horses, the cardiovascular system must not only efficiently supply blood to all parts of a large animal, but must also function well during strenuous racing or training, in many cases.

THE CARDIOVASCULAR SYSTEM IN HORSES

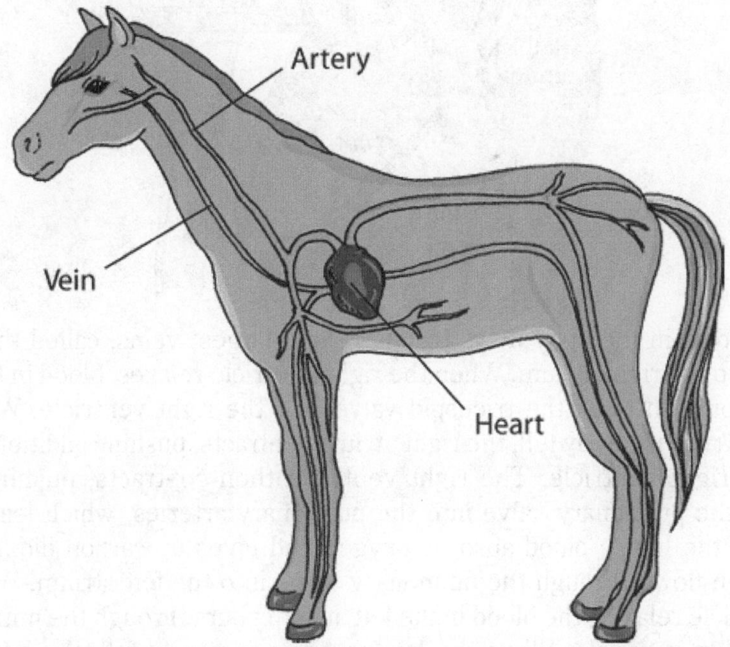

The heart is a hollow, muscular organ which, in mammals, is divided into 4 chambers. The muscular tissue is called the myocardium. There are upper chambers on both the left and right sides of the heart called the left and right atria (the plural form of atrium). There are also 2 lower chambers called the left and right ventricles.

A series of valves keep blood flowing in one direction through the heart. The atrioventricular valvesare valves between the atria and the ventricles. The semi-lunar valves are valves between the heart and the aorta and between the heart and the pulmonary artery. Each ventricle has an inlet and an outlet valve. In the left ventricle, the inlet valve is called the mitral valve, and the outlet valve is called theaortic valve. In the right ventricle, the inlet valve is called the tricuspid valve, and the outlet valve is called the pulmonary valve.

A LOOK INSIDE A HORSE'S HEART

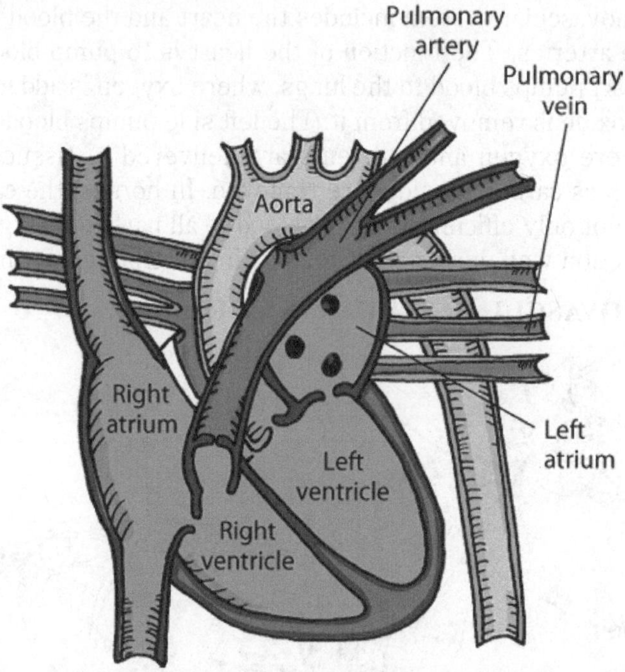

Blood from the body flows through the 2 largest veins, called the venae cavae, into the right atrium. When the right ventricle relaxes, blood in the right atrium pours through the tricuspid valve into the right ventricle. When the right ventricle is nearly full, the right atrium contracts, pushing additional blood into the right ventricle. The right ventricle then contracts, pushing blood through the pulmonary valve into the pulmonary arteries, which lead to the lungs. In the lungs, blood absorbs oxygen and gives up carbon dioxide. The blood then flows through the pulmonary veins into the left atrium. When the left ventricle relaxes, the blood in the left atrium pours through the mitral valve into the left ventricle. When the left ventricle is nearly full, the left atrium contracts, pushing additional blood into the left ventricle. The left ventricle then contracts, pushing blood through the aortic valve into the aorta, the largest artery in the body. This blood carries oxygen to all of the body except to the lungs.

Each heartbeat consists of 2 parts: diastole and systole. The first half of a heartbeat is the sound of the mitral and tricuspid valves closing. The second half is the sound of the aortic and pulmonary valves closing. During diastole, the ventricles relax and fill with blood. During systole they contract and pump blood out to the body.

The rate and force of contraction of the heart and the degree of narrowing or widening of blood vessels are controlled by different hormones and the autonomic nervous system, the part of the nervous system that controls involuntary activity.

Heart Rate

The heart beats because of a tiny electrical current that originates in the heart's pacemaker, called thesinoatrial node. Rhythmic electrical impulses or discharges cause the contraction of muscle fibers in the heart. While a horse is at rest, its sinoatrial node discharges about 15 times per minute.

Heart Sounds and Murmurs

Heart sounds are produced by the rapid acceleration and deceleration of blood and the resulting vibrations in the heart due to the circulation of blood. They can be heard using a stethoscope. In horses, 4 heart sounds can possibly be heard. An absence or abnormality of one of these sounds may indicate a heart abnormality.

Heart murmurs are vibrations that can be heard coming from the heart or major blood vessels and generally are the result of turbulent blood flow or vibrations of heart structures such as part of a valve. Murmurs are typically described by their timing (that is, whether they occur continuously or only intermittently), intensity (that is, whether they can be heard easily or with difficulty), and location. In horses, early systolic and diastolic murmurs can be noted in the absence of heart disease or anemia (having too few red blood cells). A short, high-pitched, squeaking, early diastolic cardiac murmur is sometimes seen in healthy young horses.

Arrhythmias

Arrhythmias are abnormalities of the rate, regularity, or site of heartbeat formation. An arrhythmia does not necessarily indicate heart disease. Many arrhythmias have no functional significance and require no specific treatment. Some arrhythmias, however, may cause severe signs such as loss of consciousness due to lack of blood flow to the brain or lead to sudden death. Many disorders are associated with abnormal heart rhythms.

Atrial fibrillation is a type of arrhythmia that commonly occurs in horses. In atrial fibrillation, the electric current running through the atria is not coordinated, stimulation of the atrioventricular node is frequent but random, and the heart rate is rapid and irregular.

Pulse

A pulse is the rhythmic expansion of an artery that can be felt with the fingertips during physical examination. A jugular pulse in the neck can be noted in normal animals. A pulse may be absent, increased (strong), or decreased (weak)—each of which may indicate a specific type of heart disease or defect.

HEART DISEASE AND HEART FAILURE IN HORSES

Unlike diseases of many other organ systems, cardiovascular diseases generally do not go away but almost always become more serious. In addition, cardiovascular diseases may be more difficult to detect and quantify because the heart cannot be seen and is protected so well by the rib cage. In horses, cardiovascular disease is less common than in people, dogs, or cats. However, because horses are often expected to perform work or athletic feats, the cardiovascular condition of the animal may be relatively more important.

Heart disease can be defined as any abnormality of the heart. It encompasses a wide range of conditions, including congenital abnormalities and disorders of physical structure and function. It can be classified by various methods, including whether the disease was present at birth or not (that is, congenital or acquired), causes (for example, infectious or degenerative), duration (for example, long- or short-term), clinical status (for example, left heart failure, right heart failure, or biventricular failure), or by physical structure malformation (for example, ventricular septal defect).

Heart failure is any heart abnormality that results in failure of the heart to pump enough blood to meet the body's needs. It is not a specific disease; rather, it is a condition in which congestion or an abnormal accumulation of fluid, decreased blood flow to the body, and/or abnormally low blood pressure arise as the final consequence of severe heart disease. Heart disease can be present without ever leading to heart failure. Heart failure, however, can only occur if heart disease is present because it is a consequence of heart disease.

ABNORMALITIES OF THE CARDIOVASCULAR SYSTEM

The following abnormalities of the cardiovascular system can lead to heart disease: 1) the heart valves fail to close or open properly (valvular disease); 2) the heart muscle pumps too weakly or relaxes inadequately (myocardial disease); 3) the heart beats too slowly, too rapidly, or too irregularly (arrhythmia); 4) the blood vessels offer too great an interference to blood flow (vascular disease); 5) there may be openings between chambers of the left side and right side of the heart (cardiac shunts); 6) there is too little or too much blood compared with the ability of the blood vessels to store that blood; and 7) there is parasitism of the cardiovascular system.

Signs associated with any of these diseases are due either to inadequate blood flow through the organs (signs include exercise intolerance, weakness, and fainting) or to blood damming up in organs, which causes fluid to leak from blood vessels into tissues (signs include abnormal accumulation of fluid in the lungs or abdomen). A horse showing signs of having too little blood in the tissues to sustain normal function is said to be in heart failure. An animal showing signs caused by blood damming up in poorly drained organs is said to be in congestive heart failure. When there is not enough oxygen in the blood, the mucous membranes develop a blue tinge, and often there is an increased concentration of red blood cells.

Horses with heart disease may deteriorate gradually, most often due to pulmonary failure, or they may die suddenly, due to a nearly instant stoppage of blood circulation. Pulmonary failure occurs because the lungs become too stiff to bring in adequate oxygen.

DIAGNOSIS OF CARDIOVASCULAR DISEASE IN HORSES

A veterinarian often diagnoses cardiovascular disease by reviewing the medical history and signs, conducting a physical examination, and interpreting the results of specific tests or imaging procedures. The physical examination includes using a stethoscope to listen to the sounds made by the horse's internal organs, especially the heart, lungs, and abdominal organs, xrays, electrocardiography (recording electrical activity of the heart), and echocardiography (a type of ultrasonography).

Most cardiovascular diseases can be diagnosed by physical examination and xrays. Electrocardiography is a specific test for diagnosis of arrhythmias. Echocardiography is excellent for confirming tentative diagnoses, for detecting heart tumors, or for detecting pericardial disease. Occasionally, more specialized tests such as cardiac catheterization (-using a thin flexible tube inserted and threaded through an artery into the heart) or nuclear studies (x-ray tests that include injection of radioactive isotopes) are necessary.

General Signs of Cardiovascular Disease

Horses with heart disorders or defects may have a general loss of condition, become fatigued easily (particularly after exercise), have difficulty breathing or shortness of breath, and show signs of weakness (including fainting or collapse). In addition, excess fluid that has accumulated in the chest or abdominal area may indicate heart failure. Signs may show up only after exercise, but over time they may occur even when the horse is at rest.

THE NERVOUS SYSTEM OF HORSES

The nervous system is made up of the brain, spinal cord, and several different kinds of nerves that are found throughout the body. These create complex circuits through which animals experience and respond to sensations.

A familiar type of nervous system circuit is a reflex. Reflexes are simple networks found in the nervous system of all animals. For example, when the eyelid is touched, it closes; when the toe is pinched, the foot pulls away "automatically."

Many different types of diseases can affect the nervous system, including birth defects, infections, inflammatory conditions, poisoning, metabolic disorders, nutritional disorders, injuries, degenerative diseases, or cancer.

Neurologic diseases are often more common in a particular breed or sex, or tend to occur at a certain age.

PARTS OF THE NERVOUS SYSTEM IN HORSES

The central nervous system includes the spinal cord and the brain. The brain is divided into 3 main sections—the brain stem, which controls many basic life functions; the cerebrum, which is the center of conscious decision-making; and the cerebellum, which is involved in movement and motor control. The spinal cord of horses is divided into regions that correspond to the vertebral bodies (the bones that make up the spine) in the following order from neck to tail: cervical, thoracic, lumbar, sacral, and caudal segments. Specialized tissue called the meninges cover the brain and spinal cord, and cerebrospinal fluid surrounds and protects the brain and spinal cord.

The peripheral nervous system consists of the nerves that are found throughout the rest of the body.

THE NERVOUS SYSTEM OF THE HORSE

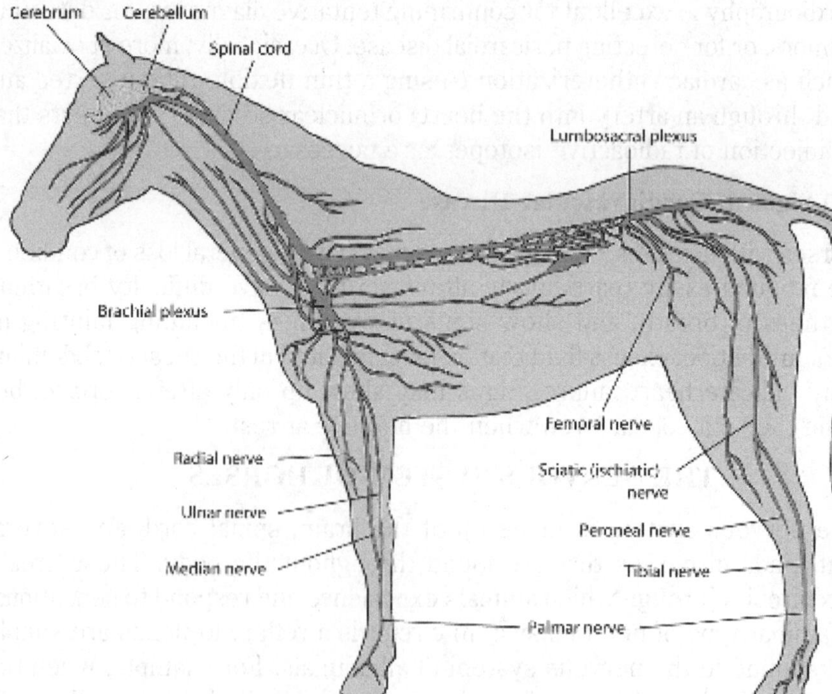

Neurons

Both the central and peripheral nervous systems contain billions of cells known as neurons. Neurons connect with each other to form neurological circuits. Information travels along these circuits via electrical signals.

All neurons have a center portion called a cell body and 2 extensions called dendrites and axons. Dendrites receive signals from other neurons and transmit electrical charges *to* the cell body. Axons transmit the electrical charges *away* from the cell body. When the current reaches the end of the axon, the axon releases chemicals called neurotransmitters. Neurotransmitters pass the signal to the dendrites of other neurons, or to muscles or glands.

Neurons in the peripheral nervous system combine to form pairs of spinal nerves and pairs of cranial nerves. The spinal nerves arise from the spinal cord and extend axons outward into the front and hind legs and to the chest, abdomen, and tail. These nerves subdivide into smaller nerves that cover the entire surface and interior of the body. The cranial nerves include sensory and motor neurons that connect the head and face to the brain.

Types of Neurons

Sensory neurons carry information from the body to the spinal cord or brain stem, and then on to the cerebellum and cerebrum for interpretation. Sensory information includes sensations of pain, position, touch, temperature, taste, hearing, balance, vision, and smell.

Motor neurons carry responses to the sensory information from the spinal cord and brain to the rest of the body. Inside the spinal cord, the axons of motor neurons form bundles known as tracts, which transmit this information to motor peripheral nerves going to muscles in the limbs. Motor neurons are important for voluntary movements and muscle control.

A specialized set of neurons controls and regulates basic, unconscious bodily functions that support life, such as the pumping of the heart and digestion. These neurons make up what is called the autonomic nervous system, which sends axons from the brain stem and spinal cord to various areas of the body such as the heart muscle, the digestive system, and the pupils of the eyes.

NERVOUS SYSTEM DISORDERS AND EFFECTS OF INJURIES IN HORSES

A change in an animal's ability to sense its environment can be caused by disease in either the central nervous system or the peripheral nervous system. The primary signs of nervous system disorders include behavioral changes, seizures, tremors, pain, numbness, lack of coordination, and weakness or paralysis of one or more legs. The effects of an injury on sensory and motor functions depend on its location and severity.

A spinal cord injury can cause loss of feeling and paralysis below the level of the injury. Mild spinal cord injuries can result in clumsy movement and mild weakness of the limbs. Moderate spinal cord injuries can cause a greater weakness of the limbs. In severe spinal cord injuries, a complete loss of movement (paralysis) and feeling can occur. However, not all spinal cord injuries cause paralysis. For example, injury to the spinal cord in the lower back can result not in limb paralysis but in loss of bladder control.

Brain injuries result in different effects, again depending on which part of the brain is affected. Injuries to the brain stem can cause a loss of balance, weakness of the limbs, hyperactive reflexes, stupor, or coma. Injuries to the cerebellum can result in a lack of coordination of the head and legs, tremors, and a loss of balance. Injuries to the cerebrum can cause complete or partial blindness, loss of the sense of smell, seizures, coma, stupor, pacing or circling behavior, and inability to recognize an owner.

Some injuries to the nervous system can cause damage that is not evident until 24 to 48 hours after the injury occurs. Longterm damage is usually caused by swelling or internal bleeding of the vessels in the brain. Strokes caused by clogged arteries or high blood pressure are rare in animals.

Mechanisms of Disease

In addition to the effects of injuries, nervous system disorders can include birth defects, infections and inflammations, poisoning, metabolic disorders, nutritional deficiencies, degenerative diseases, or cancer.

Most birth defects, often called congenital disorders, are obvious at birth or shortly after. Some genetic diseases cause the neurons to degenerate slowly and irreversibly in the first year of life. In other inherited diseases, such as epilepsy, the animal may not show any signs for 2 to 3 years.

Infections of the nervous system are caused by specific viruses or microorganisms. Other inflammations such as certain types of meningitis can be caused by the body's own overactive immune system. These are known as autoimmune disorders. Various chemicals can cause a toxic reaction in the nervous system. These include certain pesticides and herbicides, rat poisons, antifreeze, and sedatives. Botulism, tetanus, and tick bites, as well as coral and tiger snake venom, can also affect the nervous system and cause paralysis.

Some metabolic disorders affect the function of the nervous system, including low blood sugar, shortness or loss of breath, liver disease, and kidney failure. Thyroid gland abnormalities can also cause neurologic signs.

REPRODUCTIVE DISORDERS OF HORSES

The reproductive system is the group of organs that produce offspring. In both males and females, the reproductive system is composed of primary sex organs and primary regulatory centers. The primary sex organs are the testes in males and the ovaries and uterus in females. The primary regulatory centers are in the brain.

They control the production of hormones that in turn influence the function of the primary sex organs.

THE GONADS AND GENITAL TRACT OF HORSES

Both sexes have a pair of sexual organs or gonads (ovaries or testes), the main function of which is to produce eggs or sperm, respectively.

The Ovaries

Ovaries are female gonads that produce eggs and female sex hormones, including estrogen and progesterone. These hormones are necessary for the development of eggs and preparation of the uterus for pregnancy. Once puberty is reached and the horse starts its heat or estrous cycle, the size and form of the ovaries change. Within the ovary, a group of special cells form a follicle, which surrounds each egg. The estrous cycle begins when follicle stimulating hormone causes follicles to grow, leading to maturation of eggs and production of the hormone estrogen. Estrogen causes the brain to release luteinizing hormone, which stimulates the release of eggs from the ovary (a process called ovulation). After ovulation, the follicle becomes a corpus luteum, which produces progesterone. Progesterone causes the mare to go out of heat and prepares the uterus for pregnancy. If the mare does not become pregnant, the corpus luteum is destroyed and the mare returns to heat.

The reproductive system of the mare

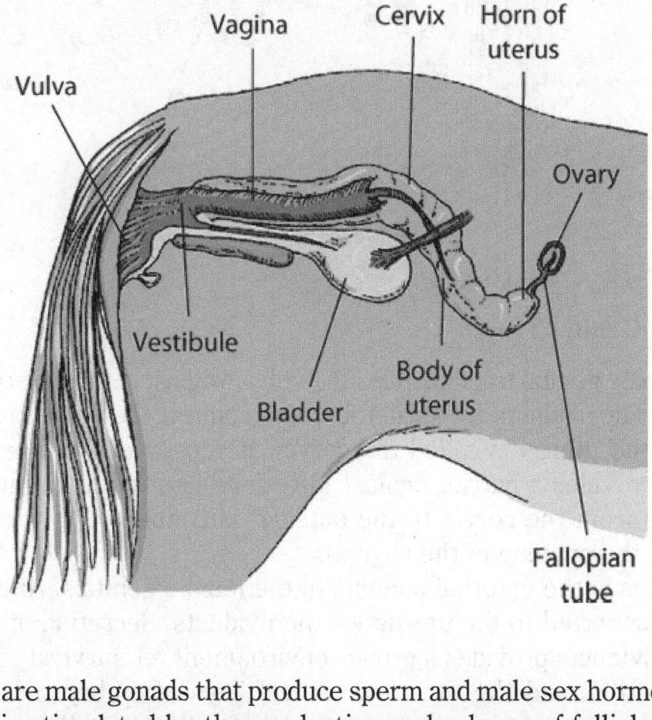

The Testes

Testes are male gonads that produce sperm and male sex hormones. Sperm maturation is stimulated by the production and release of follicle stimulating hormone and testosterone. Testosterone is required for proper function of the accessory sexual glands, male sex characteristics, and sexual behavior. For sperm production, the testes must descend into the scrotum (a pouch of skin outside the abdomen), because regular body temperature is too high for sperm to develop normally. The function of the testicles can be assessed by an evaluation of semen samples and hormonal tests. In addition, testicle examination and measurement may reveal reproductive diseases.

The reproductive system of the stallion

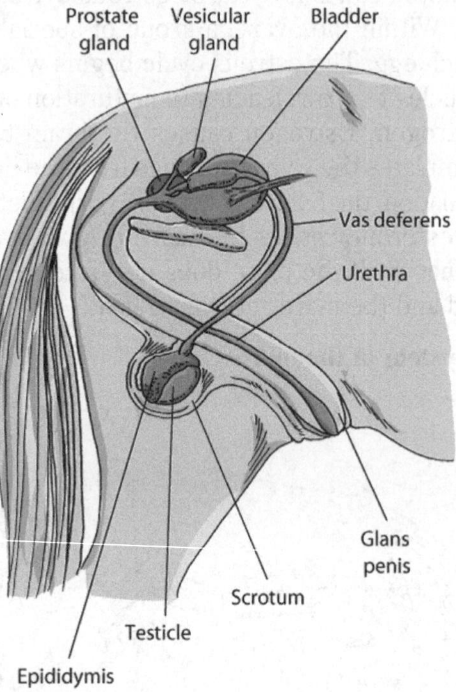

The Female Genital Tract

The female genital tract includes the vulva, vagina, cervix, uterus, oviducts, and ovaries. The oviducts are small tubes that connect the ovaries to the uterus. The end of the uterus is called the cervix. It separates the uterus from the vagina and provides a barrier against infection. The vagina (a muscular tube that extends from the cervix to the outside) is connected to the vulva (skin surrounding the opening of the vagina).

The vulva is the external opening of the female genitals. After ovulation, eggs are transported to the uterus via the oviducts. Secretion of protein-rich fluid in the oviducts provides a proper environment for survival of the mature egg, fertilization, and the first few critical days of embryonic life. Proper functioning of the uterus and cervix are also required for the establishment and maintenance of pregnancy. Infections contracted at mating are common causes of female infertility because they interfere with the proper function of the uterus. Infertility in females can be diagnosed through various means, including physical examination, ultrasonography, hormonal tests, uterine culture, and biopsy of the uterine lining. The vagina and vulva serve as the copulatory organs and as the last part of the birth canal. Mares that have poor function of the vulva and vagina may be prone to recurrent contamination of the reproductive tract, leading to the development of chronic uterine infections and infertility.

The Male Genital Tract

In males, the genital tract provides a pathway for sperm cells and semen. The testicles produce sperm and testosterone (the main male hormone). Sperm pass from the testicles into the epididymis, where they mature and are stored. The epididymis is connected to the ductus deferens, which carries sperm to the urethra. The accessory sex glands, such as the prostate and seminal vesicles, create the fluid portion of semen. Abnormalities of the genital tract can be diagnosed by physical examination or ultrasonography. A semen evaluation can provide information regarding the quality and quantity of sperm and can give some estimate of fertility.

MANAGEMENT OF REPRODUCTION IN HORSES

Reproductive Cycle and Breeding Management

Most mares only have estrous cycles during the seasons of the year when day length is long. During winter, the ovaries are inactive, and the reproductive hormones are at baseline levels in the bloodstream. As the days get longer in the spring, mares go through a transitional stage as the reproductive tract starts to prepare for the breeding season. During this transition, mares will have 3 to 4 episodes of sexual receptivity due to the development of waves of large follicles on the ovaries. Mares are not fertile at this time of year because the ovaries do not release eggs. After 8 to 10 weeks, the level of luteinizing hormone will be high enough to cause one of the follicles to ovulate. Once this occurs, the mare will then establish a normal estrous cycle with regular ovulation. Mares ovulate every 21 days throughout the breeding season, with periods of estrus (receptivity to breeding with a stallion) lasting 2 to 8 days. During estrus, follicles on the ovary enlarge. One of these will become dominant and release a mature egg. The follicle becomes a corpus luteum and produces the hormone progesterone, which prepares the uterus for implantation of the egg. If the mare breeds and the egg is fertilized, the egg passes into the uterus and begins a pregnancy. Otherwise, the corpus luteum is destroyed and another estrous cycle begins.

After the period of winter inactivity and transition, mares naturally begin estrous cycles in April in the northern hemisphere. Because the changes in cycling are stimulated by increasing amounts of daylight, it is possible to hasten the onset of cycles by exposing the mare to increased amounts of artificial light during the winter. The timing of ovarian cycles can also be manipulated by giving injections of various reproductive hormones. This is typically done to facilitate breeding appointments and to allow mares and stallions to remain in competition during much of the breeding season. Your veterinarian can advise you on how this is done.

Detecting Estrus

A successful breeding program revolves around good estrus detection. The mare should be presented to a stallion (teaser) daily or every other day during the breeding season, and an accurate interpretation and record of her response should be made. A mare in estrus (the receptive phase of the cycle) raises her tail, squats, urinates, and presents her rear to the stallion. She will also allow the stallion to mount and copulate. A mare in diestrus (the nonreceptive phase of the cycle) usually squeals, kicks, bites, and rejects the stallion's attempts to sniff the mare or mount. Adequate exposure to the teaser stallion is necessary to determine the mare's receptivity. Prolonged and irregular periods of estrus are common at the beginning and end of the breeding season.

A mare in estrus may not appear receptive at first due to nervousness or inexperience. Some mares with a foal at their side will appear less receptive because they are protective of the foal. The mare's behavior should be consistent with the condition of her reproductive tract as determined by a physical examination. The response to teasing can determine if estrus has begun and indicate when a mare should be bred. Failure to return to estrus 2 to 3 weeks after breeding may suggest that the mare is pregnant. Pregnancy can be confirmed by several methods. Ultrasonography of the uterus (through the rectum) allows the earliest detection. Examination by rectal palpation (feeling a bulge in the uterus) can detect pregnancy from as early as day 28 in some mares. Hormonal tests are most accurate after about day 60 of pregnancy.

DIGESTIVE DISORDERS OF HORSES

The digestive system includes all of the organs that are involved in taking in and processing food. It begins with the mouth and includes the esophagus, stomach, liver, pancreas, intestines, rectum, and anus.

MAJOR DIGESTIVE ORGANS OF THE HORSE

The process of digestion begins when your pet picks up food with its mouth and starts chewing. Enzymes found in saliva begin breaking down the food chemically. The process continues with swallowing, additional breakdown of food in the stomach, absorption of nutrients in the intestines, and elimination of waste. Digestion is critical not only for providing nutrients but also for maintaining the proper balance of fluid and electrolytes (salts) in the body.

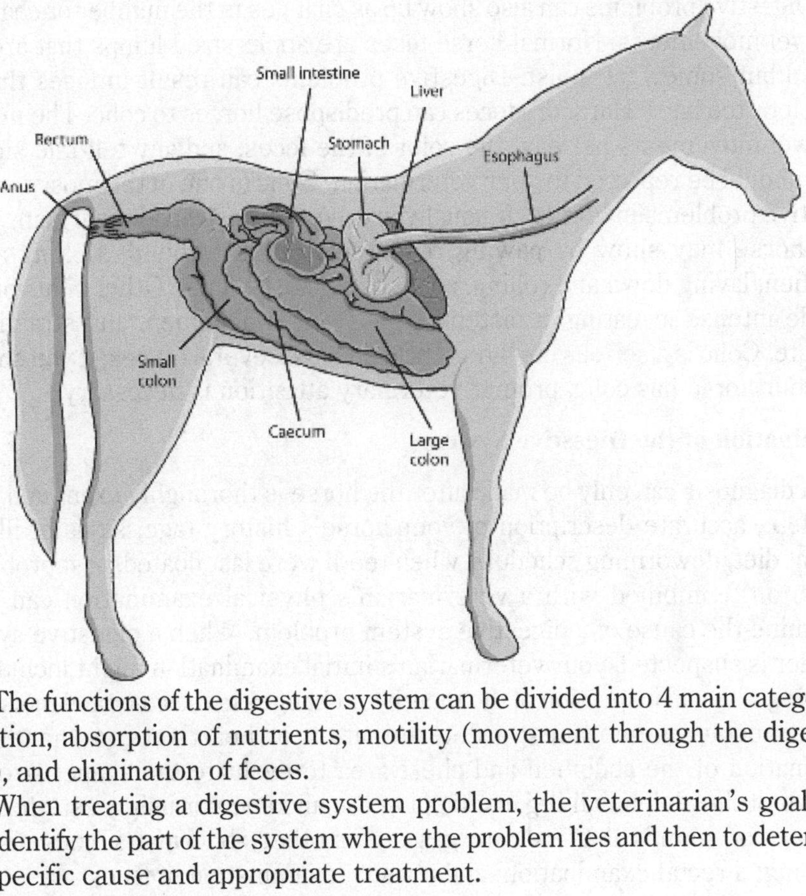

The functions of the digestive system can be divided into 4 main categories: digestion, absorption of nutrients, motility (movement through the digestive tract), and elimination of feces.

When treating a digestive system problem, the veterinarian's goal is to first identify the part of the system where the problem lies and then to determine the specific cause and appropriate treatment.

General Signs of Digestive System Disorders

Signs of digestive system disease can include excessive drooling, diarrhea, constipation, loss of appetite, bleeding, abdominal pain and bloating, shock, and dehydration. Horses cannot vomit or regurgitate.

Diarrhea is often a sign of digestive system disorders, but it can have many causes. Large-volume, watery diarrhea usually is associated with hypersecretion, a condition in which excess fluid is secreted into the intestines. This can be caused by bacterial infection.

Diarrhea can also be caused by malabsorption, the failure to properly absorb nutrients. Malabsorption is due to a defect in the intestinal cells responsible for absorption. This condition can be caused by viral infection, by a defect that limits the ability of the intestines to absorb liquids, or by defects in the pancreatic secretions needed for effective digestion. Dehydration and electrolyte (salt) imbalance, which may lead to shock, are seen when large quantities of fluid are lost (for example, from diarrhea).

Digestive problems can also show up as changes in the number or character of bowel movements. Normal horse feces are apple-sized lumps that are well formed but somewhat moist. Digestive problems can result in feces that are too soft or too hard. Hard, dry feces can predispose horses to colic. The number of bowel movements per day, the color of the feces, and any tell-tale signs of blood should be reported to your veterinarian. Colic is one of the most common digestive problems in horses. It usually involves intense abdominal pain, which your horse may show by pawing repeatedly at the ground, kicking at the abdomen, laying down and rolling, and looking at the flank. Other signs of colic include intense sweating, a distended or swollen abdomen, and straining to defecate. Colic is a serious condition that can have several causes; if you suspect that your horse has colic, prompt veterinary attention is necessary.

Examination of the Digestive System

A diagnosis can only be made after the horse is thoroughly examined. Your complete, accurate description of your horse's history (age, signs of illness, current diet, deworming schedule, when teeth were last floated, past problems, and so on) combined with a veterinarian's physical examination can often determine the cause of a digestive system problem. When a digestive system disorder is suspected, your veterinarian's initial examination might include the following: a visual inspection of the mouth for changes in the color and moistness of the mucous membranes; a determination of heart rate; a "hands on" examination of the abdomen and chest area to evaluate the shape, size, and position of the abdominal organs; tapping the abdomen or using a stethoscope to listen for sounds that indicate gas or for other abnormal sounds such as splashing; a rectal examination; and a visual examination of feces.

Depending on what the initial examination reveals, additional tests might include performing bacterial culture and virus isolation from feces; passing a stomach tube; using a long needle to collect fluid from distended abdominal organs or from the abdominal cavity for analysis; creating xray images, which may include use of special dyes, to detect blockages and other problems; performing abdominal ultrasonography; obtaining a biopsy (sampling and microscopic analysis) of liver or intestinal tissue; or conducting additional blood tests to detect possible malabsorption or maldigestion.

Infectious Diseases

Because it is easy for foreign organisms and other "invaders" to enter the digestive tract through the mouth, this body system is prone to infection by bacteria, viruses, parasites, and other organisms (see Digestive Disorders of Horses: Infections of the Digestive System in Horses). These infections spread in various ways, but the most common are by direct contact or by contamination of food or water by feces.

People and animals all have small numbers of certain intestinal microorganisms found within the digestive tract—most commonly in the intestines—that become established within a few hours after birth. These socalled intestinal flora are actually beneficial, in some cases aiding in digestion and in others helping to prevent infection. However, sometimes infections occur when these organisms, normally found in small numbers, suddenly multiply. This can occur after a period of stress, under unhygienic conditions, or in an animal whose immune system is weakened. For example, salmonellosis in horses can develop after transportation, extended anesthesia, or surgery.

Diagnosis of a specific infectious disease depends on finding and identifying the organism suspected to cause the disease. This may require one or more fecal samples, which will be submitted to a diagnostic laboratory.

Parasites are a frequent cause of digestive tract disorders in horses. Many species of parasites can infect the digestive tract and cause disease. The life cycles of some parasites are direct, which means that there is only one host. Eggs and larvae are passed in the feces, develop into an infective stage, and are eaten by the horse. No other animals or organisms are needed to complete the life cycle. Other parasites have complex life cycles that involve an intermediate host such as an insect. In this situation, infection is acquired when the intermediate host—or parasite shed by that host—is consumed by the horse. The extent to which a parasite affects an animal depends on the parasite itself, as well as the animal's own resistance, age, nutrition, and overall health.

Parasites can cause severe disease or simply decrease your horse's overall fitness. Some of these parasites also infect humans. Because parasite infection is easily confused with other illnesses, diagnosis depends on the veterinarian's knowledge of seasonal cycles of parasite infection, as well as examination of feces for evidence of parasite eggs or larvae. In some cases, blood tests can also be used to detect the presence of parasites.

Noninfectious Diseases

Many digestive system diseases are not caused by infective organisms. Their causes include overeating, eating poor-quality food, chemicals, obstruction caused by swallowing foreign objects, or injury to the digestive system. Digestive system disease can also be caused by enzyme deficiencies, damage to the digestive tract such as from stomach (gastric) ulcers, or birth defects. Digestive system signs such as diarrhea may also occur because of kidney, liver, or adrenal gland disease. The causes are uncertain in several diseases, including stomach ulcers in foals. Some conditions, such as colic, may have both infectious (such as damage from internal parasites) and noninfectious (such as excessive gas or simple obstruction) causes.

In noninfectious diseases of the digestive tract, usually only a single animal is affected at one time; exceptions are diseases associated with excessive food intake or poisons, in which multiple animals living together can be affected.

Treatment Overview of Digestive System Disorders

Elimination of the cause of the disease may involve drugs that kill bacteria or parasites, antidotes for poisons, or surgery to correct defects or displacements.

Use of drugs to correct diarrhea or constipation is done depending on the specific case. Although such drugs might seem to be a logical choice, they are not beneficial in every situation. For example, diarrhea can actually be a defense mechanism for the animal, helping it to eliminate harmful organisms and their toxins. In addition, the available drugs may not always give consistent results.

Replacement of fluids and electrolytes (salts) is necessary in cases where the animal is at risk of dehydration, such as from excessive diarrhea. Relief of distension (bloating) by stomach tube or surgery may be required if the digestive tract has become distended with gas, fluid, or food. Pain relief is sometimes provided. However, a horse being given pain medicine must be watched carefully to ensure that the pain relief is not masking a condition that is becoming worse.

Treatment and Control of Infectious Disease

Bacterial and parasitic diseases of the digestive system are often treated with medications designed to kill the infectious organisms. There are currently no specific medications for treatment of viral diseases. Antibiotics (drugs effective against bacteria) are commonly given for several days until recovery is apparent, although their effectiveness in treating digestive system disease is still uncertain. Antibiotics may be given by injection when septicemia (infection of the blood) is apparent or likely to occur. Your veterinarian will make the decision whether to prescribe antibacterial medication based on the suspected disease, likelihood of benefits, previous results, and cost of treatment.

Advances in understanding the life cycles of parasites, coupled with the discovery of effective antiparasitic drugs, have made successful treatment and control of digestive system parasites possible. Response to treatment is usually rapid, and a single treatment is often all that is needed unless reinfection occurs or the damage caused by the parasites is particularly severe.

Control of digestive diseases and parasites depends on practicing good sanitation and hygiene. This is achieved primarily by providing adequate space for your horse and by regular cleaning of its living areas. In addition, adequate nutrition and preventive medical care will minimize the stress on your horse and help it to stay healthy.

CONGENITAL AND INHERITED DISORDERS OF THE DIGESTIVE SYSTEM IN HORSES

Congenital abnormalities are conditions that an animal is born with; they are often called birth defects. Some of these conditions are inherited and tend to occur within particular families or breeds, while others are caused by chemicals or injury during pregnancy. For still others, the cause is unknown.

Mouth

A cleft palate or cleft lip (harelip) is caused by a defect in the formation of the jaw and face during embryonic development. It leads to a gap or cleft in the center of the lip, the roof of the mouth (hard palate), or both. Often this condition leaves an open space through the roof of the mouth into the breathing passages. These conditions have a wide range in severity. Usually the upper lip and palate are affected; a cleft in the lower lip is rare. Most cases are inherited, although nutritional deficiencies during pregnancy, drug or chemical exposure, injury to the fetus, and some viral infections during pregnancy have also been suggested as causes.

Cleft palate or lip will usually be noticed shortly after birth when the foal might have problems with nursing. For example, milk might be seen dripping from the nostrils or the foal might have difficulty suckling and swallowing. The veterinarian can usually identify the problem by examining the foal's mouth, although a cleft involving only the soft palate may be difficult to see. Affected foals require intensive nursing care, including hand or tube feeding and possibly antibiotics to treat secondary respiratory infections. Surgical correction is effective only in minor cases. The decision to perform surgery should be made carefully, and the affected animal should be neutered to prevent passing the defect on to future offspring.

Brachygnathia occurs when the lower jaw is shorter than the upper jaw. This condition is called "parrot mouth" in horses. It can be a minor problem or a serious defect depending on the degree of abnormality. Mild cases may cause no problems. More severe cases can cause damage to the hard palate (roof of the mouth) or restriction of normal jaw growth.

Prognathia occurs when the lower jaw is longer than the upper jaw. If a foal is badly affected, nursing may be impossible. Treatment consists of rasping or shearing the points and projections on the teeth that are causing difficulty.

Teeth

In most animals, having too few teeth is rare. Extra teeth occur in some horses; these extra teeth are commonly found in the incisor and molar regions of the mouth. The result may be crowding and rotation of the teeth. Extra teeth are either removed or periodically rasped, especially if they interfere with chewing or are irritated by a bit.

Abnormalities in placement or shape of teeth sometimes occur. This may affect the incisors and result in rotation or overlapping of adjacent teeth. Usually, this does not cause any problems, but it may require extraction of some teeth if crowding or bite abnormalities occur.

Enamel defects such as disruption of tooth enamel formation may be caused by fever, trauma, malnutrition, poisoning, or infection. The damage can vary, depending on the severity and duration of the cause, from pitted enamel to the absence of enamel with incomplete tooth development. Affected teeth are prone to plaque and tartar accumulation and subsequent tooth decay.

Cysts of the Head and Neck

Cysts (lumps) in the head and neck can be caused by imperfections during fetal development. These need to be distinguished by your veterinarian from abscesses or lumps caused by infection or other disease. Dentigerous cysts arise from abnormal tooth development and often contain tooth fragments. These cysts can be found in horses less than 3 years of age, and they may be difficult to distinguish from cystic sinuses, another condition which can result in face or jaw distortion. Surgical removal of the cysts is required. The veterinarian can then make a final diagnosis based on examination of the removed tissue.

SOME CYSTS ARISE FROM ABNORMAL TOOTH DEVELOPMENT.

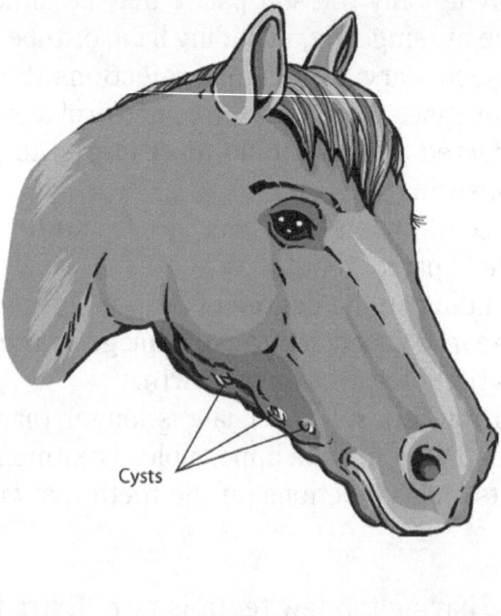

Hernias

A hernia is the protrusion of a portion of an organ or tissue through an abnormal opening. One common congenital type, called a hiatal hernia, involves an abnormal opening in the wall of the diaphragm (the sheet of muscle that separates the chest from the abdomen) or the abdomen. The defect may allow abdominal organs to pass into the chest or bulge beneath the skin. Hernias may be congenital (present at birth) or result from injury. Signs of a hernia vary from none to severe and depend on the amount of herniated tissue and its effect on the organ involved.

Hernias involving the abdominal wall are called umbilical, inguinal, or scrotal, depending on their location. Diagnosis of umbilical herniasis usually simple, especially if the veterinarian is able to push the hernia back through the abdominal wall (called reducing the hernia). If not, the hernia must be differentiated from an umbilical abscess, which is common in horses. The veterinarian may need to use a fine needle to collect tissue or fluid for confirmation of the diagnosis. These hernias are corrected by surgery. The tendency to develop hernias may be inherited.

Inguinal hernias are sections of the intestine that break into the inguinal canal located in the groin. Inguinal hernias in colts often resolve spontaneously during the first year of life, so early corrective surgery is not recommended unless the hernia is strangulated or so large that it interferes with walking.

Stomach

Besides hiatal hernias, another abnormality involving the stomach is pyloric stenosis. It can be acquired or inherited, but the inherited form is rare in horses. Pyloric stenosis results from muscular thickening of the "exit" of the stomach (pyloric sphincter). This obstructs the flow of digested food from the stomach to the small intestine. Signs may include loss of appetite and other signs, such as abdominal pain, that are usually associated with colic. Treatment is through dietary modification and medication. In more severely affected cases, surgery may be beneficial.

Small and Large Intestine

Maldigestion is a condition in which certain foods are not properly digested. Malabsorption occurs when nutrients are not properly absorbed into the bloodstream. Both of these conditions often cause chronic, persistent digestive system problems including weight loss and diarrhea. Malabsorption and maldigestion are often treated with a combination of dietary changes and medication; the exact treatment will depend on the condition being treated.

A defect in the colonic nerves called ileocolonic agangliosis is seen in white foals produced by matings of Overo horses. Although the foals appear normal at birth, they soon develop colic and die on the second day. The affected horses are white and have blue irises. Diagnosis can be confirmed by the lack of specific nerves in the colon.

Congenital defects of the rectum and anus generally result from arrested embryonic development. Inherited closures (atresias) of the small- and large-intestinal tracts are relatively common in large animals and are always fatal.

Atresia of the colon has been reported in Percheron horses. Segmental aplasia (also called rectal agenesis) is a condition in which the rectum is closed off before reaching the anus. Surgical correction is difficult because the location of the closure varies, and damage to nerves in the area may occur during surgery.

Liver

The most common congenital liver defect present at birth is a portosystemic shunt. In a healthy animal, blood coming from the intestines is processed by the liver, which removes toxins from the bloodstream before they reach the brain or other organs. In an animal with a portosystemic shunt, however, blood bypasses the liver through one or more "shortcuts" (shunts) and enters directly into the general circulatory system. Signs of a portosystemic shunt include nervous system disturbances and a failure to grow and thrive. In the late stages, protein-containing fluid may accumulate in the abdomen, a condition called ascites. Your veterinarian may also notice enlargement of the kidneys and kidney stones. A definite diagnosis is made by using an opaque dye to highlight the blood vessels, followed by x-rays. This procedure can identify the location of the shunt and determine whether it is single or multiple. It also allows the veterinarian to assess whether surgical correction is possible. Animals with multiple shunts tend to do poorly. Other liver developmental anomalies include hepatic (liver) cysts, which generally cause no signs. They are significant mainly because they must be differentiated from abscesses in the liver. A veterinarian who finds a hepatic cyst will often want to examine the kidneys because hepatic cysts often occur along with polycystic kidney disease.

HORMONAL DISORDERS OF HORSES

Hormones are chemical messengers that have many different functions. The effects of hormones in the body are wide-ranging and varied. Some familiar examples of hormones include insulin, which is important in the development of diabetes, and estrogen and progesterone, which are involved in the female reproductive cycle.

The Endocrine System

The endocrine system consists of a group of tissues that release hormones into the bloodstream for travel to other parts of the body. Most endocrine tissues are glands (such as the thyroid gland) that release hormones directly into small blood vessels within and around the tissue. Several important hormones are released from tissues other than glands, such as the heart, kidney, and liver. Some hormones act only on a single tissue, while others have effects on virtually every cell in the body. Hormones are present in the blood in very small quantities, so laboratory tests done to measure hormone levels must be very sensitive.

THE MAJOR ENDOCRINE GLANDS IN THE HORSE

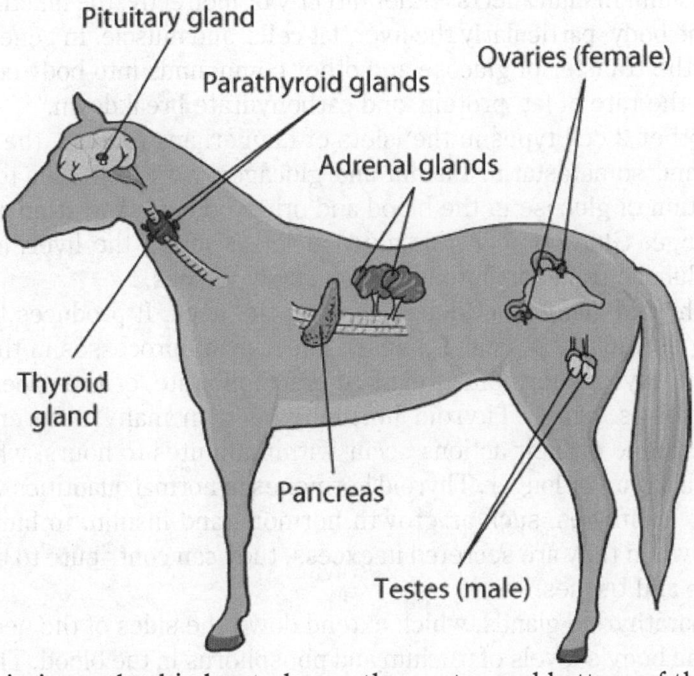

The pituitary gland is located near the center and bottom of the brain. It produces a number of critical hormones that control many parts of the body, including several other endocrine glands. For this reason, it is sometimes called a "master gland." Because large numbers of hormones are produced by the pituitary, a variety of different conditions can be caused by pituitary disease or tumors. The specific illness and signs depend on the cause and the area of the pituitary gland that is affected.

The adrenal glands are located just in front of the kidneys. The adrenal gland has 2 parts—the cortex and the medulla. The adrenal cortex consists of 3 layers, each of which produces a different set of steroid hormones. The outer layer produces the mineralocorticoids, which help to control the body's balance of sodium and potassium salts. The middle layer produces glucocorticoids, which are involved in metabolizing nutrients as well as in reducing inflammation. The inner layer produces sex hormones such as estrogen and progesterone. The adrenal medulla plays an important role in response to stress or low blood sugar (glucose). It releases epinephrine (sometimes called adrenaline) and norepinephrine, both of which increase heart output, blood pressure, and blood glucose, and slow digestion.

The pancreas is composed of several types of cells that have distinct functions involved in the production of hormones and digestive enzymes. The islets of Langerhans in the pancreas consist of 3 types of cells, each of which produces a different hormone. Most of the cells, which are called beta cells, produce insulin. Insulin affects, either directly or indirectly, the function of every organ in the body, particularly the liver, fat cells, and muscle. In general, insulin increases the transfer of glucose and other compounds into body cells. It also decreases the rate of fat, protein, and carbohydrate breakdown.

The other 2 cell types in the islets of Langerhans produce the hormones glucagon and somatostatin. Insulin and glucagon work together to keep the concentration of glucose in the blood and other body fluids within a relatively narrow range. Glucagon controls glucose release from the liver, and insulin controls glucose transport into numerous body tissues.

The thyroid gland is a 2-lobed gland in the neck. It produces the iodine-containing hormones, T_3 and T_4, which affect many processes in the body. In general, the thyroid hormones regulate metabolic rate, or the speed at which body processes "run." Thyroid hormones act on many different cellular processes. Some of their actions occur within minutes to hours, while others take several hours or longer. Thyroid hormones in normal quantities work along with other hormones, such as growth hormone and insulin, to build tissues. However, when they are secreted in excess, they can contribute to breakdown of proteins and tissues.

The parathyroid glands, which extend down the sides of the neck, help to regulate the body's levels of calcium and phosphorus in the blood. The way the body processes calcium and phosphate, the function of vitamin D (which acts more like a hormone than a vitamin), and the formation of bone are all tied together into a system that involves 2 other hormones—parathyroid hormone and calcitonin—that are secreted by the parathyroid glands.

Development of Endocrine System Disease

The body monitors and adjusts the level of each hormone by using a feedback system specifically for that hormone. Hormones function to keep factors such as temperature and blood sugar (glucose) levels within certain ranges. Sometimes, pairs of hormones with opposite functions work together to keep body functions in balance.

Endocrine system diseases can develop when too much or not enough hormone is produced, or when normal pathways for hormones to be used and removed are disrupted. Signs can develop because of a problem in the tissues that are the source of the hormone, or because of a problem in another part of the body that is affecting the secretion or action of a particular hormone.

A tumor or other abnormal tissue in an endocrine gland often causes it to produce too much hormone. When an endocrine gland is destroyed, not enough hormone is produced. Diseases caused by overproduction or excess of a hormone often begin with the prefix hyper-. For example, in hyperthyroidism, the thyroid gland produces too much thyroid hormone. Diseases caused by a lack or deficiency of a hormone often begin with the prefix hypo-. For example, in hypothyroidism, the thyroid gland does not produce enough thyroid hormone.

In many cases, the abnormal gland not only overproduces hormone, it also does not respond normally to feedback signals. This causes hormone to be released in situations in which its levels would normally be reduced. Sometimes, the over-production is caused by stimulation from another part of the body. Occasionally, a tumor outside the endocrine system can produce a substance similar to a hormone, causing the body to respond as though that hormone were being produced.

Diseases caused by not enough hormone secretion can also have multiple causes. Endocrine tissue can be destroyed by an autoimmune process, in which the body incorrectly identifies some of its own tissue as foreign and destroys the tissue cells. In early stages of tissue loss, the body may compensate by producing additional hormone from the remaining tissue. In these cases, signs of disease may be delayed until the tissue has been destroyed completely.

Disorders resulting in signs of reduced endocrine activity may also develop because tissues distant from the hormone source are disrupted. This can occur when the function of one hormone is to stimulate the production of a second hormone. For example, the pituitary gland secretes a hormone that stimulates the thyroid gland to secrete thyroid hormones. If the levels of the thyroid-stimulating hormone from the pituitary gland are abnormally low, the levels of thyroid hormones will also be low even if the thyroid gland is healthy. Another potential cause for reduced endocrine function is tissue loss caused by tumors that do not produce hormones themselves but compress or destroy the nearby endocrine gland.

Endocrine diseases and related conditions also result from changes in the response of tissues targeted by a hormone. An important example is type 2 diabetes mellitus, in which the body produces insulin but the cells no longer respond to it. This condition is often associated with obesity.

Treatment of Endocrine System Disease

Endocrine diseases caused by the presence of too much hormone may be treated surgically (tumor removal), by radiotherapy (such as the use of radioactive iodine to destroy an overactive thyroid gland), or with medication. Syndromes of hormone deficiency are often successfully treated by replacing the missing hormone, such as insulin injections to treat diabetes mellitus. Steroid and thyroid hormone replacements can usually be given by mouth.

Animals taking hormone replacement treatment must be monitored for adverse effects and periodically retested to make sure the dosage is correct. In some cases, such as after surgical removal of an endocrine tumor, the diseased gland will recover and hormone replacement will no longer be needed. However, most of the time, lifelong treatment is required.

COMMON HORMONAL DISORDERS IN HORSES

The most common hormonal disorders in horses affect the adrenal glands (Addison's disease and Cushing's disease), thyroid gland (hypothyroidism and goiter), and the pancreas (diabetes mellitus).

Cushing's Disease (Hypertrichosis)

Cushing's disease, also called hyperadrenocorticism, is the most common endocrine disease in horses. The signs are due primarily to chronic excess of the hormone cortisol. Increased cortisol levels may result from one of several mechanisms, such as destruction of a portion of the pituitary gland and overproduction of certain other hormones. Unlike Cushing's disease in dogs or people, the cause is not usually related to a pituitary tumor. However, pituitary tumors do occur, particularly later in the disease.

The disorder is seen in older horses (over 15 years of age) of any breed. Mares and geldings are most often affected. The most striking sign is development of an abnormally long or heavy hair coat (calledhypertrichosis), which can grow up to 4 to 5 inches (10 to 12 centimeters) long, and is thick, wavy, and often matted. Other signs include excessive thirst and urination, increased appetite and weight, an enlarged abdomen, and bulging eyes. Horses with Cushing's disease tend to have a weakened immune system and may be prone to infections (such as dental disease or respiratory infections) or parasites. A diagnosis is based on a history and signs, physical examination, and appropriate blood tests.

Once a horse has been diagnosed with Cushing's disease, there are treatments available that may improve its condition and, in some cases, even return it to normal health. Pergolide and cyproheptadine are the 2 most commonly used drugs for treatment of this disorder in horses. However, treatment requires that the owner administer the drug daily and schedule regular veterinary checkups, including blood tests to monitor the horse's response. If treatment is stopped, the signs can reappear within a few weeks.

Even without medical treatment, many horses respond to careful management. This includes regular dental care, dewormings, hoof care, and an appropriate diet.

Addison's Disease

Addison's disease (hypoadrenocorticism) is caused by a deficiency of adrenal gland hormones. It is seen occasionally in horses. The cause is usually not known, but an autoimmune condition in which the body destroys some of its own tissue is likely. The adrenal gland can also be destroyed by other conditions, including cancer in other parts of the body. Secretion of aldosterone, the main mineralocorticoid hormone, is reduced, which affects the levels of potassium, sodium, and chloride in the blood. Potassium gradually builds up in the blood and, in severe cases, may cause the heart to slow down or beat irregularly. Signs of Addison's disease include repeated episodes of vomiting and diarrhea, loss of appetite, dehydration, and a gradual loss of body condition. Weight loss is often severe. Although signs can be hard to identify while Addison's disease is developing, severe consequences, such as shock and evidence of kidney failure, can develop suddenly.

A tentative diagnosis is based on the history, signs, and certain laboratory abnormalities, such as very low levels of sodium and very high levels of potassium in the blood. The diagnosis is confirmed by specific evaluation of adrenal function.

This is done by measuring the level of cortisol in the blood, treating the animal with adrenocorticotropin (a hormone that stimulates the adrenal gland in healthy animals), and then measuring the level of cortisol in the blood a second time. Affected horses have low baseline cortisol levels, and there is little response to administration of adrenocorticotropin.

An adrenal crisis is a medical emergency and requires treatment with intravenous fluids to restore levels of body fluids, salt, and sugar to normal. Hormone replacement therapy can often be started while the animal is being stabilized. Laboratory values are monitored regularly to assess the response to treatment and adjust doses if needed. For longterm treatment, replacement hormones can be given by mouth or injection. Supportive treatment and rest are indicated for horses with Addison's disease.

Diabetes Mellitus

Diabetes mellitus (often called simply diabetes) is a chronic disorder of carbohydrate metabolism caused by either a deficiency of insulin or a resistance to insulin. Diabetes caused by a deficiency of insulin (also called primary diabetes mellitus) is rare in horses; however, resistance to insulin (also called secondary diabetes mellitus) is more common and tends to develop in horses with Cushing's disease. A diagnosis of diabetes mellitus is based on finding high levels of sugar in the blood and urine after a period of fasting. Treatment with insulin cannot reverse the insulin resistance seen in secondary diabetes mellitus.

Hypothyroidism

In hypothyroidism, decreased levels of thyroid hormones result in a slower metabolic rate. Adult horses rarely develop hypothyroidism. However, foals may be born with hypothyroidism if the pregnant mare grazed plants that contained goiter-producing substances, or if she was fed a diet with either not enough or too much iodine. Most commonly, these foals are affected by a specific syndrome in which they have both thyroid abnormalities and multiple congenital musculoskeletal abnormalities. This syndrome has been reported most commonly in western Canada, and it may be related to feeding a high nitrate diet (for example, greenfeed) to pregnant mares.

Because a deficiency of thyroid hormone affects the function of all organ systems, signs vary. Many signs could also be caused by other diseases. Most signs are directly related to slowing of metabolism, which results in lethargy, unwillingness or inability to exercise, and weight gain without an increase in appetite.

Diagnosis requires evaluation of signs and various laboratory tests, including demonstration of low serum concentrations of thyroid hormones that do not respond to administration of thyroid-stimulating hormone. Identifying hypothyroidism can be difficult at times because other conditions in the adult horse can make it appear that thyroid hormone levels are low when they are actually normal.

Hypothyroidism is treated using replacement with synthetic thyroid hormone. The success of treatment can be measured by the amount of improvement in signs. Serum thyroid hormone concentrations are also monitored to determine whether the dosage of thyroid hormone needs adjustment. Once the dose has been stabilized, thyroid hormone levels are usually checked once or twice a year. Treatment is generally lifelong.

Goiter

A goiter is a noncancerous enlargement of the thyroid gland that develops most often when the diet does not contain enough iodine. Goiters develop in people, all domestic mammals, and birds. In addition to iodine deficiency, other major causes of goiter include goiter-causing substances, too much iodine in the diet, and inherited defects in the body's production of thyroid hormones. Many animals with goiter appear to have normal thyroid hormone levels, but signs of hypothyroidism may develop in some, especially in newborns.

Iodine Deficiency

Goiter due to iodine deficiency was common in many areas of the world before iodized salt was routinely added to animal diets. Outbreaks of goiter caused by iodine deficiency are now sporadic and affect fewer animals, but iodine deficiency is still responsible for most goiters seen in horses.

A lack of iodine reduces the ability of the thyroid to make thyroid hormone. Because thyroid hormone levels drop, the pituitary gland secretes more thyroid-stimulating hormone, which results in the thyroid gland enlarging in an effort to make more thyroid hormone. Often, the enlarged thyroid gland can make enough thyroid hormone to bring the levels into the normal range. However, foals born of mares on iodine-deficient diets are more likely to develop severe thyroid enlargement and have signs of hypothyroidism. The neck is usually enlarged, and the skin and other tissue may be thickened, flabby, and swollen.

Many foals die before or soon after birth, but treatment with iodized salt may resolve the goiter and associated signs in mildly affected foals. Prevention is much more effective than treatment, and adding stabilized, iodized salt to the diet is recommended in areas known or suspected to be deficient in iodine.

Iodine Toxicity

Foals born of mares fed too much iodine may develop extreme thyroid enlargement and die before birth or soon after. Signs include general weakness, long hair, and limb abnormalities.

Goiter-causing Substances

Certain plants can produce goiter if eaten in sufficient amounts, especially when the diet does not contain enough iodine. Soybeans, cabbage, rape, kale, and turnips all contain goitrogens. Cooking or heating (and the usual processing of soybean meal) destroys the goiter-causing substance in these plants. Goiter-causing substances act by interfering with production of thyroid hormone. As in iodine deficiency, the pituitary gland responds to the lower levels of thyroid hormone by increasing its secretion of thyroid-stimulating hormone. In turn, the thyroid responds by enlarging in an attempt to produce more thyroid hormone. In adult animals the disease is usually not significant, but newborn foals can develop severe thyroid enlargement and signs of hypothyroidism.

LUNG AND AIRWAY DISORDERS OF HORSES

The respiratory system consists of the large and small airways and the lungs. When a horse inhales, the air travels down the trachea, which divides into the tubes known as the right and left bronchi, then into the smaller airways called bronchioles in the lungs. The bronchioles end in the small sacs called alveoli, where the barrier between the air and the blood is a thin membrane.

The most important function of the respiratory system is to deliver oxygen into the blood, which distributes it throughout the body, and to remove carbon dioxide from the blood. The exchange of oxygen and carbon dioxide occurs in the alveoli. When this exchange fails or becomes inefficient because of disease, the animal can become seriously ill. The respiratory system protects its own delicate airways by warming and humidifying inhaled air and by filtering out particles. Large airborne particles usually land on the mucous lining of the nasal passages, after which they are carried to the throat to be either swallowed or coughed up. Small particles and microorganisms are destroyed by the body's immune system.

Although the basic functions are the same, the anatomy of the respiratory tract varies among species. For example, the respiratory systems of dogs and cats are somewhat similar to each other, but differ from the respiratory systems of horses and humans. These differences explain in part why some diseases affect only certain species of animals. Horses are nasal breathers, and are not able to breathe through their mouths. Because of this, the horse's nasal passages are large and can expand somewhat during strenuous exercise in order to increase the intake of air.

THE LUNGS AND AIRWAYS IN A HORSE

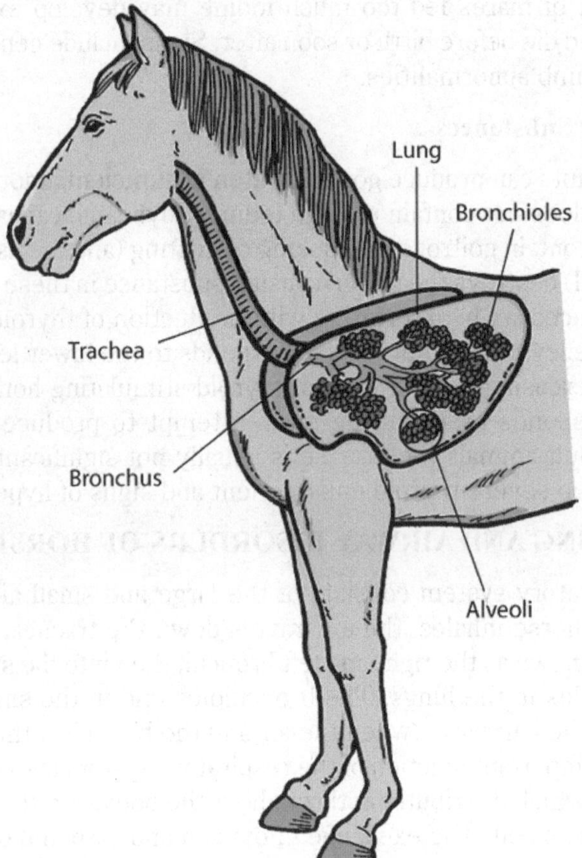

When the level of oxygen in the blood is too low (called hypoxia or anoxia), the animal will show signs of respiratory distress. Low oxygen levels can be caused by reduced oxygen-carrying capacity of the red blood cells, insufficient movement of gases in and out of the lungs, or inability of tissues to use available oxygen (a condition caused by some poisons). The animal's body attempts to compensate for low oxygen in the blood by increasing the depth and rate of breathing, increasing contraction of the spleen (to force more red blood cells into circulation), and increasing blood flow and heart rate. If the brain suffers from lack of oxygen, respiratory function may be reduced even further due to depression of nervous system activity. In addition, heart, kidney, and liver functions may be reduced, as may the normal movement and secretions of the intestine. If the body is not able to compensate for the reduced oxygen level, a "vicious cycle" may begin in which all body tissues function less efficiently.

Causes of Lung and Airway Disorders

Lung and airway disorders are often caused by direct infection with viruses, bacteria, fungi, or parasites, as well as by immune-mediated reactions or inhalation of irritants or toxic substances. Trauma (such as crashing into a fence) may lead to the collapse of a lung or airway.

Viral respiratory infections are common in horses. Most bacterial respiratory infections (with the exception of strangles) occur after an attack of a viral disease. This is because viral respiratory infections impair and/or destroy respiratory defense mechanisms, making the horse more susceptible to additional infections. The most common organisms associated with pneumonia in horses are bacteria normally present in the upper respiratory tract that take advantage of the horse's weakened state (called opportunistic bacteria). Secondary bacterial disease may result in bacterial infections of the mucous membranes (rhinitis and tracheitis), or may produce more serious invasive disease such as pneumonia and pleuropneumonia. *Streptococcus equi equi*, the organism that causes strangles, is a primary bacterial disease organism of the upper respiratory tract and is capable of causing infection without predisposing factors such as a previous viral disease. *Rhodococcus equi* is a disease organism of the lower respiratory tract of foals less than 5 months of age, which produces lung hardening and formation of abscesses.

THE UPPER PORTION OF THE HORSE'S RESPIRATORY SYSTEM

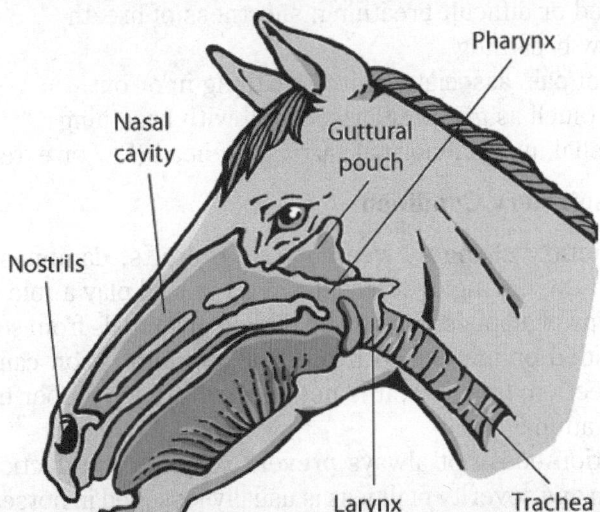

Noninfectious respiratory disease (disease not caused by a bacterium or other microorganism) may occur in adult horses of various ages and can limit their performance. Inflammatory airway disease is characterized by excessive mucus in the airways and poor exercise performance in young horses. The cause is unclear, but viral respiratory infection, allergy, and environmental factors may play a role. Reactive airway disease (heaves) is triggered by exposure to organic dusts in older horses with a genetic susceptibility to allergic airway disease. The small airways in the lungs are obstructed by constriction and excessive mucus production. The severity of signs ranges from exercise intolerance to labored breathing at rest.

Diagnosing Lung and Airway Disorders

Your horse's history and the veterinarian's physical examination will help to determine the possible cause and site of respiratory disease. Other techniques commonly used for diagnosis include xrays of the chest or neck; the use of various types of endoscopes to view the nose, throat, and airways; and examination of the contents of the lung and airways, which can be obtained through aspiration or retrieval of fluid used to "bathe" the airways.

Signs of Respiratory Disorders

- Discharge from the nose (mucus, pus, or blood, depending on the cause)
- Coughing that may be dry or may include mucus or blood
- Rapid breathing (not always a sign of disease, such as in healthy animals after exercise; however, rapid breathing at rest is never normal)
- Labored or difficult breathing; shortness of breath
- Shallow breathing
- Signs of pain associated with breathing in or out
- Noise (such as grunting) associated with breathing
- Head shaking or abnormal carriage of head (low or extended)

Control of Respiratory Conditions

Sudden dietary changes, weaning, cold, drafts, dampness, dust, poor ventilation, and the mixing of different age groups all play a role in respiratory disease in groups of animals. Stress and mixing of animals from several sources should be avoided or minimized if possible. Immunization can help control respiratory infection; however, it is not a substitute for proper environmental conditions and animal care.

Immunization does not always prevent respiratory infections in horses, but the duration and severity of illness is usually lessened in horses with regular vaccination depending on factors such as the disease and specific vaccine. The veterinarian will help weigh the cost and hazards of each vaccination against the probability of the horse's exposure and potential disease and the effectiveness of the vaccines. Vaccination recommendations and schedules will vary according to the use of the horse and its potential for exposure to contagious animals.

General Treatment of Lung and Airway Disorders

Respiratory disorders often involve the production of excess secretions in the respiratory system (for example, in the nose and lungs) that the affected animal may not be able to remove without assistance. One goal of veterinary treatment is to reduce the volume and thickness of the secretions and to make their removal easier. This can be accomplished by controlling infection, thinning the secretions, and when possible, improving drainage and mechanically removing the material. Therapies can include altering the inhaled air as well as the use of expectorants (which help an animal to cough up the secretions), cough suppressants, bronchodilators (to help open airways), antibiotics, diuretics (to reduce fluid buildup), and other drugs.

Regardless of the type of respiratory disease, environmental factors and supportive care are important to aid recovery. A dust- and ammonia-free stable environment prevents further damage to the respiratory system. Highly palatable feeds help to prevent weight loss and weakness during the treatment and recovery period. Adequate water intake will decrease the thickness of respiratory secretions, making their removal from the lower respiratory tract easier. A comfortable, dry, temperature-appropriate environment will allow the horse to rest and will lessen the need for the respiratory tract to work at regulating the horse's body temperature.

EYE STRUCTURE AND FUNCTION IN HORSES

The eyes of animals, including horses, function much like your eyes. Animals also develop many of the same eye problems that people can have, including cataracts, glaucoma, and other problems. Because sight is the way in which horses get the majority of their information about their surroundings, it is important for your horse to receive good eye care to protect its sight and allow the horse to interact comfortably with its environment. In general, horse vision is a little blurrier and a little less colorful than human vision. However, horses see movement very well throughout the 340° arc of their peripheral vision.

The eye is an active organ that constantly adjusts the amount of light it lets in and focuses on objects near and far. It produces continuous images that are quickly relayed to the brain.

The bony cavity or socket that contains and protects the eyeball is called the orbit. The orbit is a structure that is formed by several bones. The orbit also contains muscles, nerves, blood vessels, and the structures that produce and drain tears.

Structures that protect the eye

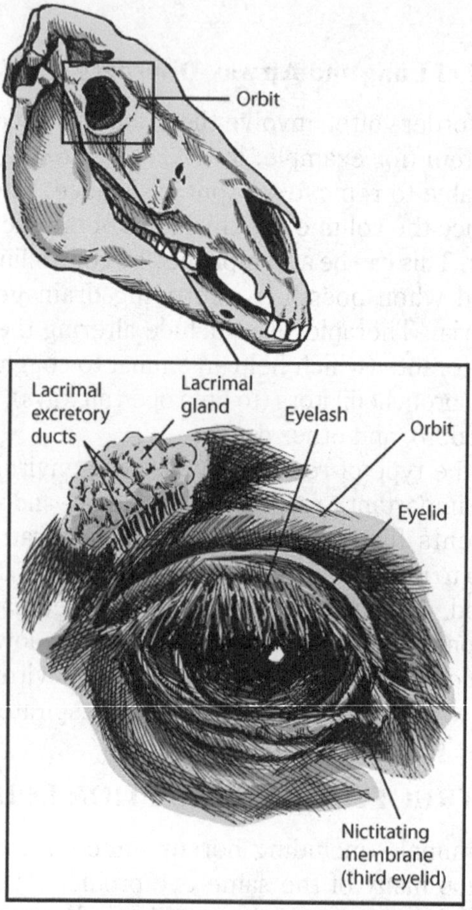

The white of the eye is called the sclera. This is the relatively tough outer layer of the eye. It is covered by a thin membrane, called theconjunctiva, located near the front of the eye. The conjunctiva runs to the edge of the cornea and covers the inside of the eyelid. The cornea is a clear dome on the front surface of the eye that lets light in.

The cornea not only protects the front of the eye, but also helps focus light on the retina at the back of the eye. The iris is the circular, colored area of the eye. It controls the amount of light that enters the eye by making the pupil larger or smaller. The pupil is the black area in the middle of the eye. It is controlled by the circular sphincter muscle. When the environment is dark, the pupil enlarges to let in more light; when the environment is bright, the pupil becomes smaller to let in less light.

The lens, which sits behind the iris, changes its shape to focus light onto the retina. Small muscles called ciliary muscles contract to cause the lens to become thicker, which allows the lens to focus on nearby objects. The ciliary muscles relax to cause the lens to become thinner when it focuses on distant objects. In horses, the very large lens appears to have limited changes. The retinacontains the cells that sense light (photoreceptors). The most sensitive area of the retina is called thevisual streak in horses. This area contains thousands of tightly packed photoreceptors that make visual images sharp. Each photoreceptor is attached to a nerve fiber. All the nerve fibers are bundled together to form the optic nerve. The photoreceptors in the retina convert the image into electrical impulses, which are carried to the brain by the optic nerve.

The upper and lower eyelids are thin folds of skin that can cover the eye and reflexively blink to protect it. Blinking also helps spread tears over the surface of the eye, keeping it moist and clearing away small particles. The eyes of a horse are protected not only by the same types of eyelids that people have, but also by the nictitating membrane, which is sometimes called the third eyelid. This additional eyelid is a whitish pink color, and it is found under the other eyelids in the inside corner of the eye (near the nose). The third eyelid extends up when needed to protect the eyeball from scratches or in response to inflammation. To function properly, eyes must be kept moist. Tears are the source of this needed moisture. Tears are produced by 2 types of glands. Lacrimal glands produce the watery portion of tears. They are located at the top outer edge of each eye. Mucus glands in the conjunctiva produce mucus that mixes with the watery portion. This creates a more protective tear that is slower to evaporate. Nasolacrimal ducts allow tears to drain from each eye into the nose. Each of these ducts has openings at the edge of the upper and lower eyelids near the nose.

Physical Examination of the Eye

When examining your horse's eyes, a veterinarian will begin by checking to see that the shape and outline of the eyes are normal and that there are no obvious abnormalities. Using light and magnification in a darkened room, the reflexes of the pupils and the front part of the eye will be examined. A test, called the Schirmer tear test, may be performed to ensure that the eyes are producing enough tears to keep them moist. This is a relatively simple test in which small paper strips are inserted under the eyelid to measure the amount of moisture produced. A small drop of fluorescein staining may be put into each eye allowing defects in the cornea of the eye to be detected.

Pressure within the eye is painlessly measured using an instrument called a tonometer. If eye pressure is too high, optic nerve damage can occur, leading to irreversible blindness. A swab may also be done to culture for bacteria or fungi. The eyelids may be turned inside out to examine the underside of the eyelids. The nasal tear duct may be flushed to evaluate the external parts of the eye. Drops may be added to the eyes to dilate the pupils so that the veterinarian may examine the internal part of the eye using an ophthalmoscope.

To ensure the safety of both horse and veterinarian, appropriate restraint during the eye examination—such as topical anesthesia, use of a nose or ear twitch, or sedation—are often necessary.

DISORDERS OF THE EYELIDS IN HORSES

Problems affecting the eyelids may be congenital (present at birth) or may occur as a result of injury, infection, or exposure to various types of irritants.

Entropion

Entropion is the turning in of the edges of the eyelid so that the eyelashes rub against the eye surface. It occurs in foals as a congenital defect. It may also be acquired in older horses as a result of chronic eye irritation or spasms. The turning in of eyelashes or facial hairs causes discomfort and irritation of the conjunctiva and cornea. Extremely long lashes can cause scarring, abnormal coloring, and possibly the formation of slow-healing sores or ulcers on the cornea. Early spasms of entropion may be reversed if the cause is removed or if pain is lessened. Turning the lid hairs back away from the eye with stitches in the lid, injections of medication into the lid close to the area where the lid is turning in, or using anesthetics to block the nerves in the eyelids are some of the methods that have been used to lessen the pain. Established entropion may require surgery to correct the defect.

Lacerations of the Eyelid

Eyelid lacerations (rips or tears in the eyelid) are common in horses. They must be repaired quickly to avoid infection, reduce swelling, and prevent further damage to the eye. In many cases, your veterinarian will clean the wound and use stitches to repair it so that minimal scarring occurs.

RIPS OR TEARS IN THE EYELID REQUIRE PROMPT ATTENTION.

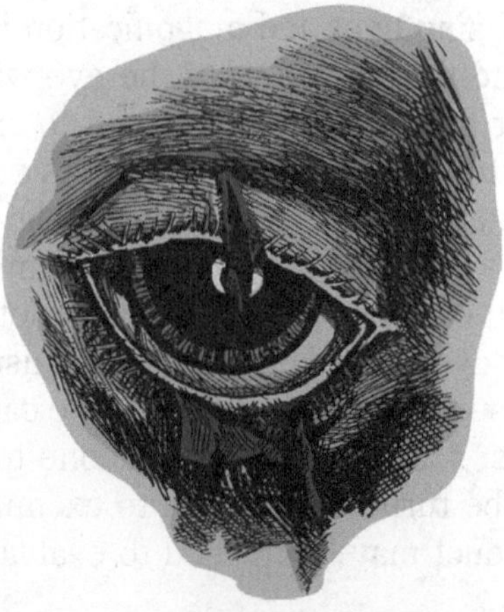

Inflammation of the Eyelids (Blepharitis)

Inflammation of the eyelids can result from the spreading of a generalized inflammation of the skin (dermatitis), conjunctivitis (inflammation of the conjunctiva), local glandular infections, or irritants such as plant oils or sunlight. Infection with certain fungi or bacteria can also lead to inflammation of the eyelids; however, this is uncommon in horses.

Parasites, such as eyeworms and the larvae of stomach worms (*Habronema* species), are a common cause of blepharitis in horses, particularly during warm-weather months. Treatment is targeted toward removing the parasites by the use of appropriate antiparasitic drugs, although eyeworms can sometimes be removed directly from the surface of the conjunctiva using forceps. If needed, topical ointments to reduce swelling may be prescribed.

STRUCTURE OF THE SKIN IN HORSES

The skin is the largest organ of your horse's body. It provides a protective barrier against the environment, regulates temperature, and gives your horse its sense of touch. Depending on the species and age, the skin may be 12 to 24% of an animal's body weight. The skin has 3 major layers: the epidermis or outermost layer, the dermis or middle layer, and the subcutis or innermost layer. Other important components include skin appendages (such as hair and hooves), and subcutaneous muscles and fat.

STRUCTURE OF A HORSE'S SKIN

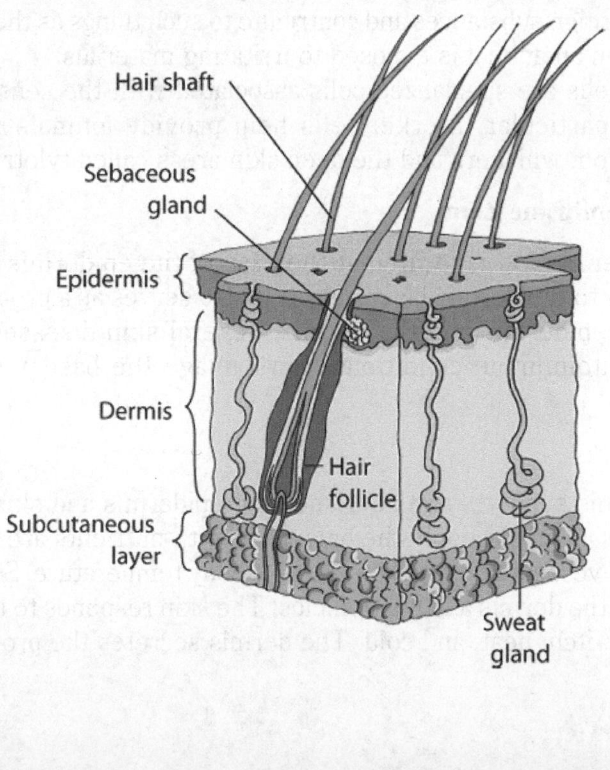

Epidermis

The epidermis is the outer layer of skin, which is composed of several layers of cells. It provides a barrier of protection from foreign substances. The epidermis is thickest in large animals like horses. It includes multiple types of cells, including keratinocytes, melanocytes, Langerhans cells, and Merkel cells. Each of these cells has special functions.

Keratinocytes provide a protective layer that is constantly being renewed in a process called keratinization. In this process, new skin cells are created near the base of the epidermis and migrate upwards. This produces a compact layer of dead cells on the skin surface. This layer keeps in fluids, electrolytes, and nutrients, while keeping out infectious or noxious agents. The top layer of dead skin cells are continuously shed and replaced by cells from lower layers. The rate of cell replacement is affected by nutrition, hormones, tissue factors, immune cells in the skin, and genetics. Disease and inflammation can also change normal cell growth and keratinization.

Melanocytes are located at the base of the epidermis, the outer root sheath of hairs, and the ducts of the sebaceous and sweat glands. The melanocytes produce the skin and hair coloring (pigment) called melanin. Production of melanin is controlled by both hormones and the genes received from parents. Melanin helps protect the cells from the damaging rays of the sun.

Langerhans cells are part of the immune system. These cells are damaged when exposed to excessive ultraviolet light and glucocorticoids (anti-inflammatory drugs). Langerhans cells play an important role in the skin's response to foreign substances and contribute to such things as the development of rashes when an animal is exposed to irritating materials.

Merkel cells are specialized cells associated with the sensory organs in the skin. In particular, Merkel cells help provide animals with sensory information from whiskers and the deep skin areas called tylotrich pads.

Basement Membrane Zone

This layer of skin is located at the base of the epidermis and connects the epidermis to the dermis layer below. It also serves as a protective barrier between the epidermis and the dermis. Several skin diseases, including a number of autoimmune conditions, can damage the basement membrane zone.

Dermis

The dermis supports and nourishes the epidermis and skin appendages. The blood vessels that supply the epidermis with nutrients are located in the dermis. Blood vessels also regulate skin and body temperature. Sensory nerves are located in the dermis and hair follicles. The skin responds to the sensations of touch, pain, itch, heat, and cold. The dermis secretes the proteins collagen and elastin that give support and elasticity to the skin. Also present are immune cells that defend against infectious agents that pass through the epidermis.

Skin Appendages

Hair follicles, oil and sweat glands, and hooves are all skin appendages that grow out of the epidermis and dermis. The hair follicles of horses are simple—the follicles have 1 hair emerging from each pore.

The growth of hair is affected by nutrition, hormones, and change of season. The size, shape, and length of hair are controlled by genetics and hormones. Disease, drugs, nutrition, and environment also affect the health of hair.

The hair coat protects the skin from physical and ultraviolet light damage, and helps regulate body temperature. Trapping air between secondary hairs conserves heat. This requires that the hairs be dry and waterproof. The cold-weather coat of many animals is often longer and finer to facilitate heat conservation. The hair coat can also help cool the skin. The warm-weather coat has shorter thicker hairs and fewer secondary hairs. This anatomic change allows air to move easily through the coat, which facilitates cooling.

Oil glands (also called sebaceous glands) secrete an oily substance called sebum into the hair follicles and onto the skin. They are present in large numbers near the hooves, back of the neck, rump, mouth, and tail area. Sebum is important for keeping the skin soft, moist, and pliable. It gives the hair coat sheen and has antimicrobial properties.

Sweat glands are part of the horse's system to regulate body temperature. They are found over most of the body except the legs. The evaporation of sweat from the skin is the primary cooling mechanism of the body for horses.

Subcutis

The subcutis is the innermost major layer of skin. It contains the subcutaneous fat and muscles. (The word subcutaneous means "beneath the skin.") The twitch muscle is the major subcutaneous muscle. The subcutaneous fat provides insulation; a reservoir for fluids, electrolytes, and energy; and a shock absorber. Nerves and blood vessels that supply the skin are also found in the subcutis.

BONE, JOINT, AND MUSCLE DISORDERS IN HORSES

The musculoskeletal system includes the bones, cartilage, muscles, ligaments, joints, tendons, and other connective tissue. It supports the body, permits movement, and protects the vital organs. Because many other body systems (including the nervous system, blood vessels, and skin) are interrelated, disorders of one of these systems may also affect the musculoskeletal system.

Musculoskeletal disorders are common in horses, in part due to the fact that they are often expected to work, perform, or carry a rider, which places additional stress on the muscles and bones. Horses also have a complex anatomy of the front and hind limbs with a number of bones, tendons, and joints. Disorders of any of these musculoskeletal components can lead to lameness and other problems.

COMPONENTS OF THE MUSCULOSKELETAL SYSTEM OF HORSES

Bones provide rigid structure to the body and shield internal organs from damage. They also house bone marrow, where blood cells are formed, and they maintain the body's reservoir of calcium. Old bone tissue is constantly replaced with new bone tissue in a process called remodeling. This helps keep the bones healthy.

Bones come together to form joints. The type of joint formed determines the degree and direction of motion. For example, joints with a ball-and-socket formation allow for rotation, while hinge joints only allow bending and straightening. Some joints do not move at all. In a joint, the ends of the bones are covered with cartilage, which is a smooth protective tissue that helps reduce friction as joints move.

The skeleton of the horse

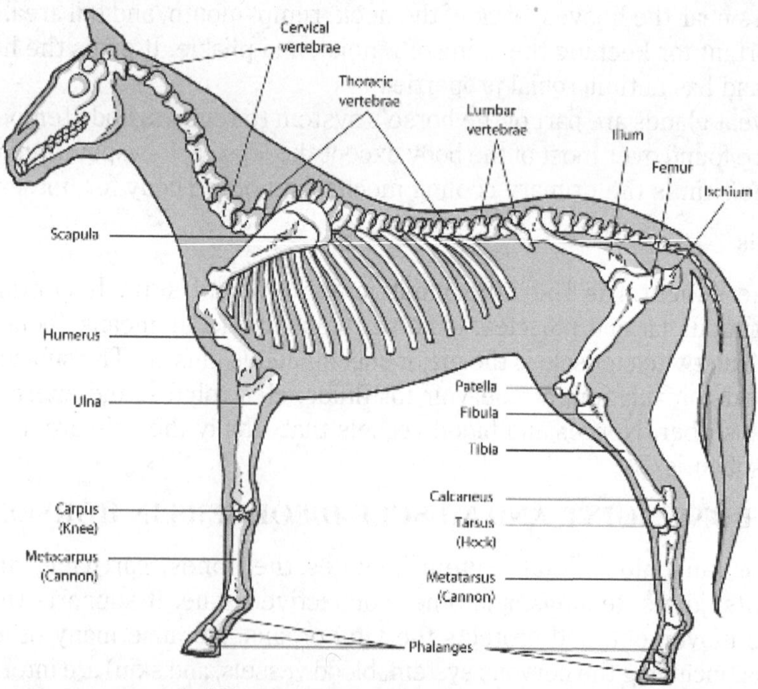

There are several different types of muscles in the body. Two of these, skeletal muscle and smooth muscle, are part of the musculoskeletal system. Skeletal muscles are responsible for posture and movement. They are attached to bones and arranged around the joints. Smooth muscle helps facilitate many processes in the body such as the flow of blood (by surrounding arteries) and the movement of food along the digestive tract. Tendons are tough bands of connective tissue made up mostly of a protein called collagen. They stretch very little. Tendons attach each end of a muscle to a bone. At selected points, they are located within sheaths that allow them to move easily. Ligaments are also tough cords formed of connective tissue. Unlike tendons, they can stretch to some extent. Ligaments surround joints and help to support and stabilize them. They can also connect one bone to another.

MUSCULOSKELETAL DISORDERS IN HORSES

Disorders of the musculoskeletal system most often affect the horse's ability to move. How severely movement is impaired depends on the type and severity of the problem. Skeletal and joint disorders are the most common, but problems in the musculoskeletal system can also indicate diseases of the muscles, neurologic problems, toxins in the body, hormonal abnormalities, metabolic disorders, infectious diseases, blood and vascular disorders, poor nutrition, and birth defects.

Many different systems in the body rely on the muscles. A horse's ability to see, breathe, urinate, breed, and even chew and swallow may be affected by a muscular condition. With many disorders, however, the musculoskeletal system is simply the location where signs of other underlying conditions show themselves. Veterinarians often trace the real cause of musculoskeletal trouble back to the nervous system.

Disorders that primarily affect the muscle membrane or muscle fibers are called myopathies. Muscle membrane disorders may be hereditary or acquired. Myopathies involving the actual muscle fiber components include exertional myopathy. Diagnosing a myopathy usually requires laboratory tests.

Tendons stretch very little, so they are prone to injury and may become torn if a large amount of force is applied to them. Such injuries lead to tendinitis, which is inflammation of the tendons. Because tendons and ligaments are relatively poorly supplied with blood, they heal slowly and sometimes imperfectly. Injuries to ligaments and tendons require patience and careful longterm rehabilitation.

Bone diseases are generally present at birth or the result of nutritional deficiencies or injuries. An imbalanced level of minerals in the diet, particularly of trace minerals such as copper, zinc, and magnesium, is a common dietary cause of bone defects. Growing animals that are fed too much protein can also develop nutritional disorders affecting bones. Getting either too much or too little of certain vitamins, particularly vitamins A and D, can influence bone growth and development.

Most bone disorders stem from some sort of trauma, such as fractures or cracks. Bone fractures are classified as simple (the bone is broken into 2 pieces) or compound (there are 3 or more pieces). They are also divided into closed (the skin is not broken) or open (the skin is broken). Infections that cause bone tissue to break down and die can lead to bone disorders. In other situations, diseases of the ligaments or tendons may cause secondary bone troubles.

Movable joints are vulnerable to joint diseases or disorders affecting their membranes, as well as related ligaments, cartilage, and bone. Joint disorders may be caused by trauma to the joint, longterm inflammation, developmental problems, or infections. Traumatic injuries may produce short-term consequences such as dislocation, fracture, or the distortion of a joint. More longterm effects may include arthritis or the rupture of nearby ligaments or membranes.

Chronic or longterm inflammation is most commonly seen in joints associated with movement. The effects of longterm inflammation can be complicated. Any joint injury changes the composition and amount of fluid inside the joint, which affects the amount of pressure on the connecting bones. Injuries also raise the white blood cell count in a joint, which can break down cartilage over time.

Recent years have seen great advances in techniques for diagnosing and healing musculoskeletal disorders. When detected early, the disorders often can be corrected, allowing the horse a full return to healthy life.

LAMENESS IN HORSES

A horse is said to be lame when its normal stance or gait is changed by a problem in one or more of the limbs, the neck, the trunk, or the quarters. Lameness is not a specific disease but may indicate a disorder in the musculoskeletal system. When unaccompanied by pain, the lameness may be mechanical. A horse with mechanical lameness may develop an abnormal gait as a result of scarred and thickened connective tissues or of an abnormality of movement produced by a syndrome such as stringhalt. It is important that a veterinarian properly distinguish the cause of mechanical lameness.

Physically immature horses that are subjected to repetitive stress on bones are prone to lameness. Immature bones may be anatomically normal but weak due to the age of the horse. Such bone weakness may also be caused by developmental orthopedic disease. An example of repetitive stress on bones is the continuous training of racehorses around left-handed bends. This can produce shin soreness, stress fractures, or imbalances of the feet that abnormally distribute body weight among the limbs. Other factors that cause lameness include direct or indirect trauma to a limb, fatigue in racehorses racing over long distances, or inflammation—more often than not without infection—of joints, tendons, and ligaments. Lameness in one part of a limb will often produce soreness in another area of the limb as well. It can also lead to a secondary lameness in the fore- or hindlimb on the opposite side of the body.

The Lameness Examination

A thorough investigation of a lame horse is necessary in order to ensure a correct diagnosis and appropriate treatment. The examination begins with a full medical history. The horse's type, age, and training regimen may give important clues to the lameness. Your veterinarian will ask how much time has passed since the onset of lameness and how it has been managed. The length of time since the last shoeing will be noted, as well as any indication that the lameness improves with either rest or exercise. The horse's response to anti-inflammatory or pain-relieving medications may provide useful information. Results of laboratory tests may reveal other problems that influence overall performance.

The feet are thoroughly examined, inspecting any compression of the walls or soles with hoof testers. Wear patterns of shoes and feet are noted. In addition to the legs and feet, the back and neck should be thoroughly examined with the horse restrained and standing on a level surface.

Examination during exercise is done only if the degree of lameness is minor and longterm. If lameness is major and severe (for example, a suspected fracture), additional exercise could catastrophically injure the horse. It is important to tell your veterinarian if your horse was given pain-relieving medication before a lameness examination. In some cases, a ridden assessment of the horse may be necessary, particularly with a subtle lameness or a horse that is unwilling to perform certain movements (for example, a dressage horse).

Because lameness may indicate a nerve dysfunction, a neurologic examination should always be part of the lameness examination. This might include observing the horse execute "complicated" movements such as turning short, backing, "hopping" on one forelimb (with the other forelimb held up), and negotiating a small step. These tests help identify reduced response to stimuli, weakness, or spasms, as well as abnormalities in the motor function of the major muscle groups that flex and extend the limbs.

Imaging Techniques

Imaging techniques provide important information necessary to diagnose and treat specific conditions. Imaging can be divided into anatomic and physiologic methods. Anatomic methods are used to view physical structures, and physiologic methods assess how the anatomy functions. Anatomic imaging methods include x-rays and ultrasonography. Physiologic imaging methods, including scintigraphy and thermography, evaluate metabolism or circulation, completing the picture of the disorder. Imaging techniques may help narrow the problem to a specific region when other diagnostic methods fail. Imaging may also be necessary if the lameness is very subtle, or if the horse resists handling or an injection. Imaging may also help prevent injury by allowing early detection of the physiologic changes associated with injuries.

Imaging Techniques for Evaluation of Lameness

- *X-rays:* Plain x-rays allow evaluation of bony tissues and reveal longterm changes. Contrast x-rays (x-rays taken after a dye has been administered to the horse either intravenously or by mouth) provide information about joint cartilage and surfaces and are of particular value in determining whether cysts below the cartilage interfere with the joint.
- *Ultrasonography:* Ultrasonographic examination can be used to evaluate most soft tissues. It is most useful in the evaluation of tendons and ligaments but can also be used to evaluate muscle and cartilage. Used together, x-rays and ultrasonography provide a complete picture of bony tissues and the soft tissues that connect and support them.
- *Thermography:* This noninvasive technique creates a picture of the surface temperature of an object. It measures emitted heat and is useful for detecting inflammation that may contribute to lameness. Disease and injury nearly always affect blood circulation, which in turn affects the temperature of an injured area. Due to the increased blood flow to an affected area, thermography can identify "hot spots" of local inflammation. If a disease reduces blood supply to a particular area, that area of decreased heat is usually surrounded by a warmer rim, which thermography can also identify.
- *Scintigraphy:* During scintigraphy, a small amount of a radioactive substance is injected intravenously. Because inflammation causes a local increase in blood flow, the radioactive substance will accumulate in the inflamed area. A special camera that is sensitive to the injected radioactive substance is then used to locate the affected area.

Arthroscopy

Arthroscopy can be used to both evaluate and treat musculoskeletal disorders, particularly those involving the soft tissues of the joint. It is a minor surgical technique in which an endoscope (a long, flexible tube with a tiny camera in the tip) is inserted through a small surgical incision. Most arthroscopies take place with the horse under general anesthesia. Because it is a minimally invasive surgery, arthroscopy allows for rapid healing of soft tissues.

With diagnostic arthroscopies, an attached camera transmits images to a monitor allowing the veterinarian to see inside the joint. Corrective surgeries (such as removal of bone fragments or cartilage, or repair of fractures) can often be performed at the same time. Arthroscopy also enables biopsies within the joint.

Endoscopes may also be used to explore and correct problems involving the tendons or bursae, particularly in cases of infection. For a number of disorders, these techniques have revolutionized treatment and dramatically improved recovery rates.

Regional Analgesia

The use of regional, or local, painkillers (analgesics) to help diagnose lameness is an important part of the examination. This procedure is done before x-rays are taken. It is used when physical examination fails to identify the affected area (for example, in cases such as superficial foot pain, inflammation of the navicular bone and forefoot of the horse, traumatic joint disease, or inflammation of a ligament). A horse given diagnostic analgesia should be lame enough that any improvement can be detected. In conditions of severe lameness, the analgesia should be used with great care. Suddenly removing the protective effect of pain can cause a horse to further damage an injured area.

In most cases, interpreting the results of diagnostic analgesia is fairly straightforward. Pain reduction lessens the lameness, helping your veterinarian isolate the location of the injury. But various factors may complicate the procedure. Severe pain can be difficult to block fully. Also, mechanical and neurologic lameness, which are not necessarily painful to the horse, may be undetectable with analgesia. Sometimes a local anesthetic diffuses more widely than planned, reducing diagnostic accuracy. The arrangement of nerves in a given area may vary from horse to horse, and visible lameness may lessen as a horse "warms up." Overall, however, regional analgesia remains very useful for diagnosing lameness.

7

Tumors in Cow

BOVINE GRANULOSA CELL TUMORS

Granulosa cell tumors are the most common neoplasm involving the ovaries of cattle. Even though they are the most common ovarian neoplasm, the incidence has been reported to be approximately 0.5 per cent. Granulosa cell tumors affect all breeds of cattle but appear to occur more often in dairy cattle rather than beef cattle. Of the dairy breeds, Guernseys and Holsteins are most frequently affected; however, these breeds also represent the majority of the dairy breeds in absolute numbers.

The higher incidence of these neoplasms occurring in dairy cows may also be partially due to more intense reproductive management practiced on most dairy operations compared to beef cattle operations. The reproductive status of a cow on a dairy farm is evaluated more frequently through repeated rectal examinations than beef cattle on average. It has been proposed that this intense management results in a higher percentage of these tumors being diagnosed rather than dairy cattle being predisposed to developing them. These tumors have been diagnosed in every age of cow, ranging from virgin heifers to very old cows. Granulosa cell tumors have also been reported in pregnant cows, but the occurrence is rare.

These tumors vary in size and structure and are rarely malignant. They have been reported to range from 11.9 grams – 12.3 kilograms. Granulosa cell tumors are most often unilateral and may suppress the function of the contralateral ovary. When this occurs, the contralateral ovary becomes atretic while the ovary with the tumor may continue to grow. These tumors vary in size from relatively small, solid, yellow to white structures to large structures composed of multiple cysts (some being as large as 7.5 cm in diameter), a single large cystic structure, or a combination of solid and cystic structures. The surface of these tumors may be smooth or lobulated. They may be so highly vascularized that one can palpate fremitus in the middle uterine artery. The occurrence of malignant granulosa cell tumors is rare and metastasis is even more rare. It has been reported, however, that an aged cow experienced metastases to the liver. Cows with granulosa cell tumors may present with a variety of clinical signs ranging from anestrus to nymphomania to male-like behavior. Other clinical signs that may be present include abnormal estrous cycles, deepening of the voice, virilism, and mammary gland development. Mammary gland development and colostrum-like secretions have been reported in virgin heifers. The presence of vaginal discharge and enlargement of the vulva and clitoris may also be present. Cows have also been presented for chronic weight loss due to persistent heat or male-like behavior. These cows may be noticed constantly riding other cows within the herd expressing bull-like behavior. The presence of each of the above clinical signs depends on how much functional luteal tissue is present within the tumor. The amount of functional tissue relates to the dominant hormone being produced within the tumor. In the majority of cases, the tumor continues to grow and may cause affected cows to progress through a series of various stages beginning with nymphomania and ending with virilism.

Diagnosis of granulosa cell tumors is most frequently made through rectal palpation. Most often one will palpate one ovary that is abnormally enlarged while the contralateral ovary will be small. As a rule, granulosa cell tumors should be suspected when the affected ovary is greater than 10 cm in diameter. In some cases, due to the size and/or weight of these tumors, it may be impossible to palpate if located deep in the abdominal cavity. Using a board to pry the abdomen from ventral to dorsal may be one way to make rectal palpation possible. The use of cervical forceps may also be an option to retract the uterus and ovaries to make rectal palpation effective. Ultrasound examination is another means of diagnosing granulosa cell tumors. Upon ultrasound examination the affected ovary may appear to have a honeycomb appearance. If the majority of the tumor tissue is solid the tumor will appear solid and may be misdiagnosed as a large corpus luteum. A granulosa cell tumor should also be suspected when a chronic cystic structure does not respond to conventional treatment regimens. Cystic ovaries may produce the same clinical signs as a granulosa cell tumor. Monitoring a cow's response to exogenous hormones or prostaglandins may aid in differentiating between the two conditions. Another more practical way to differentiate between a granulosa cell tumor and cystic ovaries is serial rectal palpations. Most often a granulosa cell tumor will continue to enlarge, whereas cystic ovaries tend to be more static. However, the only way to have a definitive diagnosis for a granulosa cell tumor is through histopathological examination of the tumor sections.

Treatment of granulosa cell tumors is limited. Unilateral ovariectomy is the only treatment option available. Most often, this is only used in valuable breeding stock. It is not recommended to perform this on animals that exhibit changes in secondary sexual characteristics such as enlargement of the vulva and clitoris. Animals showing these changes have a reduced fertility post surgical removal of the tumor. It has also been reported that removing the ovary in cows exhibiting these changes resulted in cystic follicles on the contralateral ovary. As a rule, cattle have reduced fertility after a unilateral ovariectomy when compared to the mare. Economically it is often not feasible to perform surgery to remove the tumor on animals with average genetics. The surgical approach for removing the ovary depends on the size and location of the ovary and the demeanor of the cow. Most dairy cows will stand for the procedure, whereas beef cattle often need to be cast into lateral or dorsal recumbency. If the ovary fits in the palm of one's hand, most often the ovariectomy via colpotomy approach is successful. In order to safely remove large tumors the surgical approach recommended is through a laparotomy incision in the paralumbarfossa or on ventral midline.

BOVINE LEUKOSIS

Lymphosarcoma in cattle may be sporadic or result from infection with bovine leukemia virus (BLV); the latter is often referred to as an enzootic bovine leukosis. Sporadic lymphosarcoma in cattle is unrelated to infection with BLV. Despite the lack of association, animals with sporadic lymphosarcoma may possibly be infected with the virus. Sporadic lymphosarcoma manifests in three main forms: juvenile, thymic, and cutaneous. Juvenile lymphosarcoma occurs most often in animals <6 mo old, thymic lymphosarcoma affects cattle 6–24 mo old, and cutaneous lymphosarcoma is most common in cattle 1–3 yr old.

Etiology, Transmission, and Epidemiology

Enzootic bovine leukosis is caused by BLV, an exogenous C-type oncogenic retrovirus of the BLV-human T-lymphotropic virus group. BLV has a stable genome, does not cause chronic viremia, and has no preferred site of proviral integration. Despite the lack of preferred proviral integration sites, the tumors generated by the virus in a single individual are typically monoclonal and have a single integration site. The virus escapes the immune response by low levels of viral replication. It appears that replication is blocked at the transcriptional level, but the mechanism is not completely understood.

The prevalence of BLV infection varies from country to country. Many European countries, Australia, and New Zealand have eradication programs in place that have led to negligible rates of BLV infection. Although voluntary control programs are in place in the USA, prevalence is high compared with much of the rest of the world. The most recent surveys in the USA estimate that 44% of dairy and 10% of beef cattle are infected with the virus. Prevalence tends to increase on dairies with increasing herd size, while the converse is true in beef cattle. In general, the prevalence of viral infection increases with age. Cattle are infected with BLV through the transfer of blood and blood products that contain infected lymphocytes. Once infected, cattle develop a lifelong antibody response, primarily to the gp51 envelope protein and the p24 capsid protein. B lymphocytes harbor the integrated provirus but rarely express viral proteins on their cell surface. The exact site of viral replication and expression that drives the immune response remains elusive.

Under experimental conditions, most routes of viral exposure can successfully transmit infection. However, many of these settings are unlikely to be encountered naturally. Many bodily fluids, including urine, feces, saliva, respiratory secretions, semen, uterine fluids, and embryos, have been examined for their ability to transmit BLV and are considered to be noninfectious. Only on rare occasion has virus been found in these fluids. Colostrum from BLV-positive cows contains virus and has been found to be infectious experimentally. However, colostrum also contains large amounts of antibody, and it is believed that the protective effects of colostral antibody outweigh the infectious potential when colostrum is administered in a normal fashion.

Most BLV transmission is horizontal. Close contact between BLV-negative and BLV-positive cattle is thought to be a risk factor. Many common farm practices have been implicated in viral transmission, including tattooing, dehorning, rectal palpation, injections, and blood collection. Vectors such as tabanids and other large biting flies also may transmit the virus. Vertical transmission may occur transplacentally from an infected dam to her fetus, intrapartum by contact with infected blood, or postpartum from the dam to the calf through ingestion of infected colostrum. Any material that is blood contaminated or lymphocyte rich has the potential to infect animals with BLV.

Pathogenesis

There are three main outcomes in cattle infected with BLV. Most animals remain persistently infected with no outward signs of infection. Approximately 29% of BLV-infected cattle develop persistent lymphocytosis, while <5% of BLV-infected cattle develop lymphosarcoma.

Persistent lymphocytosis is sometimes referred to as a preneoplastic syndrome, but there is no convincing evidence that affected cattle have an increased risk of developing lymphosarcoma. The lymphocytes present in persistent lymphocytosis are not neoplastic, although they may have mild reactive changes consistent with normal blood smears in cattle. Persistent lymphocytosis is considered a benign condition associated with BLV infection. For this reason, it is often overlooked. However, these cows may serve as a reservoir of infection.

The increased lymphocyte count is attributed to a 45-fold increase of infected $CD5^{+}$ and a 99-fold increase in infected $CD5''$ B cells. It has been suggested that cows with persistent lymphocytosis may be at greater risk of passing BLV infection on to their calves in utero and may show decreased milk production and alteration of milk components.

Lymphosarcoma is rarely seen in animals <2 yr old and is most common in the 4- to 8-yr-old age group. Less than 5% of BLV-infected cattle develop lymphosarcoma. Lymphosarcoma, including both sporadic and enzootic forms, is one of the main causes of condemnation of adult dairy cows at slaughter.

Clinical signs associated with development of lymphosarcoma are highly variable, because the affected organ(s) will dictate the predominant clinical signs.

Juvenile lymphosarcoma is often characterized by a sudden onset of diffuse lymphoid hyperplasia with or without visceral organ involvement. Weight loss, fever, tachycardia, dyspnea, bloat, and posterior paresis have all been described with this form of lymphosarcoma. Profound lymphocytosis (>50,000/ìL) often accompanies this fatal form of bovine lymphosarcoma. Thymic lymphosarcoma may involve the cervical or intrathoracic thymus, or both. Clinical signs associated with this form of lymphosarcoma depend heavily on the location and size of the tumor. A cervical swelling may be evident. Dyspnea, bloat, jugular distention, tachycardia, anterior edema, and fever have been documented. The affected cell population is an immature, poorly differentiated lymphocyte. Cutaneous lymphosarcoma presents as cutaneous plaques, 1–5 cm in diameter, on the neck, back, rump, and thighs. Regional lymph nodes may also be enlarged. This form of lymphosarcoma may undergo spontaneous remission; however, relapses may occur.

Lesions

Animals with BLV-associated lymphosarcoma commonly show lesions in the central or peripheral lymph nodes, leading to lymphadenopathy. Lesions of the abomasum may lead to signs of cranial abdominal pain, melena, or abomasal outflow obstruction. Pelvic limb paresis progressing to paralysis can occur in animals with extradural spinal lesions. Retrobulbar lesions cause protrusion of the globe, resulting in exposure keratitis and eventually proptosis. Lesions of the right atrium may be mild and undetectable clinically, or may cause arrhythmias, murmurs, or heart failure. Uterine lesions may lead to reproductive failure or abortion. Lesions of the internal organs typically involve the spleen, liver, or kidneys and ureters. Lesions of the spleen are often initially asymptomatic but may result in rupture of the spleen and exsanguination into the peritoneal cavity. Lymphosarcoma of the liver is often asymptomatic but may lead to jaundice and liver failure. Disease of the kidney and ureter can lead to abdominal pain and the subsequent development of hydroureter or hydronephrosis and clinical signs associated with renal failure.

Lymphosarcoma may appear as yellow-tan, discrete nodular masses or a diffuse tissue infiltrate. The latter pattern results in an enlarged, pale organ and can be easily misinterpreted as a degenerative change rather than neoplasia. Histologically, the tumor masses are composed of densely packed, monomorphic lymphocytic cells.

Diagnosis

Lymphosarcoma is often included on the differential diagnosis list for many diseases because of the wide range of clinical findings. Viral infection is diagnosed by serology or virology, persistent lymphocytosis is identified by hematology, and neoplastic tumors are identified by histologic examination of biopsies. Positive serology or virology for BLV confirms viral infection but not the presence of lymphosarcoma.

Serology is the most common and reliable way to diagnose infection with BLV. Agar gel immunodiffusion is still recognized by most countries as the official import/export test, but ELISA is the most common test for routine diagnostic use. Serology is unreliable in calves that have ingested colostrum from BLV-positive cows because of the passive acquisition of maternal antibodies that typically wane by 4–6 mo of age. PCR is a sensitive and specific assay for diagnosis of BLV infection in peripheral blood lymphocytes. This test can identify proviral DNA of BLV in the lymphocytes of infected animals and differentiate positive from negative calves in the presence of maternal antibodies. The diagnosis of lymphosarcoma must be made by cytology or histopathology. Cytologic diagnosis is sometimes difficult because of the frequency of blood contamination of the aspirates.

Treatment and Control

There is no treatment for viral infection or for lymphosarcoma in cattle, although parenteral corticosteroids can transiently decrease the severity of clinical signs. Eradication programs have been developed but success has been variable, primarily because of the expense and high prevalence of infection among cattle in the USA relative to the economic cost of disease. The most commonly recommended eradication protocol is as follows: 1) identify infected animals using a serologic test, 2) cull seropositive animals immediately, 3) retest the herd in 30–60 days, 4) use PCR to test young calves and as a complementary test to clarify test results in herds with a low prevalence of infection, and 5) repeat testing and cull until the entire herd tests negative. Testing is then repeated every 6 mo. The herd is declared free when there have been no positive tests for 2 yr. Additions to the herd should have two negative tests 30 and 60 days before arrival.

When test and cull programs are economically untenable, test and segregation programs have been recommended but are rarely implemented. These programs necessitate running two completely separate operations and require additional resources, including money, time, and available workforce.

Prevention

Eliminating the movement of blood from infected animals to naive animals is the cornerstone of prevention protocols. In calves, feeding colostrum from seronegative cows is often advocated. However, most epidemiologic evidence suggests that the protective effect of colostral antibody outweighs the risk of infections, particularly in high prevalence herds. The replacement of whole milk feeding with high-quality milk replacer may also be considered. Bloody milk should never be fed to calves.

Cautery or other bloodless methods of dehorning should be used. Equipment used for castration, tattooing, ear tagging, or implanting should be adequately cleaned and disinfected between animals.

Transmission can be decreased in adult cattle by changing rectal sleeves in between cows. Artificial insemination or embryo transfer (using negative recipients) may limit transmission. In beef herds, the use of a negative bull may limit transmission, but natural service is an uncommon method of viral transmission unless breeding is traumatic.

Additional recommendations include disinfection of equipment that has come in contact with blood or body tissue. Single use, disposable needles should always be used for blood collection and IM injections. It is preferable to use single-use disposable needles for vaccination, but the risk of transmitting BLV virus via SC vaccination is low. Handling facilities that become contaminated with blood should be cleaned between animals. Fly control helps minimize the potential for tabanid-associated transmission. Blood transfusions and vaccines containing blood, such as those used for babesiosis and anaplasmosis, are particularly potent ways to spread the disease, and donors must be carefully screened.

BOVINE OCULAR SQUAMOUS CELL CARCINOMA (BOSCC) AND ITS SURGICAL MANAGEMENT.

Animals also battle with cancer. This segment is on Eye cancer in cattle

Eye cancer in cow is a very painful and a fatal disorder which start as small nodular growth in the adnexa of eye. Squamous cell carcinoma is regarded as the most frequently diagnosed cancerous tumour in the bovines. Bovine ocular squamous cell carcinoma (BOSCC) represents the most economically important neoplasm in large animals. The malignant tendencies of this disease make early recognition critical.

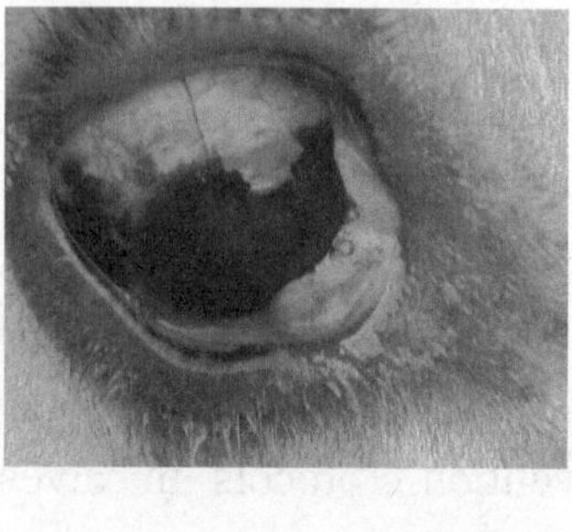

I got opportunity to see a wide variety of Ocular Squamous cell carcinoma cases in working bullocks and dairy cattle. UV rays from scrotching sunlight could be the main source of eye cancers in working bullocks and free grazing dairy cattle especially in states like Maharashtra and northern Karnataka.

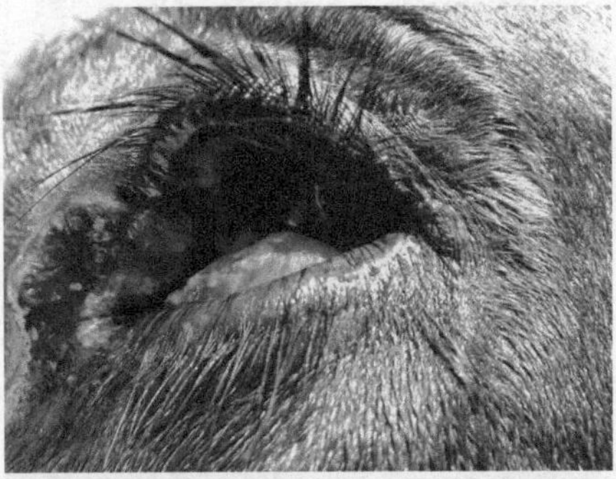

Prolonged exposure to ultraviolet light, lack of pigment and lack of hair at the affected site are common factors associated with the development

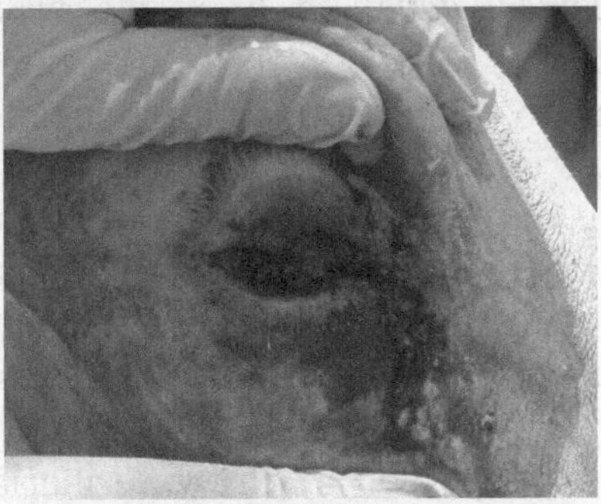

of a squamous cell carcinoma.These tumours in all species develops through a series of premalignant stages before proceeding to carcinoma. Squamous cell carcinomas are solitary lesions, however multiple tumours may develop in conjunction with solar injury.Human negligence and callous attitude can lead to severe inflicting injuries to the eye which ultimately culminates in eye cancer.

I would like to share my experience during my tenure in Animal Rahat, PETA, the largest working animal welfare organization dedicated for working bullocks. We witnessed many animals in a northern Maharashtra village with eye cancer on the right eye. When enquired, it was a shocking finding that the abuse of the eyes by rubbing a combination of pepper and salt to make the animals run during bullock races and sugarcane season had led to high incidence of eye cancer.

I am of the opinion that Willful persuasive extension education should be part and parcel of the veterinary practice and I strongly believe that it is the duty of every vet to educate the animal owner/ farmer about what the animal is going through. Thus eye cancer not only drastically reduce the performance of the animal and considerable economic importance to the farmer, but also has serious welfare constraints in animal ethology point of view.The cancer growth may start at the medial canthus, lateral canthus or on eyelid. Third eyelid is the most common site of origin of eye cancer. The stage of presentation of the case decides the chances of saving an affected eye.

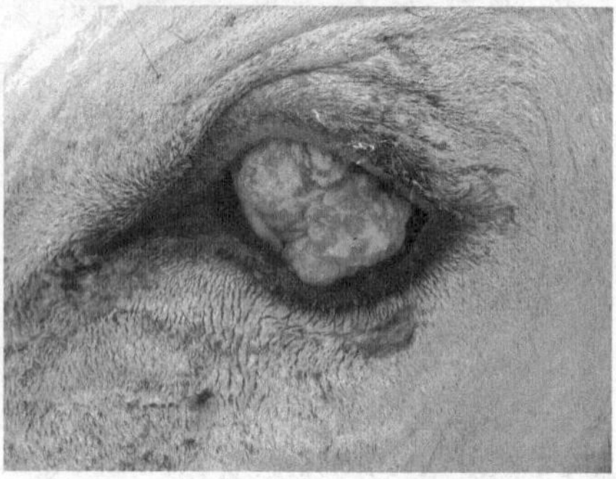

Cases presented in initial stages can be fully recovered and the eyes can be saved. But many cases come in advanced chronic stage with extensive tumor mass eating up the entire globe.

Treatment of cancer eye depends on its location and the degree of involvement of ocular structures. Veterinarians easily remove lesions on the third eyelid by cutting out the free border of the third eyelid using local anesthesia and tranquilization. Lesions removed in this way are very unlikely to recur. Applying hyperthermia (high temperature) or cryotherapy (freezing) to kill the remaining tumor cells in the tissue of the lids but the procedure can not be employed for all cancers.

Three terms are widely used by the veterinarian in this context are Extripation or Exenteration, Evisceration, Enucleation

Enucleation is removal of the entire globe, including the cornea, sclera, and a portion of the optic nerve.

Evisceration surgery involves removal of the contents of the globe while leaving the sclera and optic nerve in place. The cornea can be retained in some evisceration cases.

Exenteration of Eye or Extripation of Eye ball is the removal of the globe, as well as the complete or partial removal of the soft tissues of the orbit.

The ultimate goals of these surgeries are to safely and effectively remove the diseased eye or orbital contents using advanced surgical techniques, eliminate the severe underlying ocular pathology

Removal of the entire eyeball is indicated when tumors have spread to the extent that the eyeball is blind; the tumor has invaded deeper structures surrounding the globe; or the eyelid is involved to the extent that it cannot be repaired after removal of the tumor. It may appear to be a drastic procedure, but success of the extirpation is to accomplish it quickly and humanely.

ANETSHESIA AND RESTRAINT

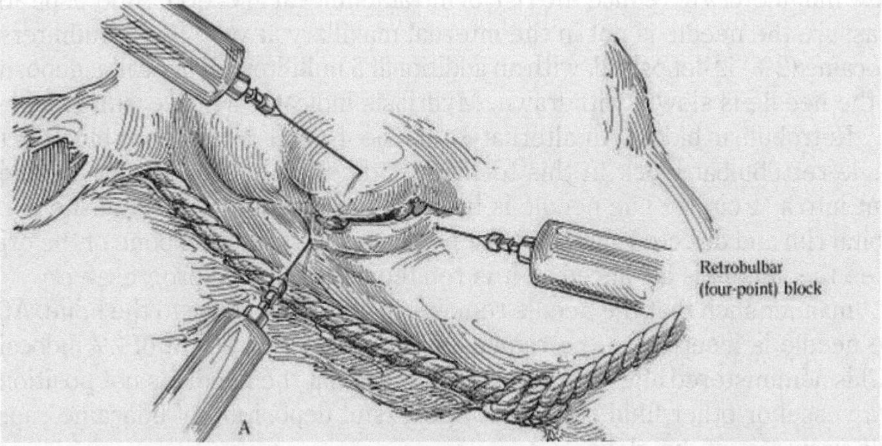

4 Point Retrobulbar Nerve Block: The 4-point retrobulbar block is technically easier and can be done more rapidly as compared with the Peterson eye block. In this technique, an 18 gauge, 9-cm long needle is introduced through the skin on the dorsal, lateral, ventral and medial aspects of the eye, at 12, 3, 6, and 9 o´clock positions, respectively. Introduction of the needle through the

conjunctiva should be avoided to reduce the occurrence of ocular contamination. The needle is directed behind the globe using the bony orbit as a guide. When the needle is introduced into retrobulbar sheath, the eye will move slightly with the tug of the needle. After this location is reached and aspiration is performed to assure that the needle is not in a vessel, 5-10 milliliters of lidocaine (2%) is deposited at each site. Mydriasis indicates a successful block.

Other ocular nerve blocks in cattle are

Auriculopalpebral Nerve Block: The auriculopalpebral nerve can be palpated as it crosses the zygomatic arch, roughly 5-6 centimeters behind the supraorbital process. Inject 5 milliliters of 2% lidocaine HCl subcutaneously on the dorsal aspect of the zygomatic arch at this location.

Peterson Nerve Block: This is a complicated one.

After performing a small local skin block over the intended site of puncture, a 3.8-cm long 14 gauge needle is inserted through the skin as a cannula for introduction of an 18-gauge 9-cm long needle for the nerve block. The cannula is inserted caudal to the junction of the supraorbital process and zygomatic arch and is introduced through the skin. Then, the 18-gauge, 9-cm long needle is introduced through the cannula needle and is directed in a horizontal and slightly dorsal direction until the coronoid process is encountered. The needle is "walked off" the rostral aspect of the coronoid process and advanced in a ventromedial direction along the caudal aspect of the orbit until the needle encounters the bony plate encasing the foramen orbitorotundum. Once the needle is advanced to the foramen, it is advised that the needle be drawn back a few millimeters to reduce the risk of intrameningeal injection. After aspirating to assure the needle is not in the internal maxillary artery, 10-15 milliliters of lidocaine (2%) is deposited, with an additional 5 milliliters of lidocaine deposited as the needle is slowly withdrawn. Mydriasis indicates a successful block.

Retrobulbar block: An alternative to the 4-point retrobulbar block is the single retrobulbar block In this technique, the 9-cm long 18-gauge needle is bent into a ½ circle. The needle is inserted immediately ventral to the dorsal orbital rim and directed such that the needle impacts into the bone of the orbit. Then the needle is advanced as it is rotated ventrally in a progressive

manner such that the needle remains in close proximity to the bone. After the needle is inserted to the caudal aspect of the eye, 20 ml of 2% lidocaine HCl is administered after aspiration to ensure that the needle is not positioned in a vessel or other fluid structure. Successful deposition of lidocaine causes mild proptosis of the globe.

Ring Block: Additional local anesthesia of the eyelids is recommended as the Peterson and retrobulbar blocks typically result in incomplete analgesia of the eyelids. Five to ten milliliters lidocaine (2%) is infiltrated subcutaneously 2.5 centimeters from the eyelid margins as a ring block.

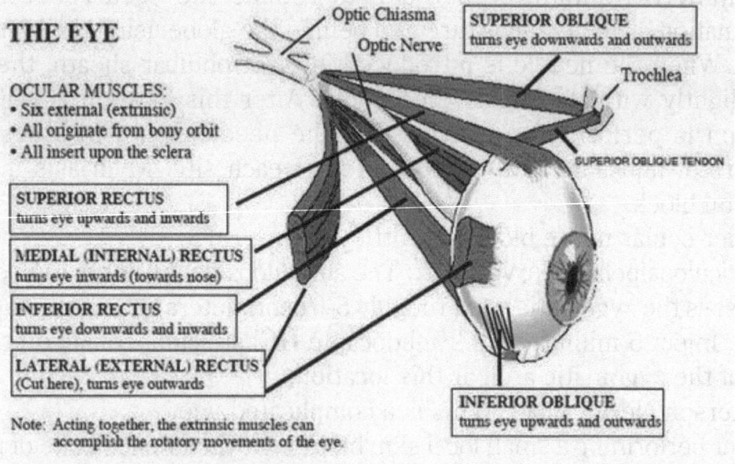

Surgical procedure of Exenteration

A transpalpebral ablation technique is utilized to remove the eye. The upper and lower eyelids are sutured closed or alternatively, eyelids can be closed using multiple towel. A circumferential skin incision is made approximately 1 centimeter from the edges of the eyelids. Using a combination of blunt and sharp dissection, Mayo scissors are used to dissect through the orbicularis oculi muscle, fascia, and subcutaneous tissue surrounding the eye. The interior of the bony orbit is used as a guide. The medial and lateral canthal ligaments are sharply transected to allow access to the caudal aspect of the orbit.

As there is a large vessel associated with the medial canthus and optic stalk which need to be ligated after the resection. A vascular clamp can aid in hemostasis while additional excision of remaining orbital tissue is undertaken. In cases where neoplastic infiltration of the bony orbit has occurred, affected areas of ocular periosteum should be thoroughly excised. I had not encountered any bony lesions extending from BOSCC so far.

The skin incision can be closed in a variety of patterns with a nonabsorbable suture such as No. 3 nylon. Common patterns include the Ford interlocking, cruciate or simple continuous after sufficiently obliterating the dead space with iodine impregnated gauze rolls. This needs to be changed for 5 dyas till the cavity dries and heals up.

Post-operative Care: The animal should be kept in a confined area for several days after surgery to allow for appropriate hemostasis to occur. Daily observation of the surgical site and assessment of general well being is recommended until suture removal. Sutures should be removed in 14 to 21 days to allow for complete healing of the skin.

EFFICACY OF COW URINE THERAPY ON VARIOUS CANCER PATIENTS

In cow blood, there is pran shakti (Life Force). Cow urine is cow's blood that is filtered by kidney. Kidneys filter blood. Whatever elements are present in blood are present in cow urine also. Cow, as Kamdhenu is the provider of all happiness 'Matrah Sarva Bhutanam, Gavah Sarv Sukh Prada'. Meaning 'Cow being mother of all living entities gives all pleasures to everyone'.

Cow is a mobile dispensary. It is the treasure of medicines. The cow urine therapy is capable of curing several curable and incurable diseases. The holy texts, such as Atharva Veda, Charak Samhita, Rajni Ghuntu, Vridhabhagabhatt, Amritasagar, Bhavprakash and Sushrut Samhita, contain beautiful description about these things. Cow Urine Treatment and Research Center has conducted a lot of research in the past few years and reached at the conclusion that it is capable of curing diabetes, blood pressure, asthma, psoriasis, eczema, heart attack, blockage in arteries, fits, cancer, AIDS, piles, prostrate, arthritis, migraine, thyroid, ulcer, acidity, constipation, gynecological problems, ear and nose problems, abortion and several other diseases.

The analysis of cow urine has shown that it contains nitrogen, sulphur, phosphate, sodium, manganese, carbolic acid, iron, silicon, chlorine, magnesium, maleic, citric, tartaric, succinic acids and calcium salts, vitamin A, B, C, D, E, minerals, lactose, enzymes, creatinine, hormones and gold acids. A person falls ill when there is deficiency or excess of the substances inside the body. The cow urine contains those substances, which are present in the human body.

Therefore, consumption of cow urine maintains the balance of these substances and cures incurable diseases.

Persons suffering from cancer had been told that they would not survive for long; cow therapy was done on such cases. Out of the many patients, who were suffering with cancer for the last 4 years are now leading a healthy life. In the same way, the diabetes patient who was taking insulin and having a sugar level of 488 or 420 mg/ dl did not have the necessity to take insulin after this treatment. In the same way AIDS, asthma, psoriasis, eczema, blood pressure, heart disease, prostrate, piles were also cured with this treatment. Subject 1 was suffering from asthma and eosinophilia. He had palpitations and cough along with phlegm. He is now enjoying a healthy life after taking cow urine therapy. The cough and phlegm reduced. Subject 2 was having varicose veins and dismenorrhoea. She used to have swollen nerves, pain and swelling as soon as she used to stand up and it was very painful at the time of menses. She was advised operation for both the problems. However, she went for cow urine therapy and is now healthy. Subject 3 suffered a heart attack; he had high levels of cholesterol and chest pain. After taking cow urine therapy, he is able to walk up to 4 km, he does not get chest pain and cholesterol level is also normal. Subject 4 suffered from cancer of the food pipe 4 years back; he could not take liquids also. Now he is able to eat and drink properly, he is healthy and working in the fields. There are thousands of people who have recovered from serious diseases after taking this therapy.

Today, many AIDS patients are taking cow urine therapy. People who were suffering with migraine and headache for the past 15 years have recovered within 6 months of taking this therapy. It is an important measure of relieving tension or improving the memory power. In the past few years the Cow Urine Treatment and Research Center has treated 150,000 of people. Out of the total patients, 85- 90% were patients of constipation; there is an old saying that if the stomach is clean half of the diseases get cured automatically. The patient taking cow urine therapy says that he is enjoying sound heath and passing bowels easily within 1 month of this therapy.

In Charaka-Samhita, Sushruta Samhita and Vangbhatt, we can find there are eight types of animal urines that can be used in medicine. These can be divided in two groups:

Male: Elephant, camel, horse

Female: Cow, goat, sheep, buffalo

However, according to Sushruta Samhita and Ayurveda Samgraha for medicinal and other purposes cow urine is beneficial and better than all other animals. Cow, *Bos indicus*, is most valuable animal in all *Veda* and it is called as the 'Mother of all'. The use of gomutra (cow urine) in India can be traced back to the Vedic times and probably it was used in the prevedic period also. In Vedas and Dharmashastras, it has been well documented as a purifying substance and as an ingredient of the Panchayagavya (Milk, Urine, Dung, Ghee and Curd of Cow). Essentially, gomutra is a Kshara dravya and finds place in chemical preparation and also as a substance for purification. In India, gomutra is used by the majority of rural population as folklore remedy in almost all the states. As per Ayurvedic literature, gomutra has katu, tikta, kashaya rasa, ushna-veerya (potency) kshra, laghu and deepana prabhav. It is useful in number of diseases particularly in gulma and kusta, ascites, elephenthsis, aburda (cancer), etc.

Traditional Uses of Gomutra

Cow urine have been used for lot of diseases, some of the uses are:

Uses of cow urine in purification and detoxification process

1. Purification of guggal
2. Detoxification of aconite
3. Detoxification of dhatura
4. Detoxification and purification of silver
5. Purification of loha and bhallataka

The composition containing cow's excretions urine, dung, cow's milk, curd and ghee, the five ingredients together known as 'Panchagawya' is given to women after she delivers a baby. Panchagawya is the main ingredient of many of our Ayurvedic preparations. Cow urine is believed to have therapeutic value and used in many formulations.

From time immemorial, we have the practice of mopping the floors (clay buildings) in our villages with cow dung. Even today, we have the practice of sprinkling water dispersed with cow dung at the entrance of our houses and the belief is that it repels microbes from entering into the house. On religious occasions (ceremonial occasions), our grandparents use to rinse the utensils with the diluted cow dung solution/dispersion. All these suggest that our forefathers believed that cow dung has anti-microbial effect and cow urine works as a purifier. It is even believed that textiles especially the new ones treated with the ashes of cow dung before they are worn for the first time would have a beneficial effect, probably a hygiene benefit.

Ancient Indian Vedic Scriptures including Manu Smriti, Charaka Samhita and Sushruta Samhita quoted the rational use of these animal products for diseases of the liver, gastro- intestinal systems, heart, lungs, cancer and many others.

COW URINE WINS OVER DISEASES

Cow urine has amazing germicidal power to kill varieties of germs. All germ-generated diseases are thus destroyed. Cow urine balances the tredosh (mucous, bile and air), thus diseases are cured. Cow urine corrects functioning of liver. So, liver makes healthy pure blood.

It gives disease resistance power to the body. Cow urine has all elements, which compensate for deficiency of nutrients in our body, which are required for healthy life. Cow urine contains many minerals especially Copper etc. It compensates for bodily mineral deficiency. The presence of gold salts protects body against diseases. Mental tension hurts the nervous system. Cow urine is called madhya and hridya, which means it, gives strength to brain and heart. Thus, cow urine protects heart and brain from damages caused by mental tension and protects these organs from disorders, and excessive use of any medicine leaves some residue in our body, this residue causes diseases.

Cow urine destroys the poisonous effects of residues and makes body disease free. Electric currents (rays) keep our body healthy. These currents (rays) are present in environment. These rays in form of extremely small currents enter our body through copper. We get copper from cow urine. To attract these electric waves is a quality of copper. Thus, we become healthy.

The nutrients, which are less available in food, are compensated by nutrients present in cow urine and thus help to regain health by acting against the voice of soul (immoral and sinful action), the heart and mind become narrow minded. Due to this, the functioning of body is affected and causes diseases. Cow urine provides mode of goodness. Thus helps us to perform correct activities by mind and protects from diseases. In scriptures, some diseases are said to be due to actions performed in previous lives, which we have to bear.

Ganga resides in cow urine. Ganga is destroyer of sins, thus cow urine destroys such previous sins and so diseases are cured. The diseases caused by entrance of ghosts in body are cured by the intake of cow urine. The Master of ghosts is Lord Shiva. Lord Shiva holds Ganga on his head. Ganga is in cow urine also. Thus, by taking cow urine, the ghosts get to see Ganga over their master's head. So, they are calmed and become peaceful. So, they do not trouble the body. Thus, diseases caused by entrance of ghosts are also destroyed. By regularly taking cow urine before sickness, we get so much immunity that any attack of diseases is repulsed. Cow urine being miraculous poison destroyer destroys the disease caused by poison (Toxin). Extremely dangerous chemicals are purified by cow urine.

Cow urine provides immunity power by increasing resistance power against diseases in human body. It contributes to immunity power. It is not only non-poisonous but also destroys poison. It is anti-toxin. There are many formulations of cow urine in market that have potential therapeutic uses.

MATERIALS AND METHODS

Materials

Patient data relevant to the study were obtained from the following sources:

- Patients data collection form;

- Patient's prescriptions;
- Patient's pathology reports;
- Patient interview;
- A questionnaire was developed to assess the efficacy of cow urine therapy.

There was a high proportion (30.87%) of throat cancer and the other prevalent cancer was breast cancer (14.70%) followed by cervix and uterine cancer (5.88%), buccal cavity cancer and sinus (4.41%) lung cancer, lymphoma and bone cancer (2.94%), both throat and buccal (5.88%) and other cancer (8.82%), respectively.

The symptoms (pain, inflammation, burning sensation, difficulty in swallowing, irritation etc.) of cancer patients were categorized in severe, moderate and mild categories, respectively.

Persons suffering from cancer were told that they would not survive for long; cow urine therapy was performed on such cases. Out of the many patients, who were suffering from cancer for the last 4 years are now leading a healthy life.

Intensive study of the data of cancer patients revealed that the degree of severe, moderate and mild symptoms were 82.16%, 15.8% and 1.58% on the first day and 7.9%, 55.3% and 36.34% on the seventh day, respectively.

It was evaluated that patients who were receiving cow urine therapy since 2-3 months were most benefited. Hence, this traditional therapy is really a boon to cancer patients.

CANCER EYE IN CATTLE

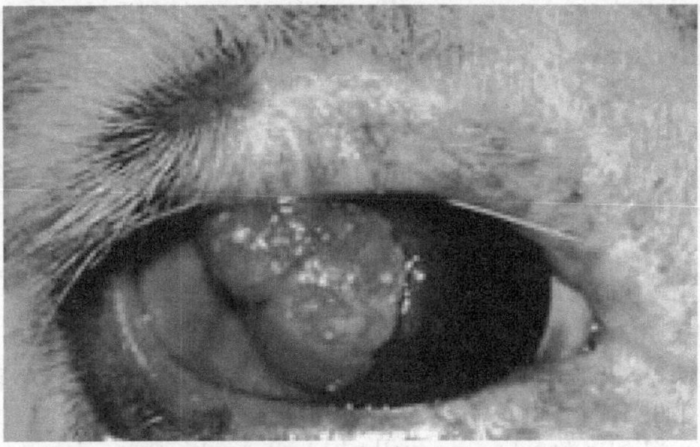

A typical cancer of the eyeball, characterised by a pink, fleshy growth. Animals with such cancers should be disposed of promptly.

Cancer eye (ocular squamous cell carcinoma) is the most common form of cancer in cattle. It is a skin cancer occurring on the eye and eyelids. Estimates of its prevalence vary from 1% to 20% in some herds. Some of the predisposing factors are heritable, and in certain lines or families nearly all cattle may be affected during their lifetime. Australian Quarantine Inspection Service (AQIS) abattoir figures (from August 2002 to August 2003) indicate that 22% of total condemnations in mature cattle, and 7.3% of condemnations in young cattle (less than 8 teeth, or younger than approximately 4 years) are due to cancer eye. Overall, cancer eye represents 47% of all cancers that result in condemnation.

FACTORS AFFECTING INCIDENCE OF CANCER EYE

BREED

The disease occurs in cattle that have areas of unpigmented skin or conjunctiva. Breeds such as Hereford, Poll Hereford and white-faced Friesian are most susceptible, with Hereford crosses being less susceptible than purebreds. Cancer eye is rare in breeds with fully pigmented skin, such as Angus and Santa Gertrudis.

STRAIN AND SIRE

Susceptibility to the disease varies with strain and sire. The heritability of cancer eye has been estimated as moderate at 0.4.

PIGMENTATION

Cattle with unpigmented skin around the eyes are more susceptible to cancer eye. Cancers usually start on unpigmented skin, but can then spread to pigmented areas. Eyelid pigment is present at birth and is easily assessed. Eyeball pigment usually takes more than 5 years to develop fully. Pigmentation of the eyelids and pigmentation of the eyeball are highly heritable and appear to be genetically related.

CONFORMATION

It is likely that protruding eyeballs are more susceptible than 'hooded' eyes, which are more protected from sunlight.

AGE

Older cattle are more prone to cancer eye. The disease is uncommon in cows under 5 years of age. Males and females are equally susceptible, but the cancer is seen less frequently in males because steers are sent for slaughter at an early age, and bulls tend to be culled at an earlier age than cows.

NUTRITION

It has been reported that a high level of nutrition and growth rate increases the risk of cancer eye developing.

Anecdotal evidence suggests that iodine deficiency may be associated with an increased risk of cancer eye.

Cancer eye is more common during and after droughts due to longer hours spent grazing in the sun, more sunlight, dust and flies, and because stress lowers immunity.

PREGNANCY

Cancer eye often seems to develop or progress more rapidly in the latter half of pregnancy. This may be due to increased stress or the immunosuppression associated with pregnancy.

It may simply be that farmers become more aware of a cancer eye when faced with the decision of whether it is humane to keep an affected cow long enough to calve and raise the calf.

VIRUS

Papillomavirus has sometimes been isolated from animals with cancer eye, but advanced virological techniques have so far failed to reveal any definite association between papillomavirus and the development of these tumours.

ULTRAVIOLET SOLAR RADIATION

Increased ultraviolet (UV) radiation is thought to predispose cattle to cancer eye. Therefore cattle living at low latitudes and high altitudes may be at greater risk, because latitude, altitude and daily hours of sunlight all contribute to the amount of UV radiation.

CANCER EYE LESIONS

Cancer eye growths can occur on the eyeball, eyelid or third eyelid.

All eye cancers develop from precursor lesions, but less than half of these lesions develop into cancer. The precursor lesion can be:

- A plaque (a slightly elevated, flat, opaque area on the eyeball);
- A papilloma (a wart-like growth protruding from the eyeball);
- A keratoma (a horn-like projection attached to the eyelid);
- A small ulcer on the eyelid.

As the disease progresses from the precursor stage to the cancer stage, the tumour becomes ulcerated. Bleeding and weeping are common. Bacteria invade the lesion, which usually develops externally into a festering, foul-smelling growth.

The cancerous tissue also grows inwards, invading the deeper tissues behind the eye. It can progress to the lymph nodes of the head and then to body organs such as the lungs and liver.

Cancers that begin in the third eyelid or outer eyelids usually invade the deeper tissues more quickly than do those that start on the eyeball.

Untreated cattle may live for 2–5 years after the first appearance of a cancerous lesion, although such cattle may become weak and emaciated within 6 months, indicating the involvement of internal organs.

It is more common for only one eye to be affected, but occasionally the condition may occur in both eyes.

A plaque on the edge of the eyeball. Only some of these types of lesions will progress to become tumours, and some will regress spontaneously, but it is advisable to cull such animals.

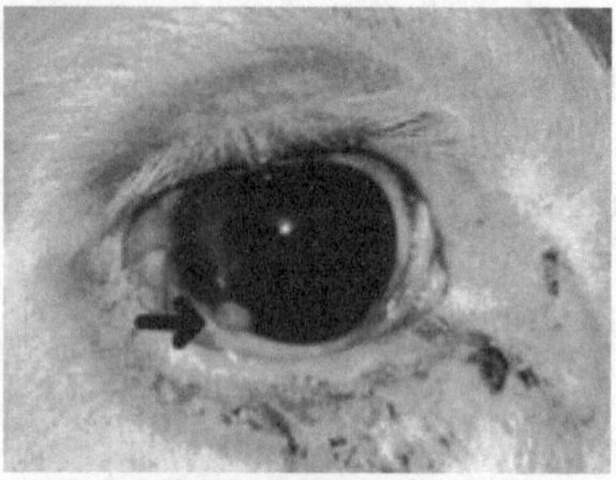

Two plaques on the third eyelid. Note the poor pigmentation of the eyelids and third eyelid.

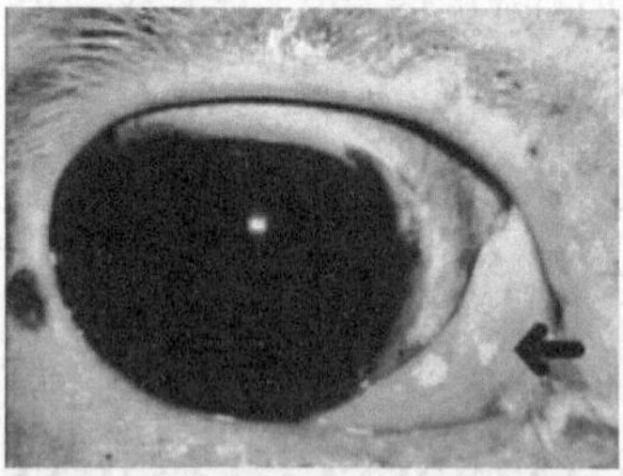

A well-developed papilloma. Such lesions can be removed with a cautery by a vet, but these animals should be disposed of as soon as practicable, and not be used for breeding.

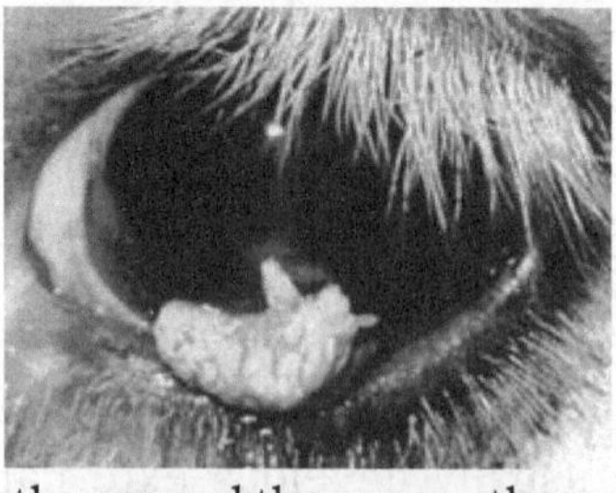

Watch for horny growths around the eye — they may also turn into cancers.

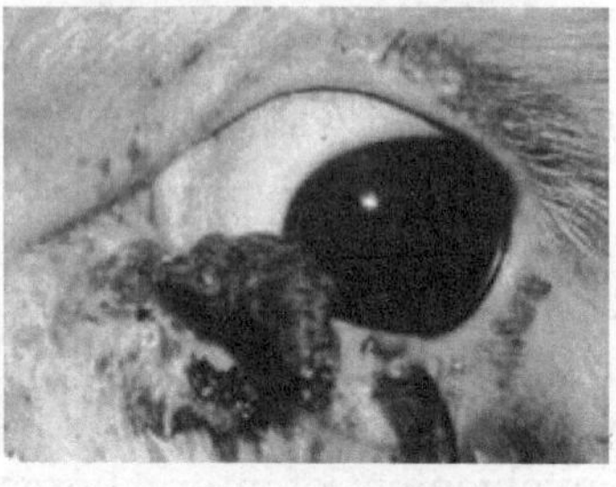

WELFARE AND SALE

It is regarded as an offence under the *Prevention of Cruelty to Animals Act* to allow cancer eye to develop to the advanced stages, and owners failing to cull such animals early enough may face prosecution. On extensive properties, where disease may become well advanced before it is detected, severely affected stock should be destroyed for humane reasons as soon as they are found.

Stock with moderate to severe disease (e.g. those with tumours that are large and ulcerated or have a detectable odour) are unlikely to be accepted for transportation. They may be judged unfit for sale at the saleyards, and condemned. Such cancers can be knocked during transport or at the saleyards, and because they bleed easily, can become a gruesome sight. Animal welfare and negative public perceptions need to be considered. Animals with all but very early, non-protruding lesions should be sent directly to the abattoir if regarded as still saleable, rather than to public saleyards.

If there is any evidence of spread of the cancer to nearby lymph nodes, such as lumps under the ear or jaw, the animal will be condemned at abattoir inspection. Such animals should be humanely destroyed.

A good manager will inspect stock regularly, recognise the early stages of cancer eye and cull affected animals immediately. This way the animals suffer little discomfort and the carcase will usually be passed for human consumption, thus minimising financial loss. Occasionally, however, an apparently small cancer may have already spread internally, in which case the carcase will be condemned.

PREVENTION

Cancer eye can be controlled by selective breeding. Selection for pigmented eyelids is most effective and can result in rapid improvement in the herd. This selection is recommended for all Hereford herds.

Selection for pigmentation around the eyes in Herefords will help prevent cancer eye.

Eyeball pigmentation develops slowly, but apparently is genetically linked to eyelid pigmentation, so selection for one, in effect, selects for both.

Lower-lid pigmentation is more important than pigmentation of the upper lid. The intensity of pigmentation is not important as long as it is darker than just a pale brown.

Unlike eyeball pigment, third-eyelid pigment is not affected by selection for external eyelid pigment. Third-eyelid cancers account for about 10% of cancers in Herefords. They are easily removed by surgery in the early stages. Treated cows and their progeny should be culled as soon as practicable to select against this form of cancer eye.

Breeders not committed to producing pure Herefords should consider crossbreeding if cancer eye is a problem.

There are strain and sire differences. Breeders who identify families that are more susceptible, or sires whose progeny are more susceptible, should move away from these lines in their breeding program.

TREATMENT

Surgical treatment of cancer eye is possible, particularly in the early stages, and is warranted if, for example, a cow has a young calf. Small lesions can be removed by electrocautery (burning) or cryosurgery (freezing). A veterinarian may also surgically remove lesions from the eyelids or third eyelid, scrape a lesion from the eyeball, or remove the whole eye.

In some cases, removal of very early precursor lesions may result in a permanent cure. However, more than 60% of cases recur and it is not good practice to continue to breed from treated cattle or from their progeny.

An experimental vaccine made from cancer tissue to elicit an immune response has been reported to cause regression and sometimes complete disappearance of existing eye tumours, even in some moderately advanced cases. However, the vaccine is quite difficult to make, and requires large amounts of tumour material to be collected from other animals. Development of a vaccine for treatment purposes is therefore not considered to be a commercially viable proposition.

Since it is undesirable to keep animals that are predisposed to cancer eye, vaccination is not recommended as a management tool, as it would make selection against the disease more difficult.

IMMUNOHISTOCHEMICAL EVALUATION OF VASCULAR URINARY BLADDER TUMORS FROM COWS WITH ENZOOTIC HEMATURIA

Chronic ingestion of bracken fern (*Pteridium spp.*), the only plant known to cause cancer naturally, produces a neoplastic syndrome in cows known as bovine enzootic hematuria (BEH), which is characterized by the occurrence of hematuria and urinary bladder tumors. Moreover, in approximately 62% of the cases, multiple tumors occur in the same bladder. This syndrome occurs worldwide and is very common in cows on São Miguel Island, Azores, Portugal, where bracken grows endemically. As a complexing factor, bovine papillomavirus type 2 (BPV-2) is frequently associated with bracken-induced tumors of the urinary bladder, and 84 of 200 bladder tumors collected at the slaughterhouse on São Miguel Island displayed BPV-2 DNA.

Exposure to the BEH carcinogen(s) causes the development of bladder cancer with diverse morphologies. Metastases are known to occur, namely to the lumboaortic lymph node, but no extensive study on the frequency of metastatic disease has been performed so far. Histologically, among the epithelial-derived tumors, transitional cell carcinoma is the most common neoplasm. Endothelial-derived urinary bladder tumors are also quite common in BEH, ranging from well-differentiated cavernous hemangiomas to solid hemangiosarcomas. This is a very heterogeneous group of tumors, and although some tumors present as clear-cut hemangiomas and hemangiosarcomas, others possess nearly all of the features of a hemangioma but also display increased cellularity and mitotic activity or contain well-formed vascular channels not well-demarcated and confined to the lamina propria, as is the case in most hemangiomas. Hence, in BEH, some authors use the designation "hemangioendothelioma" for those vascular tumors that histologically are intermediate in appearance between a hemangioma and a conventional hemangiosarcoma. Two of the antibodies most commonly used in the diagnosis of endothelial tumors are directed to FVIIIra and CD31. FVIIIra, also referred to as von Willebrand factor, is a soluble plasma protein, carrier of factor VIII, and present in the Weibel-Pallade bodies of endothelial cells. It is commonly used as a marker for endothelial cells in humans but was seldom expressed in some poorly differentiated vascular tumors. The platelet/endothelial cell adhesion molecule CD31 (PECAM-1) is a transmembrane glycoprotein, expressed in platelets, monocytes, and endothelial intercellular junctions.

Angiosarcomas are among the rarest malignancies in human pathology, and only 12 cases have been described in the bladder. Hemangioendothelioma is a rare vascular tumor of low-grade or borderline malignancy that arises in several organ systems, skin, and soft tissue, but to our knowledge, in humans no cases have been described in the urinary system. Hemangioma is the most common vascular lesion described in the World Health Organization classification of urinary bladder tumors in humans and is supposedly of congenital origin. Spontaneous bladder tumors of urothelial type are common in dogs but rare in all other species. In dogs, hemangiomas, "hemangioendotheliomas," and hemangiosarcomas are mostly limited to other organ systems, including the spleen, right atrium, skin, and subcutis; they are less common in other species.

Bracken fern carcinogens, often acting synergistically with BPV-2, interfere with multiple pathways encompassing DNA damage/repair mechanisms, activation of specific oncogenes, and growth factor receptors,5 which eventually perturb control of cell proliferation. Cellular proliferation is, in part, regulated by D-type cyclins, and cyclin D1, one of its members, is specifically involved in the G1/Sphase transition. Two recently published studies report cyclin D1 expression in human breast hemangiomas and hemangiosarcomas and in hemangiosarcomas of the spleen, liver, kidney, and soft tissue of dogs. In the former study, no correlation was seen between cyclin D1 immunoreactivity and malignancy grade. In 2007, Yonemaru et al. found, using immunohistochemistry, that all canine hemangiosarcomas tested displayed cyclin D1 expression, while none of the canine subcutaneous hemangiomas were found to express this protein.

Furthermore, control of cell growth and maintenance of genome stability are controlled by the nuclear phosphoprotein p53. Mutations of *p53* occur in 50% of human cancers, including soft-tissue angiosarcoma, and usually lead to p53 protein accumulation in the nucleus.

The current study of BEH endothelial urinary bladder tumor specimens included the investigation of 1) the expression levels of structural proteins, 2) the possible correlation of structural proteins with pathologic parameters, and 3) the identification of altered expression of the cell-cycle regulatory factors cyclin D1 and p53.

MATERIALS AND METHODS

SAMPLE COLLECTION

From December 2002 to March 2003, a total of 4,026 Friesian cows (after calving and >2years of age) were slaughtered in the abattoir of São Miguel Island. Rejection due to bladder lesions occurred in 612 animals (15.2%), from which 26 displaying vascular tumors were selected, at random, for the present study.

Up to 90% of the animals had hematuria at or before slaughter, and all were known to have grazed on bracken-infested pasture. Their ages ranged from 3 to 12years (mean, 6). Multiple samples were collected from bladders exhibiting more than 1 tumor (18/26), each sample corresponding to discrete, nonconfluent tumor. Hence, 56 bovine urinary bladder tumors of both endothelial origin were available for immunohistochemical studies. Regional lymph nodes and internal organs were not accessed for metastasis. Tumors were formalin fixed and paraffin embedded, and 3-μm sections were cut for routine histologic diagnosis (hematoxilin and eosin, HE) and immunohistochemical studies. One histologic section was performed in tumors <3cm in size, and 2 sections in tumors >3cm.

HISTOLOGIC TYPING AND IMMUNOHISTOCHEMICAL STUDIES

Histologic typing of vascular tumors was based on World Health Organization established criteria for the diagnosis of urinary bladder tumors in domestic animals, recent publications on BEH-associated bladder lesions, and diagnostic histopathology criteria for human blood vessel tumors. Endothelial tumors were classified as hemangiomas, hemangioendotheliomas, and hemangiosarcomas. Although the designation hemangioendothelioma is infrequently used in veterinary pathology, this category was adopted in the present study to deal with the heterogeneity within this group and assist the correlation between the expression levels of selected proteins and tumor grade. This category includes lesions with microscopic features intermediate between hemangiomas and hemangiosarcomas in terms of cellularity, mitotic activity, presence of well-formed vascular channels, and degree of invasion.

Hemangioendotheliomas show focal invasion of deeper layers of the lamina propria and increased celullarity, opposed to hemangiomas, but have well-differentiated vessels and lack moderate to high mitotic activity, anastomosing vascular spaces, and increased cellular atypia, characteristics of the hemangiosarcoma. Classification was based on HE staining for all endothelial tumors except 5 solid hemangiosarcomas, in which CD31-positive immunohistochemical staining was contributory to the final diagnosis.

To determine possible cross-reaction of the anti-UPIII antibody with nonurothelial tumors, 8 hemangiomas, 6 hemangioendotheliomas, and 8 hemangiosarcomas were subject to immunohistochemistry with this antibody. Normal blood vessels served as internal positive controls for FVIIIra/CD31, and urothelium served as internal positive controls for UPIII.

All tumors were tested for each antibody within a short period. Hence, variations in antigen retrieval technique overtime were minimal and unimportant to the outcome of the technique.

Human mantle-cell lymphoma was included as a positive control for human cyclin D1 antibody. Heat-induced epitope retrieval was essential for immunodetection of cyclin D1 and was carried out by placing sections in a high pH buffer solution (pH = 8.0) in a microwave (20minutes, 750W). The aggressiveness of the technique occasionally compromised cell/tissue morphology, and for this reason it was not possible to assess cyclin D1 staining in 9 endothelial tumors. Bovine ocular squamous cell carcinoma was used as a positive control for p53 immunodetection. The DO-1 antibody utilized in the study recognizes an epitope present in the N terminus of p53 and cross-reacts with both wild-type and mutant human protein (amino acids 20–25). In this study, a 20% cutoff value (>100 labeled tumor cells per 500 tumor cells randomly counted) was accepted as positive staining for p53 and cyclin D1. Negative controls for immunohistochemistry were processed identically to test slides, but the primary antibody was omitted.

STATISTICAL ANALYSIS

Association between presence/absence of immunoreactivity for the antibodies tested and pathologic parameters, namely tumor grade/stage, was evaluated using 2-sided Fisher's exact test. Statistical analyses were performed using SPSS Version 14.0 (Chicago, IL, USA). The associations were considered significant when P values were less than 0.05.

RESULTS

Based on evaluation of the HE-stained tissue sections, the 56 endothelial-derived bovine urinary bladder tumors were diagnosed as follows: 21 hemangiomas, 23 hemangiosarcomas, and 12 neoplasms that had microscopic features suggestive of a low-grade tumor that were classified as hemangioendotheliomas. Growth patterns of vascular tumors ranged from vasoformative to solid, and tumor vessels varied from well formed with a well-defined lumen in hemangiomas to more poorly defined in hemangiosarcomas. The proportion of vasoformative and solid components varied, and only 6 hemangiosarcomas were found to display a full solid growth pattern and epithelioid morphology. In 3 hemangioendotheliomas and 2 other hemangiosarcomas, neoplastic cells were also predominantly epithelioid, with rare spindled areas. Muscle invasion was only seen in 14 hemangiosarcomas. Lymphoplasmacytic and neutrophilic inflammatory infiltrates (often found together, present within the tumor stroma and tumor margins) varied from prominent to slight or absent.

CD31 AND FVIIIRA

Vascular tumors were further characterized through the use of endothelial markers. CD31 expression was detected in all endothelial tumors examined, including muscle-invasive hemangiosarcomas. All neoplastic cells within the tumor consistently displayed CD31 immunostaining, which was intense at the cell membrane level. FVIIIra expression was seen in all hemangiomas (20/20) and hemangioendotheliomas (12/12), and in 15 of 20 hemangiosarcomas (67%). The hemangiosarcomas negative for FVIIIra (6 solid and 1 poorly vasoformative) were all muscle-invasive lesions, and the endothelial origin of these tumors was confirmed by their positive reaction with CD31. Undifferentiated endothelial tumor cells, generally found in areas of solid growth, did not express FVIIIra. The staining was diffusely granular within the cytoplasm of endothelial cells, with occasional staining of plasma within vascular space. It was not possible to assess FVIIIra and CD31 expression in 2 and 6 endothelial tumors, respectively, because of high general background within the vascular space.

CYCLIN D1

Normal urothelium surrounding the tumor lesion demonstrated nuclear staining for cyclin D1 in <5% of basal cells. Moderate to strong nuclear staining was identified in 53% hemangiomas (9/17), 82% hemangioendotheliomas (9/11), and 100% hemangiosarcomas (19/19). Cyclin D1 expression positively correlates with malignancy grade in vascular tumors. The staining was mainly nuclear, although occasional weak cytoplasmic immunoreactivity was detected.

P53

Immunoreactive p53 was detected in 30% of the hemangiosarcomas (7/23, 6 solid and 1 poorly vasoformative) and 25% of the hemangioendotheliomas (3/12). Nuclear p53 immunoreactivity was seen in more than 90% of hemangiosarcoma tumor cells. All p53-positive hemangiosarcomas were muscle-invasive lesions. Staining intensity was moderate to strong, and cytoplasmic p53 staining was never observed. Nuclear p53 positivity graded high in 90% of cells was also seen in 1 hemangioma. In this particular case, not only the tumor but also all capillary endothelial cells exhibited p53 immunoreactivity. All p53-positive tumors were also positive for cyclin D1, except for 1 hemangioendothelioma.

UPIII

UPIII expression was not detected in the neoplastic endothelial cells of the 22 vascular tumors evaluated for this marker. A noteworthy finding in this series was that atypical UPIII expression pattern was seen specifically in the urothelium, covering 25% hemangiomas (2/8), 100% hemangioendotheliomas (6/6), and 88% hemangiosarcomas (7/8). In contrast with the normal UPIII staining pattern, very intense on the apical aspect and cytoplasm of superficial urothelial cells, this atypical staining was characterized by discontinuity of the apical pattern, heterogeneous staining of cells in the deeper layers, and presence of clustered intermediate cells with irregular membranous staining and a diffuse or patchy stain in the cytoplasm. Only 1 of these tumors displayed altered uroplakin distribution within the urothelium distant from the vascular tumor. No overlap in staining pattern was observed between p53-positive urothelial cells and atypical UPIII expression.

DISCUSSION

The current study documents the use of immunohistochemistry as a tool to characterize endothelial-derived urinary bladder tumors developing as a result of bracken fern carcinogens. No loss of CD31 expression was registered in endothelial tumors, and generally all neoplastic cells within these lesions were CD31 positive, as has previously been reported by other authors. CD31 has an extracellular domain that mediates endothelial cell-to-cell contact and a cytoplasmic domain with potential sites for phosphorylation after cellular activation, and it has been proposed to be involved in interactive events during angiogenesis. CD31 expression by tumor cells may therefore represent a selective advantage for tumor progression and invasion. In contrast to CD31, FVIIIra expression was lost in high-grade, muscle-invasive hemangiosarcomas of solid phenotype, negatively correlating with tumor grade and stage.

Through ultrastructural studies in soft-tissue angiosarcomas, Weibel-Palade bodies, which store FVIIIra, were not observed in neoplastic endothelium, whereas they were readily identified in the endothelium of capillaries of the surrounding fibrous stroma. This is in agreement with our findings. Previous reports also noted that FVIIIra is seldom expressed in poorly differentiated vascular tumors. Hence, although FVIIIra and CD31 are both specific for the characterization of a vasoformative phenotype, only the latter was found to be consistently expressed by neoplastic endothelial cells, even in poorly differentiated/solid neoplasms. FVIIIra immunoreactivity was occasionally seen in epithelial tumor cells, in a subset of urothelial carcinomas. This was unexpected because this rabbit-anti human FVIIIra antibody is supposed to react exclusively with endothelial cells, megakaryocytes, and platelets in both human and bovine tissue; production of this glycoprotein by cells other than endothelial cells and megakaryocytes has not been described. It is possible that an FVIIIra-like molecule is being coexpressed in bladder carcinoma cells. Alternatively, partial cross-reactivity with another epitope in the urothelium may explain the spurious staining.

Urothelium overlaying endothelial malignancies frequently displayed atypical UPIII expression, quite different from that of normal bladder urothelium and urothelium adjacent to the endothelial tumors. According to other authors, this pattern of expression is diagnostic of severe dysplasia/carcinoma in situ. It is unclear whether urothelial lesions are carcinogen induced and develop together with the endothelial tumor or if a specific microenvironment in the underlying suburothelial space favors the onset of dysplastic changes in the superimposed urothelium.

Evans and Mason demonstrated, for the first time in 1965, undeniable evidence for bracken carcinogenicity. Since then, several studies have shown increased numbers of chromosome aberrations in cattle maintained on natural bracken fern or with overt enzootic hematuria.

From the various toxic compounds that have been isolated from bracken fern, unequivocal urinary bladder carcinogenesis has been proved only for ptaquiloside. At sublethal or subtoxic levels, ptaquiloside causes genotoxicity through the formation of a number of different DNA adducts, the main labile adducts occurring at the N-3 of adenine, and to a minor extent with the N-7 of guanine. *H-ras* activation is also known to be an early event in a rat model of ptaquiloside-induced carcinogenesis. Overexpression of this protein has been documented in cases of bladder neoplasia and cystitis in animals naturally exposed to bracken fern. Cyclin D1 and p53 overexpression was seen in endothelial-derived bovine urinary bladder tumors. Regarding cyclin D1, this overexpression was positively associated with high tumor grade and occurred in 53% of hemangiomas (9/17), 82% of hemangioendotheliomas (9/11), and 95% of hemangiosarcomas (18/19). Cyclin D1 expression was recently reported in breast vascular tumors in humans, but no correlation with malignancy grade was seen. On the other hand, expression of this protein was seen in 39 of 39 canine hemangiosarcomas of the spleen, liver, kidney, and soft tissue, but none of the subcutaneous hemangiomas tested (n = 10) displayed cyclin D1 immunoreactivity. In our study, cyclin D1 expression positively correlates with malignancy grade occurring in benign, low-grade, and high-grade vascular tumors. It is possible that this pathway is being differentially regulated in endothelial tumors arising in distinct organ systems.

Abnormal p53 expression was seen in bovine urinary bladder cancer, being present only in muscle-invasive hemangiosarcomas. P53 nuclear accumulation and consequent detection by immunohistochemistry can be due to gene mutation or activation of mechanisms that influence p53 stability. Mutation and deletion of the *p53* tumor suppressor gene are the most frequent genetic alterations of epithelial-derived bladder cancer in humans and are associated with advanced disease. Within soft-tissue angiosarcomas, also in humans, 20–50% display p53 overexpression. The results presently obtained are in agreement with previous studies, and a significant correlation occurred between strong positive reactions with p53 and an invasive phenotype in both endothelial and urothelial bladder cancer. BPV-2 status of the tumors.

In humans, within the spectrum of hemangioendothelioma, several distinct entities are recognized, differing not only histologically but also in terms of epidemiology and biologic behavior. Here, although this term refers exclusively to a histologic pattern, it was seen that tumors within this category are also immunohistochemically distinct from hemangiomas and hemangiosarcomas. FVIIIra expression is maintained, similar to hemangiomas, but the percentage of tumors with cyclin D1 and p53 overexpression resembles that of hemangiosarcomas. It seems that, phenotipically, these lesions are well differentiated, but the cellular pathways involving certain oncogenes and tumor suppressor genes are already impaired.

In BEH no association between histologic grade/stage of bladder lesions and their biologic behavior can be made, because samples are obtained at slaughter and there is, therefore, no treatment or follow up. Nevertheless, there is unquestionable merit in the use of animal models for studying the pathogenesis of endothelial tumors, and, in fact, molecular studies in bovine urinary bladder tumors already provided evidence, for the first time in spontaneous tumors, of the importance of the angiogenesis/chemokine axis in endothelial and urothelial malignancies. Hence, results presented here can be used to the benefit of other species (other domestic animals and man) with the goal of further predicting the outcome of the host.

In conclusion, the data presented in this study show that while CD31 expression was retained by all endothelial tumors examined, FVIIIra was lost in hemangiosarcomas of solid phenotype and was only seldom expressed in poorly differentiated vascular tumors, being therefore less sensitive than CD31 for the characterization of bovine endothelial tumors. The cell cycle regulatory pathways involving cyclin D1 and p53 seem to be altered in urinary bladder tumors in BEH.

As an extention of this work, immunohistochemical studies will further be applied to epithelial-derived bovine urinary bladder tumors, and the distinctive features between the 2 different tumor histotypes will be further addressed.

Bibliography

A J Grove and G E Newell : *Animal Biology*, Shubhi Publication, Delhi, 2007.

Amita Sarkar : *A Textbook of Human Genetics*, Wisdom Press, Delhi, 2011.

Andrew L. Winton and Kate Barber Winton : *Milk Butter Cheese Ice Cream Eggs Meat Meat Extracts Gelatin Animal Fats Poultry Fish*, Agrobios Publication, Delhi, 2013.

Asok Mukhopadhyay : *Animal Cell Technology*, IK Publication, Delhi, 2009.

B.D. Pandey and B.D. Joshi : *Bioresources for Rural Livelihood, Vol. I. Genetics*, Narendra Publication, Delhi, 2010.

C D N Singh : *Advanced Pathology and Treatment of Diseases of Domestic Animals*, International Book Distributing Co., 2008.

C. LLoyd Organ : *Animal Behaviour*, MJP Publishers, Delhi, 2012.

C.H. Waddington : *An Introduction to Modern Genetics*, Vandana Publication, Delhi, 2006.

E A Jayson and P S Easa : *A Field Guide to Animal Signs* , Kerala Forest Research Institute, Delhi, 2004.

Elle Matthews : *Animal Adventures in Africa : Serengeti Safari,* Unicorn Books, Delhi, 2010.

Florence Periera Raja : *Animal Biotechnology*, Wisdom Press, Delhi, 2013.

Fr. S. Ignacimuthu,s.j. and S. Vincent : *Applied Genetics : Recent Trends and Techniques*, MJP Publication, Delhi, 2006.

James F. Crow and Motoo Kimura : *An Introduction to Population Genetics Theory*, Scientific Publication Delhi, 1998..

Lata Bhattacharya : *Animal Biochemistry*, Discovery Publication, Delhi, 2010.

Laurence M. Winters : *Animal Breeding*, Greenworld Publication, Delhi, 2002.

P Ramadass : *Animal Biotechnology : Recent Concepts and Developments*, MJP Publication, Delhi, 2008.

P. Kanakaraj : *A Text Book of Animal Genetics*, International Book Distributing Company, Delhi, 2007.

P.K.G. Nair and K. Prabhakar Achar : *A Textbook of Genetics and Evolution*, Konark Publication, 1998.

R. Spier and J. Griffiths : *Animal Cell Biotechnology*, Academic Press, Delhi, 1998.

Ram Bramha Sanyal : *A Hand-Book of the Management of Animals in Captivity*, Natraj Publication, Delhi, 1998.

S. Nandi : *Animal Cell Culture and Virology*, New India Publication, Delhi, 2009.

S.K. Dwivedi and S. Gupta : *Agro-Animal Resources of Higher Himalayas*, Satish Serial Publication, Delhi, 2010.

Shambhu Dhir : *A Complete Book on Fish Genetics and Aquatic Environment*, Cyber Tech Publication, Delhi, 2010.

Shivangi Mathur : *Animal Cell and Tissue Culture*, Agrobios Publication, Delhi, 2006.

Swarnaprabha Pradhan : *Basics of Genetics*, Anmol Publication, Delhi, 2008.

T. Ramanathan : *Applied Genetics of Oilseed Crops*, Daya Publication, Delhi, 2004.

Vaibhav Suri : *A Modern Book on Biological Science—Cell/Genetics*, Cyber Tech Publication, Delhi, 2009.

William J.A. Payne and R. Trevor Wilson : *An Introduction to Animal Husbandry in the Tropics*, Wiley Publication, Delhi, 2013.

Yougesh Kumar and Rajeev Tyagi : *Aquaculture Fisheries Biotechnology and Genetics*, Manglam Publishers, Delhi, 2013.

Index